Intracellular Calcium-Dependent Proteolysis

Editors

Ronald L. Mellgren, Ph.D.
Department of Pharmacology-Therapeutics
Medical College of Ohio
Toledo, Ohio

Takashi Murachi, D. Med. Sc.
(Deceased)
Department of Clinical Science and
Laboratory Medicine
Kyoto University Faculty of Medicine
Kyoto, Japan

CRC Press
Boca Raton Ann Arbor Boston

Library of Congress Cataloging-in-Publication Data

Intracellular calcium-dependent proteolysis/editors, Ronald L.
Mellgren, Takashi Murachi.
p. cm.
Includes bibliographical references.
ISBN 0-8493-6570-8
1. Calpain. 2. Calpastatin. I. Mellgren, Ronald L., 1946-
II, Murachi, Takashi, 1926-1990.
[DNLM: 1. Calcium-binding Proteins--physiology. 2. Calpain-
-physiology. 3. Cells--metabolism. 4. Proteins--metabolism. QU
55 I677]
QP609.C26I57 1990
599.087′61--dc20
DNLM/DLC
for Library of Congress 90-1577

Direct all inquiries to CRC Press, Inc., 2000 Corporate Blvd., N.W., Boca Raton, Florida 33431.

International Standard Book Number 0-8493-6570-8

Library of Congress Card Number 90-1577
Printed in the United States

DEDICATION

While this volume was in preparation, two researchers who made outstanding contributions to our understanding of the calcium-dependent proteolytic system died.

Judith E. Schollmeyer, who was to contribute a chapter to this volume, died in an automobile accident on June 20, 1989. Dr. Schollmeyer was involved in some of the first studies of calpain immunolocalization in collaboration with Dr. William R. Dayton at the University of Minnesota. She continued her studies on calcium-dependent proteolysis at the Roman L. Hruska U.S. Meat Animal Research Center in Clay Center, Nebraska, and made important contributions in defining the intracellular functions of calpains I and II. Just prior to her death, she published a paper in *Science* suggesting that calpain II may have a specific role in cell division (*Science,* 240, 911, 1988).

Takashi Murachi died on May 12, 1990, just as this volume, which he co-edited, was going to press. It is a truly sad irony that he will not be with us to see this volume on calcium-dependent proteolysis published. Professor Murachi was a pioneer in this area of research and his seminal contributions are too numerous to list here. The best appreciation of the impact of the work produed by Dr. Murachi and his colleagues in Kyoto will be gained by noting the number of times his studies are cited by the diverse group of contributors to this volume.

Drs. Schollmeyer and Murachi will be missed not only by their many friends and colleagues, but by the scientific community at large. The contributors dedicate this volume in memory of Judy Schollmeyer and Takashi Murachi.

PREFACE

The last ten to fifteen years have witnessed a renaissance in the study of intracellular protein catabolism. Prior to the 1980s, most research in this field centered on the importance of lysosomal mechanisms of protein breakdown and the development and utilization of methodologies to study overall proteolysis within various cells and tissues. While there was a clear indication from several sources that there were extralysosomal pathways of intracellular proteolysis, little was known of the specific intracellular proteolytic enzymes involved. Happily, things are now different, and there is overwhelming evidence for the existence in animal cells of regulated proteases that which are not sequestered within organelles and that may therefore directly act on cytoplasmic protein substrates.

One of the well-characterized intracellular proteolytic systems is energy dependent and requires, for at least some of its actions, the small polypeptide ubiquitin. This pathway is composed of many different enzymes and factors which appear to interact in a highly organized fashion to facilitate the proteolytic removal of intracellular proteins. There is some evidence that this system is important for the proteolytic processing of polypeptide chains that are foreign to the cell, or incorrectly translated, or damaged by oxidation.

The other extensively characterized system is made up of the calcium-dependent cysteine proteinases (calpains) and their endogenous inhibitor protein (calpastatin). The structural and regulatory properties of the components of this system and the potential physiologic and pathologic importance of the calpains and calpastatin are the subjects of this volume.

The purpose of this book is to provide an overview of current research on the calpains and calpastatin. It will become apparent to the reader that this research has drawn contributions from investigators with diverse backgrounds and interests and with different ideas about the roles of the calcium-dependent proteolytic system in cell function. In part II of this volume, evidence will be presented for the importance of calpain function in everything from hemostasis to memory formation to cataract development. It is fair to say that our understanding of how calpain functions in any of these systems is far from complete. However, the molecular and enzymologic studies in Part I strongly support the notion that these enzymes and their inhibitor comprise a highly regulated system designed to hydrolyze protein substrates in a controlled manner and in response to changing Ca^{2+} concentrations.

The very fact that this system appears to exist in virtually all mammalian cells, and in at least some nonmammalian species, argues for its overall importance in some aspects of cell function. That the calpains show some selectivity toward the substrates they recognize, both *in vitro* and *in vivo*, further suggests that there is a limited set of proteins that are susceptible to their action, and this argues for a more specific role than gross protein turnover. Thus, it is appropriate to search for specific actions of the calpains in different physiologic and pathophysiologic states.

As might be expected with any scientific investigation in its beginnings, there are still many basic questions about the calcium-dependent proteolytic system that remain to be answered. In addition, some of the observations that have led to the initial formulation of hypotheses about the activation and functions of calpains have been challenged. It is clear that more information is needed in order to establish the importance of such phenomena as calpain autolysis and membrane association in the physiologic regulation of this proteolytic system.

Perhaps the most important question related to calpain function concerns the substrates for the calpains in various cells. In some cases, as in the blood platelet system, the major intracellular substrates have been defined. However, in the case of most cells, we must still guess at the physiologic substrates. Other important questions center around the mechanisms for overcoming the high Ca^{2+} requirement of the enzymes, especially the calpain II isozyme, and the mechanisms for overcoming calpastatin inhibition, which could potentially shut down the system in many cell types. While there is no overwhelming evidence at this time for any particular model that will resolve these constraints on calpain activity, there are a number of exciting possibilities presented in this volume, and testable hypotheses have been generated suggesting how this proteolytic system may function within the cell. It is reasonable to expect that with so many investigators in different areas of research working on the calcium-dependent proteolytic system, within the next several years major advances will be made in understanding their physiologic significance.

THE EDITORS

Ronald L. Mellgren, Ph.D., is Associate Professor of Pharmacology and Therapeutics at Medical College of Ohio, Toledo, Ohio.

Dr. Mellgren received an A.A. degree in 1966 from Grand View College in Des Moines, Iowa; a B.A. in chemistry in 1969 from Drake University in Des Moines; and his Ph.D. in biochemistry in 1976 from Iowa State University, Ames, Iowa.

He is a past Established Investigator of the American Heart Association and holds membership in the American Society for Biochemistry and Molecular Biology, the American Society for the Advancement of Science, the International Society for Heart Research, and Sigma Xi, Honorary Research Society.

Dr. Mellgren has received nationally funded grants from the American Heart Association and the U.S. Public Health Service, and has published over 50 scientific papers on the regulation of protein kinases and phosphatases and on the regulation of intracellular proteases.

Takashi Murachi, M.D., D.Med.Sc., is Professor and Chairman of the Department of Clinical Science and Laboratory Medicine at the Kyoto University Faculty of Medicine and the Director of the Central Clinical Laboratory of the Kyoto University Hospital at Kyoto, Japan.

Dr. Murachi graduated in 1949 from the Kyoto University Faculty of Medicine with an M.D. degree. He obtained his D.Med.Sc. degree in biochemistry in 1957 from the graduate School of Kyoto University. He was appointed a Rockefeller Foundation Research Fellow at the Department of Biochemistry of the University of Washington (Seattle) from 1957 to 1959. He served as an Associate Professor of Biochemistry at the Nagoya City University School of Medicine (Nagoya) from 1959 to 1965 and as Professor and Chairman of the latter from 1965 to 1973. He assumed his present position in 1974.

Dr. Murachi is a member of the Japanese Biochemical Society, the American Chemical Society, The Biochemical Society (London), the Japan Society of Clinical Pathology, and Japan Society of Clinical Chemistry. He is a Foreign Honorary Member of the American Society for Biochemistry and Molecular Biology (elected in 1988). He served as President (1985 to 1986) of the Japanese Biochemical Society and as President (1987 to 1989) of the Federation of Asian and Oceanian Biochemists.

Dr. Murachi received the Chunichi Culture Award in 1970, the Shimadzu Award in 1990, and the Chester Beatty Lectureship of the International Union of Biochemistry in 1988.

Dr. Murachi is the author of more than 200 papers and has been the author or coauthor of four books. His current major research interests relate to the biological regulatory system involving proteolytic enzymes and their inhibitors. His research interests also include the methods for application of enzymes to clinical analysis.

CONTRIBUTORS

Katsuhiko Ase, Ph.D.
Department of Biochemistry
Kobe University
School of Medicine
Kobe, Japan

Robert W. Colman, M.D.
Sol Sherry Professor of Medicine
Director, Thrombosis Research Center
Temple University School of Medicine
Philadelphia, Pennsylvania

Jinyang Cong, Ph. D.
Department of Chemical Engineering
Tsinghua University
Beijing, China

Catherine Crawford, D.Phil.
Laboratory of Molecular Biophysics
University of Oxford
Oxford, England

Dorothy E. Croall, Ph.D.
Associate Professor
Department of Biochemistry
University of Maine
Orono, Maine

L. L. David, Ph.D.
Assistant professor
Departments of Biochemistry and Ophthalmology
Oregon Health Sciences University
Portland, Oregon

George N. DeMartino, Ph.D.
Associate Professor
Department of Physiology
Southwestern Medical Center
University of Texas
Dallas, Texas

Peter Friedrich, M.D., Ph.D.
Deputy Director
Institute of Enzymology
Biological Research Center
Hungarian Academy of Sciences
Budapest, Hungary

T. Glaser, Ph.D.
Research Associate
Department of Human Genetics
Sackler School of Medicine
Tel Aviv University
Tel Aviv, Israel

Darrel E. Goll, Ph.D.
Professor
Departments of Animal Sciences and Biochemistry
University of Arizona
Tucson, Arizona

M. Hatanaka, D. Med. Sc.
Professor
Institute for Virus Research
Kyoto University
Kyoto, Japan

Keisuke Hashimoto, M.D.
Department of Biochemistry
Kobe University School of Medicine
Kobe, Japan

David R. Hathaway, M.D.
Professor
Department of Medicine and Pharmacology
Indiana University School of Medicine
Indianapolis, Indiana

Ian Hoffman, M.D.
Pulmonary Section
Department of Medicine
Temple University
Philadelphia, Pennsylvania

Hans-P. Kapprell, Ph.D.
Technical Product Development
Abbott GmbH
Wiesbaden, F.R.G.

Akira Kishimoto, M.D., Ph.D.
Associate Professor
Department of Biochemistry
Kobe University School of Medicine
Kobe, Japan

William C. Kleese
Muscle Biology Group
University of Arizona
Tucson, Arizona

Nechama S. Kosower, M.D.
Professor
Department of Human Genetics
Sackler School of Medicine
Tel Aviv University
Tel Aviv, Israel

Toshihide Kumamoto, M.D.
The First Department of Internal Medicine
Kumamoto University Medical School
Kumamoto, Japan

Tsukasa Kuroda, M.D.
Department of Biochemistry
Kobe University
School of Medicine
Kobe, Japan

Richard D. Lane, Ph. D.
Associate Professor
Department of Anatomy
Medical College of Ohio
Toledo, Ohio

Gary Lynch, Ph.D.
Professor
Center for the Neurobiology of Learning and Memory
University of California
Irvine, California

Massatoshi Maki, Ph. D.
Associate Professor
Institute for Virus Research
Kyoto University
Kyoto, Japan

Phadungchom McClelland, Ph.D.
Research Associate
Department of Medicine
Indiana University School of Medicine
Indianapolis, Indiana

Ronald L. Mellgren, Ph.D.
Associate Professor
Department of Pharmacology and Therapeutics
Medical College of Ohio
Toledo, Ohio

Edon Melloni, Ph.D.
Professor
Department of Biochemistry
University of Genoa
Genoa, Italy

Katsuya Mikawa, M.D.
Department of Biochemistry
Kobe University
School of Medicine
Kobe, Japan

Takashi Murachi, D. Med.Sc.
Professor
Department of Clinical Science and Laboratory Medicine
Kyoto University Faculty of Medicine
Kyoto, Japan

Donald L. Mykles, Ph.D.
Associate Professor
Department of Biology
Colorado State University
Fort Collins, Colorado

Akihiro Okitani
Department of Food Science and Technology
Laboratory of Food Chemistry
Nippon Veterinary and Zootechnical College
Tokyo, Japan

Marianna Pintér
Institute of Enzymology
Biological Research Center
Hungarian Academy of Sciences
Budapest, Hungary

Sandro Pontremoli, M.D.
Professor
Department of Biochemistry
University of Genoa
Genoa, Italy

William W. Schlaepfer, M.D.
Professor
Department of Pathology
University of Pennsylvania
Philadelphia, Pennsylvania

Peter Seubert, Ph.D.
Department of Enzymology
Athena Neurosciences
South San Francisco, California

Thomas R. Shearer, Ph.D.
Professor
Departments of Biochemistry and Ophthalmology
Oregon Health Sciences University
Portland, Oregon

Robert L. Shoeman, Ph.D.
Research Fellow
Abteilung Traub
Max Planck Institute for Cell Biology
Ladenburg, F.R.G.

Dorothy M. Skinner, Ph. D.
Biology Division
Oak Ridge National Laboratory
Oak Ridge, Tennessee

Koichi Suzuki, Ph.D.
Director
Department of Molecular Biology
Tokyo Metropolitan Institute of Medical Science
Tokyo, Japan

Kenji Takahashi, Ph.D.
Professor
Department of Biophysics and Biochemistry
Faculty of Science
University of Tokyo
Tokyo, Japan

Emiko Takano, D. Med. Sci.
Department of Clonical Science and Laboratory Medicine
Kyoto University
Faculty of Medicine
Kyoto, Japan

Peter Traub, Ph.D.
Director
Abteilung Traub
Max Planck Institute for Cell Biology
Ladenburg, F.R.G.

Un-Jin P. Zimmerman
Department of Medicine
Institute for Environmental Medicine
Philadelphia, Pennsylvania

TABLE OF CONTENTS

I. Structural and Regulatory Properties of the Calpains and Calpastatins

Chapter 1

HISTORICAL BACKGROUND AND CURRENT STATUS OF THE Ca^{2+}-DEPENDENT PROTEINASE SYSTEM

Darrel E. Goll, William C. Kleese, Akihiro Okitani, Toshihide Kumamoto, Jinyang Cong, and Hans-Peter Kapprell

TABLE OF CONTENTS

I. HISTORICAL BACKGROUND

A. DISCOVERY OF A Ca^{2+}-DEPENDENT PROTEASE: THE Ca^{2+}-ACTIVATED SARCOPLASMIC FACTOR (1969—1972)

We first learned that Ca^{2+} would induce specific degradation of skeletal muscle myofibrils late in April 1969, when a graduate student, Wayne Busch, and I (DEG) were studying the effects of Ca^{2+} on tension development in strips of rabbit psoas muscle. We worked until 2:00 a.m. one night, and rather than cleaning the instrument at that late hour, we left, and the muscle strips remained in a 1-m*M*-Ca^{2+}-containing solution overnight. When Wayne removed the strips the next morning, he noticed they were unusually fragile and broke easily into small pieces. I suggested that he prepare myofibrils from the strips and examine them with the phase contrast microscope to learn whether there was any obvious reason for their fragility. He saw at once that the Z-disks were completely gone in the fragile strips.

Because our laboratory at that time was working on α-actinin, a major protein constituting the Z-disk, the Ca^{2+}-induced removal of Z-disks was of immediate interest to us. We quickly learned that loss of Z-disks was associated with Ca^{2+}; myofibrils in strips incubated in EGTA for 3 or 4 d retained their Z-disks (see Figures 1 and 2). By purifying myofibrils and incubating purified myofibrils with Ca^{2+} in the absence of the muscle cell cytoplasm, we found that Ca^{2+} itself had no direct effect on Z-disks (Figure 3). Hence, Ca^{2+}-induced Z-disk degradation was due to some "Ca^{2+} activated sarcoplasmic factor" (CASF). With the assistance of Dr. Marvin Stromer, Wayne Busch received a crash course in electron microscopy. Electron microscope studies showed that the Ca^{2+}-induced Z-disk removal was remarkably specific and occurred with very little additional ultrastructural changes in the myofibril (Figures 1 and 2). I reported these preliminary findings at a conference held in Madison, Wisconsin on 14 to 16 July, 1969[2] and at an American Chemical Society Symposium on Muscle Proteins, held in New York, 7 to 12 September, 1969. Dr. E. G. Krebs was in the audience at the ACS symposium, and after the meeting had ended, he mentioned that his group at Seattle had earlier worked with a Ca^{2+}-activated protease that activated phosphorylase kinase. Although they had not tested the effects of their protease, which they called KAF (kinase-activating factor), on myofibrillar proteins, it was activated by Ca^{2+} and therefore was a potential cause of the Ca^{2+}-induced Z-disk removal we had noticed.

By the summer of 1970, we had repeated the procedure Huston and Krebs[3] had used to partly purify KAF and found the Ca^{2+}-activated sarcoplasmic factor (CASF) that caused Z-disk removal was also purified by this same procedure, up to the step involving alumina precipitation, when much of our activity was lost. Although the principal effort in our laboratory at that time involved characterization of α-actinin, we continued a smaller effort to purify CASF. In retrospect, these efforts were thwarted because we could purify significant quantities of α-actinin from 800 to 1000 g of rabbit skeletal muscle, but when we used similar quantities of rabbit skeletal muscle to attempt to purify CASF, our active fractions would decrease until less than a mg of protein remained, and the preparation was still impure. I left in September 1971 to spend a year in Dr. John Pringle's laboratory at Oxford, and we published the results we had obtained up to that point on CASF.[1]

B. PURIFICATION OF A Ca^{2+}-DEPENDENT PROTEASE (CAF) AND SOME OF ITS PROPERTIES (1972—1976)

When I returned from Oxford in June 1972, I discussed with Bill Dayton, a beginning graduate student, the possibility of characterizing CASF as his Ph.D. research problem. He agreed, and we began an intensive effort to purify this protein (that we were then calling CAF for Ca^{2+}-activated factor). This effort evolved into gradually scaling up our preparations until we were processing 12,000 to 15,000 g of skeletal muscle per preparation and changing from rabbits to pigs, because we could much more easily obtain such quantities of muscle

FIGURE 1. Electron micrographs of rabbit psoas muscle sampled immediately after death (upper) and after the muscle had been incubated for 9 hr at 37°C in 80 m*M* KCl, 60 m*M* Tris-HCl, pH 7.2, 5 m*M* $MgCl_2$, 1 m*M* $CaCl_2$, 1 m*M* NaN_3, and 1 m*M* deoxycholate (lower). The at-death muscle displays the usual banding pattern of mammalian skeletal muscle. Note particularly the dense fibrillar Z-disks. Z indicates Z-disks, I the I-band area, and A the A-band area of the myofibril. After 9 hr in a Ca^{2+}-containing solution, the Z-disks are completely gone but both thick and thin filaments seem intact. Z indicates the area where the Z-disks normally are present. The strips of psoas muscle incubated for 9 hr were held isometrically, and after the Z-disks were degraded, shortening to the extent of double overlap of thin filaments in the center of the A-band resulted in gaps between the sarcomeres. (Reproduced from Busch, W. A., Stromer, M. H., Goll, D. E., and Suzuki, A., *J. Cell Biol.*, 52, 367, 1972, by copyright permisssion of the Rockefeller University Press.)

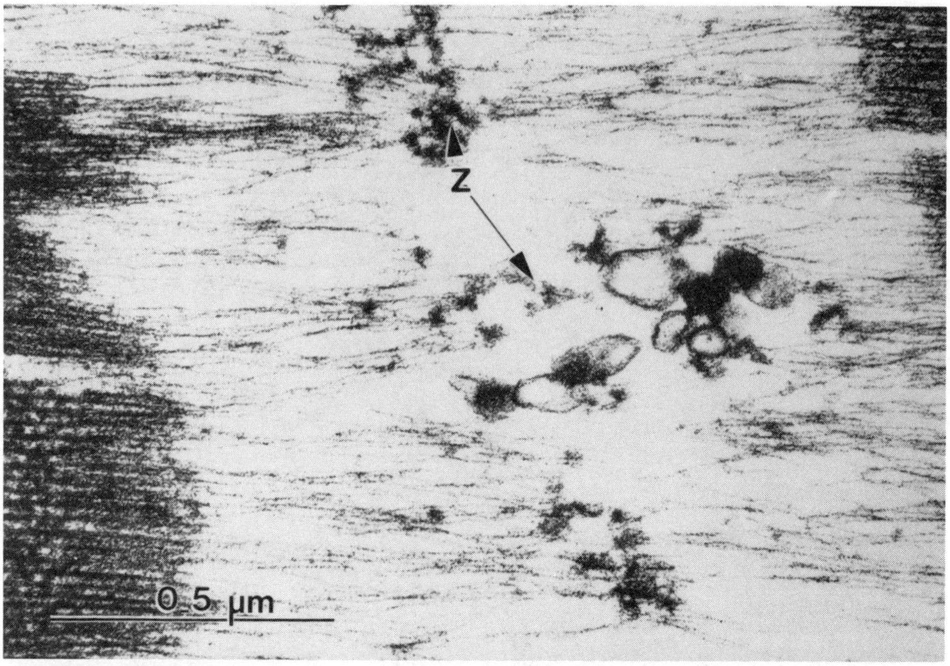

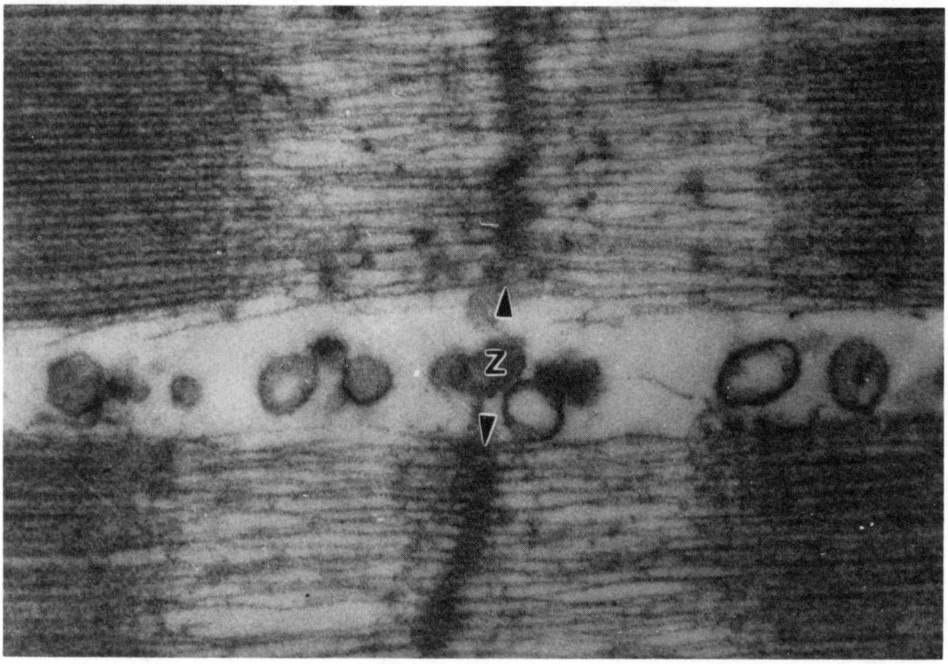

FIGURE 2. High magnification electron micrograph of the I-band region from a rabbit psoas muscle strip incubated for 9 hr at 37°C in 80 m*M* KCl, 60 m*M* Tris-HCl, pH 7.2, 5 m*M* $MgCl_2$, 1 m*M* $CaCl_2$, 1 m*M* NaN_3, and 1 m*M* deoxycholate (upper), and an electron micrograph of rabbit psoas muscle after incubation for 9 hr at 37°C in 80 m*M* KCl, 60 m*M* Tris-HCl, pH 7.2, 5 m*M* $MgCl_2$, 1 m*M* EGTA, 1 m*M* deoxycholate (lower). The high magnification micrograph shows that incubation in a Ca^{2+}-containing solution completely removes Z-disks with little or no effect on the thin filaments. Z indicates the area where Z-disks normally are present. When EGTA is substituted for Ca^{2+} in the incubation medium, Z-disks (indicated by Z) remain intact after 9 hr at 37°C and appear structurally identical to the Z-disks in at-death muscle (cf. Figure 1). (Reproduced from Busch, W. A., Stromer, M. H., Goll, D. E., and Suzuki, A., *J. Cell Biol.*, 52, 367, 1972, by copyright permission of the Rockefeller University Press.)

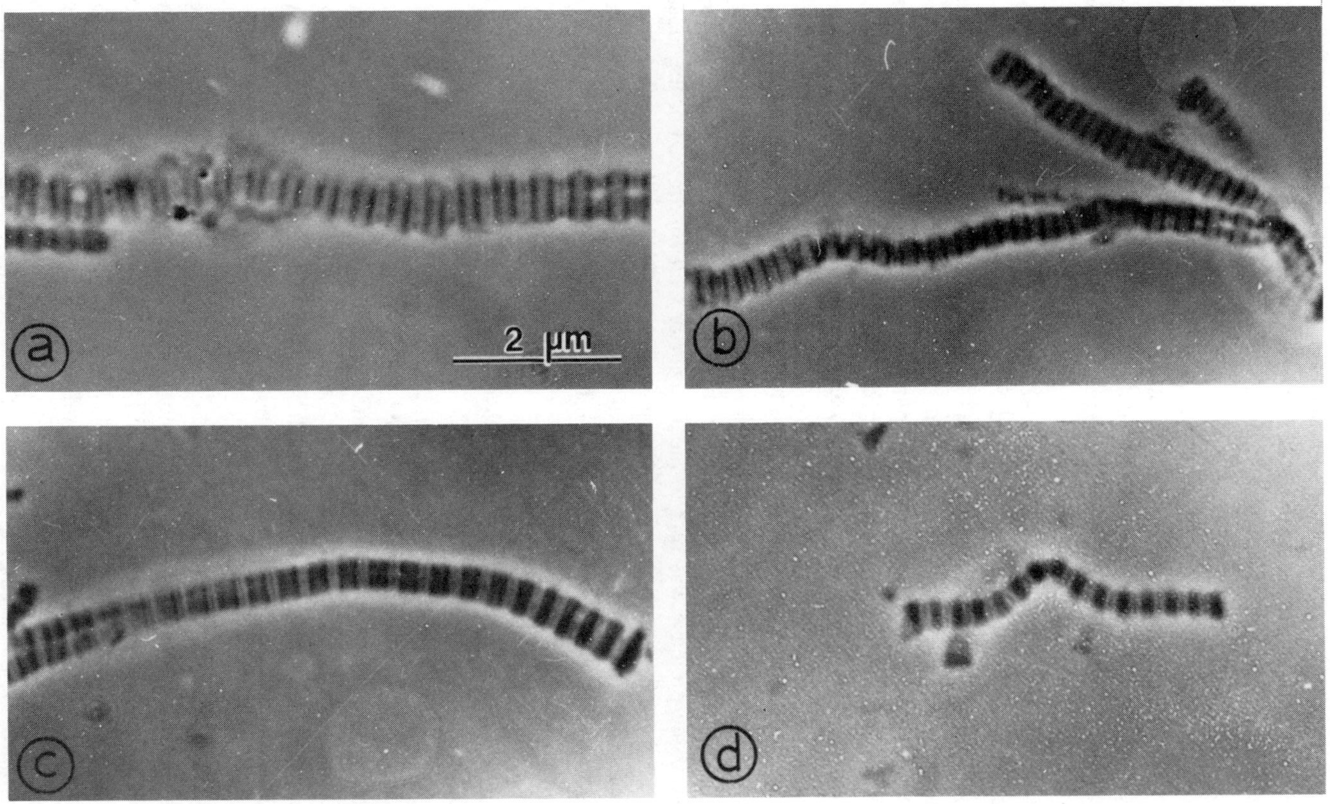

FIGURE 3. Phase contrast micrographs of myofibrils that had been purified from rabbit skeletal muscle and then incubated for 9 hr: (b) in 80 m*M* KCl, 60 m*M* Tris-acetate, pH 7.1, 5 m*M* $MgCl_2$, 1 m*M* EGTA, 1 m*M* NaN_3, and 1 m*M* deoxycholate; (c) in 80 m*M* KCl, 60 m*M* Tris-acetate, pH 7.1, 1 m*M* $MgCl_2$, 1 m*M* $CaCl_2$, 1 m*M* NaN_3, and 1m*M* deoxycholate; or (d) in 80 m*M* KCl, 60 m*M* Tris-acetate, pH 7.1, 1 m*M* EDTA, 1 m*M* NaN_3, and 1 n*M* deoxycholate. Panel (a) is a phase contrast micrograph of the myofibrils before incubation. Incubation of purified myofibrils with Ca^{2+} has no effect on Z-disks, so Ca^{2+} causes Z-disk removal by activating something in muscle cells and not by a direct effect on Z-disks.

from pigs, and pigs were plentiful in Iowa. By March 1975, we knew that CAF was a proteolytic enzyme that had two subunits of 80 kDA and 30 kDa, but we had not purified more than a milligram of it. This finding was first reported at a seminar I gave at Michigan State University in April 1975.

We continued to improve our purification procedure and began to obtain quantities of the protease sufficient to permit its characterization. Because the protease seemed so specific for Z-disk removal, we were very interested in learning what myofibrillar proteins it degraded and to what extent these proteins were affected. After a series of preliminary experiments, Bill Dayton and I decided to monitor the time course of porcine skeletal muscle myofibril degradation by CAF using SDS-PAGE. Unfortunately, we did not get the experiments started until late afternoon, and I was having a party at my house for one of the members in our laboratory, who had accepted a new position and was leaving. Bill and I loaded the SDS-gels (in tubes) at 8:00 p.m., set the current at a low level, and went to the party. We returned after 2:00 a.m. to remove the gels from the tubes and put them in stain. Even at that late hour and considering our condition, we did not break any gels, and they all eventually appeared as Figure 9 in the proceedings of a Cold Spring Harbor Conference, held in September 1975, where we reported our findings thus far and suggested that CAF may be responsible for initiating myofibrillar protein turnover.[4]

Although we could still only obtain 6 to 8 mg of purified CAF from 12,000 g of porcine skeletal muscle, Dr. Willie Reville had joined our efforts to characterize this protease, and we began a schedule of doing preparations every three weeks. As our characterization of the protease proceeded, it became clear that CAF was the same enzyme that Guroff[5] had detected in rat brain in 1964. Bill Dayton finished his Ph.D. thesis in the summer of 1976, and early in 1976, we prepared three manuscripts describing the purification and characterization of CAF. Two of these manuscripts were sent to *Biochemistry*[6,7] and one on subcellular location of the protease was sent to the *Journal of Cell Biology*[8]. For the most part, the findings in these papers have stood the test of time, although we have now significantly improved our procedure to purify the protease. Also, it is now known that the protease exists in most, if not all, animal cells and therefore has a broader role than initiating turnover of myofibrillar proteins.

C. IDENTIFICATION OF A SECOND, MICROMOLAR Ca^{2+}-REQUIRING PROTEASE IN STRIATED MUSCLE

Somewhat to their chagrin, I had been asking Bill Dayton and the others working on the CAF study to assay every fraction that we obtained during our column chromatographic procedures, both for proteolytic activity on casein and for ability to remove Z-disks from porcine skeletal muscle myofibrils. Bill was highly skilled in the laboratory, and late in 1975, he mentioned to me that a second but much smaller peak of proteolytic activity consistently eluted from our DEAE-cellulose columns between 70 and 125 m*M* KCl. He assured me that, although this activity was very small, it was consistently present, and that perhaps another Z-disk-removing protease existed in skeletal muscle. Because we were being stretched to the limit in our efforts to purify the protease responsible for the main activity eluting off the DEAE-cellulose column, I advised Bill to forget the smaller activity and to focus our efforts on purification of the larger activity. We did, however, faithfully show the smaller peak of proteolytic activity in the elution profile from the DEAE-cellulose column that we used to purify CAF (Figure 2 in Reference 6). When Ron Mellgren published his paper in 1980, describing the presence of a 2nd Ca^{2+}-dependent protease eluting at 60 m*M* KCl from a DEAE-Sephadex column of canine cardiac muscle,[9] I (and, I am sure, Bill Dayton, who was then at the University of Minnesota) immediately recognized that this was the small peak of proteolytic activity we (I) had earlier ignored. Both my laboratory (now at the University of Arizona) and Bill Dayton's laboratory independently purified this second peak of proteolytic activity,[10,11] and showed that it too had 80-kDa and 30-kDa subunits, just like the CAF we had earlier purified, but that its Ca^{2+} requirement was in the micromolar range rather than the millimolar Ca^{2+} concentration required for CAF activity. It was obvious to both Bill and me that the newly purified micromolar Ca^{2+}-dependent proteinase was very closely related to CAF, because it was inhibited by the protein inhibitor that was specific for CAF,[11] and because its effects on myofibril structure were limited to Z-disk removal,[10] just as the effects of CAF were. Therefore, we both felt at that time that this new protease was a different form of CAF that had been modified, either through limited autolysis, or through some form of covalent modification such as phosphorylation. Subsequent cDNA cloning studies, summarized in more detail elsewhere in Chapter 2 of this monograph, have now shown that the micromolar Ca^{2+}-requiring protease is not derived from CAF but is a distinct molecule that shares the same 30-kDa (now shown to be 28-kDa) subunit as CAF, but has a different 80-kDa subunit, sharing approximately 50% amino acid sequence homology with the 80-kDa subunit of CAF.[12-16] These two proteases have been given several different names, including calpain I, μ-calpain, μ-CANP and CDP I for the micromolar Ca^{2+}-requiring form, and calpain II, m-calpain, m-CANP, and CDP II for the millimolar Ca^{2+}-requiring form that we initially called CAF. In this article, the terms μ-calpain and m-calpain will be used for the two proteinases.

At this same time, Koichi Suzuki and colleagues,[17] who had been working with a m-calpain (that they called CANP for Ca^{2+}-activated neutral proteinase) from chicken skeletal muscle, found that incubation of their CANP with 0.5 m*M* Ca^{2+} for 2 min at 0 °C reduced the Ca^{2+} concentration required for half maximal activity from 400 to 30 μ*M*. The autolysis that occurred during this period decreased the mass of the large 80-kDa subunit, which was the only polypeptide present in Suzuki's preparations, from 82 to 79 kDa. Because the autolyzed protease required only micromolar Ca^{2+} for activity, Suzuki and co-workers named this protease, μCANP. Hathaway et al.[18] subsequently reported that similar autolysis of chicken gizzard m-calpain reduced the Ca^{2+} required for half maximal activity from 150 to 50 μ*M* and reduced mass of the 80-kDa subunit to 76 kDa, and the mass of the 28 kDa subunit to 18 kDa. Because of these findings, the relationship between the micromolar Ca^{2+}-requiring protease (μ-calpain) purified by Dayton et al.[10] and Szacenko et al.,[11] and the micromolar Ca^{2+}-requiring protease produced by autolysis was initially unclear. It was soon shown, however, that μ-calpain also underwent rapid autolysis in the presence of Ca^{2+}, and that autolysis of μ-calpain reduced its Ca^{2+} requirement to less than 1 μ*M* and reduced the mass of its 80-kDa subunit to 76 kDa, and the mass of its 28-kDa subunit to 18 kDa.[19,20] Consequently, it is now possible to identify four Ca^{2+}-dependent proteinases:

1. Autolyzed μ-calpain having 76-kDa and 18-kDa subunits and requiring 0.6 μ*M* Ca^{2+} for half maximal proteolytic activity;
2. μ-Calpain having 80-kDa and 28-kDa subunits and requiring 5 to 50 μ*M* Ca^{2+} for half maximal activity;
3. Autolyzed m-calpain having 78-kDa and 18-kDa subunits and requiring 10 to 300 μ*M* Ca^{2+} for half maximal activity; and
4. m-Calpain having 80-kDa and 28-kDa subunits and requiring 150 to 1000 μ*M* Ca^{2+} for half maximal activity

Amino acid sequence studies, done both by traditional Edman methods and derived from cDNA clones, have now clarified the structural relationships among these four forms of the Ca^{2+}-dependent proteinases. These studies, done largely by Suzuki and colleagues, have shown that autolysis of either μ- or m-calpain removes approximately 19 to 20 amino acids from the N-terminus of the 80-kDa subunit and 95 to 105 amino acids from the N-terminus of the 28-kDa subunit.

TABLE 1
Summary of Total Protein Yields, Specific Inhibitor Activity, Total Activity, Percentage Recovery, and Purification Factor at Each Step in Purification of the Ca^{2+}-Dependent Inhibitor (Calpastatin) from Porcine Skeletal Muscle[a]

Fraction	Total protein (mg)	Specific inhib. activity (units/mg)[b]	Total inhib. activity (units)[b]	% Activity recovered	Purification (-fold)
pH 4.9 supernatant	406,000	0.0213	8650	100	1
P_{0-50} ammonium sulfate	64,000	0.113	7230	83.6	5.3
33—55% acetone	15,500	0.218	3380	39.1	10.2
DEAE-cellulose	764	3.87	2960	34.2	182
CM-cellulose	154	15.2	2340	22.1	714
Hydroxyapatite	6.02	26.5	160	1.8	1244

[a] Figures are for a typical preparation beginning with 15,560 g of fresh porcine skeletal muscle.
[b] Expressed as a decrease in absorbance units (278 nm) of m-calpain activity using a casein substrate per mg of inhibitor protein.

The amino acid sequences of both the 80-kDa and the 28-kDa subunits of calpains purified from different species are highly homologous. Hence, the calpains seem to be highly conserved molecules. The physiological significance of autolysis is unclear and is under active investigation in a number of laboratories. These studies will be discussed in several of the chapters that follow in this monograph.

D. A PROTEIN INHIBITOR OF THE Ca^{2+}-DEPENDENT PROTEASE

One of the last things Bill Dayton and I did during our efforts to purify CAF in 1974-1975 was to construct a table summarizing the yields, recoveries, specific activities, and -fold purification of CAF at each one of the purification steps we used. As a graduate student, I had been taught that this was an essential part of any study reporting purification of an enzyme (although such tables seem to appear infrequently now). Bill immediately found late in 1975 that he could not detect CAF (m-calpain) activity in any of our initial fractions until after isoelectric precipation at pH 4.6.[6] Traditional enzyme biochemistry decrees that such a finding indicates the presence of an inhibitor. Assay of the pH 4.6 supernatant indeed indicated the presence of a calpain inhibitor in this fraction. I asked Dr. Akihiro Okitani, who had just joined my laboratory, to attempt to purify this inhibitor. He made considerable progress toward this purification in a short period of time, and we presented his results at the meeting of the Federation societies held in San Francisco in June 1976.[21] Some of his results are summarized in Tables 1 and 2 and Figures 4 and 5. Although Okitani learned that the protein inhibitor of m-calpain eluted between 50 and 100 m*M* KCl from a DEAE-cellulose column (Figure 4) and that this inhibitor was stable to pH (Figure 5) and temperature, the 1244-fold purification that he achieved was not sufficient to produce a purified protein. In August 1976, I left Iowa State University to accept the position of chair of a new department at the University of Arizona. As already indicated, Bill Dayton also left in 1976 for a position at the University of Minnesota, and Dr. Okitani returned to Japan in October 1976. Consequently, the group that had purified the Ca^{2+}-dependent proteinase and nearly purified its inhibitor disbanded, and one can only speculate what would have happened had they remained as a unit for several more years.

It required 3 years before I could reassemble a laboratory at the University of Arizona, and then with the help of another postdoctoral associate, Dr. Yuzuru Otsuka, we resumed the effort to purify the protein inhibitor of m-calpain, calpastatin. By 1980, we had succeeded in purifying calpastatin from bovine cardiac muscle and realized that it was a multiheaded inhibitor with a relative molecular weight of 110 kDa on SDS-polyacrylamide gels. We

TABLE 2
Effect of Trypsin or RNase A Treatment on Activity of the Ca^{2+}-Dependent Proteinase Inhibitor, Calpastatin[a]

Treatment	Remaining activity
Trypsin plus soybean trypsin inhibitor	100%
Trypsin	4.1%
RNase A at the end of the incubation	100%
RNase A	100%

[a] Conditions of the experiments: 8 μg of porcine skeletal muscle calpastatin was incubated for 30 min at 25°C with 100 μg trypsin (Sigma, bovine pancreas, twice crystallized) or with 100 μg RNase A (Sigma, pancreas) in 300 μl of 33 m*M* Tris-acetate, pH 7.0. Trypsin action was stopped by adding 200 μg of soybean trypsin inhibitor (Sigma). Controls were done by mixing soybean trypsin inhibitor and trypsin before incubating with calpastatin or by adding RNase A at the end of the 30-min incubation period.

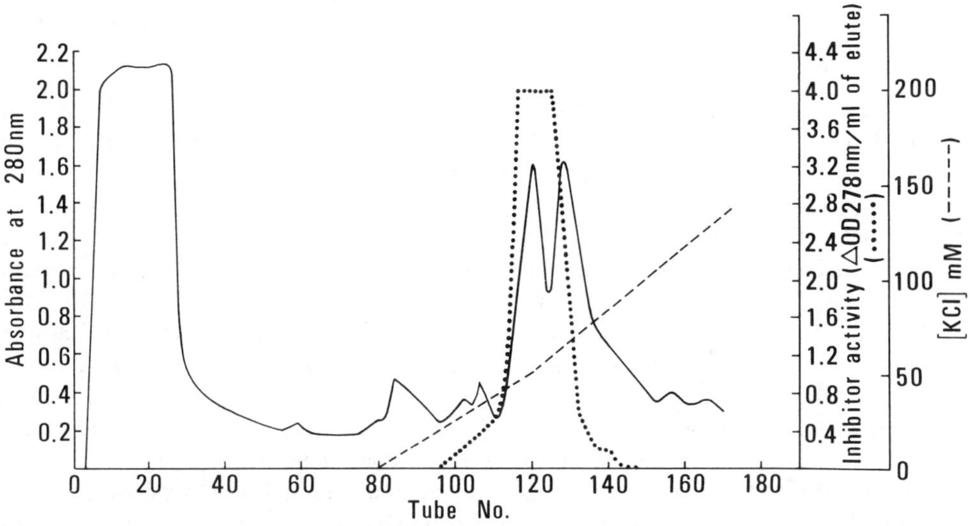

FIGURE 4. Elution profile of a crude acetone fractionated calpastatin fraction off a DEAE-cellulose column. Conditions: 6716 mg of a crude calpastatin fraction that had been precipitated between 33 and 55% acetone and dissolved in 20 m*M* Tris-acetate, pH 7.5, 1 m*M* EDTA, 14.26 m*M* 2-mercaptoethanol in a final volume of 243 ml were loaded onto a 2.5 × 42 cm column that had been equilibrated with the same buffer. The column was eluted with 600 ml of 20 m*M* Tris-acetate, pH 7.5, 1 m*M* EDTA, 14.26 m*M* 2-mercaptoethanol, followed by a linear gradient made of 700 ml each of 20 m*M* Tris-acetate, pH 7.5, 1.0 m*M* EDTA, 14.26 m*M* 2-mercaptoethanol, and this plus 200 m*M* KCl. The flow rate was 39 ml per tube, and fractions were collected at 20 min per tube.

published these results in abstract form,[22] but the press of administrative duties delayed completion and full publication of this work until some time later.[23] In the meantime, Takashi Murachi and his colleagues in Kyoto had been working on both the calpains and calpastatin, and a full report describing purified human erythrocyte calpastatin was published by Murachi in 1982.[24] Human erythrocyte calpastatin migrated with a relative molecular mass of 70 kDa on SDS-polyacrylamide gels, whereas the calpastatins from cardiac and skeletal muscle and liver migrated with a relative molecular mass of 105 to 110 kDa on SDS-polyacrylamide gels. These differences caused considerable concern during the period 1982 to 1987, and it was unclear whether some forms of calpastatin might be proteolytically degraded by the heating steps and denaturing solvents being used at that time to purify them. These questions

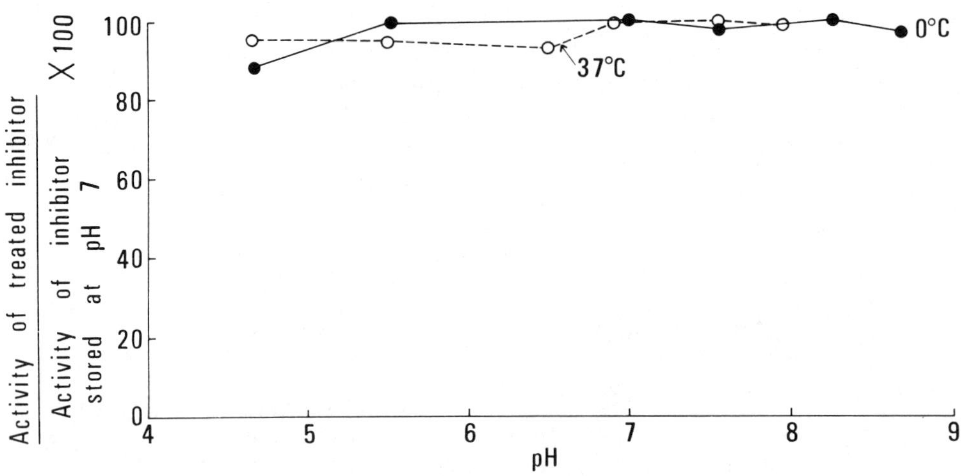

FIGURE 5. Activity of calpastatin after incubation at different pH values and two different temperatures. Conditions: 8 μg of calpastatin were incubated in 200 μl of 60 m*M* Tris-acetate at the designated pH for 22 hr at 0°C or for 2 hr at 37°C, and activity of calpastatin was assayed by measuring its ability to inhibit m-calpain digestion of casein.

were resolved when the cDNAs for calpastatin from pig[25,26] and rabbit[27,28] were cloned and sequenced. These studies showed that at least two kinds of calpastatin molecules exist; one having four repetitive domains of approximately 140 amino acids each with a N-terminal domain of approximately 150 amino acids; and a second having three repetitive domains of approximately 140 amino acids but without a 150-amino acid N-terminal domain. The first kind is found in liver and muscle cells and the second kind is found in erythrocytes.[29,30] More information on the structure of calpastatin may be found in Chapter 3 of this volume.

Thus, the Ca^{2+}-dependent proteinase system has in 13 years evolved from a single proteolytic enzyme to a three-protein system with the amino acid sequences and many of the physical properties of the three proteins known. The names, calpain and calpastatin, for these proteins were first introduced by Takashi Murachi at the 11th International Congress of Biochemistry held in Toronto, in July 1979, and these are now the names that have been approved by the Nomenclature Committee of the International Union of Biochemistry. Professor Murachi has provided an interesting summary of how he came to propose these names.[31] This proposal was a remarkable insight at the time, because the amino acid sequences and many of the other properties of the proteases were not known, and their relationship to the papain family of cysteine proteases was unclear. It seems likely that the extraordinary progress made during the last 13 years will be paralleled by an equally intensive effort during the next 13 years to learn the physiological function of the calpain system and how its activity is regulated in living cells.

II. CURRENT STATUS AND FUTURE OUTLOOK OF THE CA^{2+}-DEPENDENT PROTEINASE SYSTEM

During the last 8 years, it has become clear, both from the isolation and purification of the calpains and calpastatin from a number of different cell types and from immunohistochemical reaction with antibodies specific to these three proteins, that the Ca^{2+}-dependent proteinase or calpain system is widely distributed and exists in most, if not all, vertebrate animals.[32-34] A proteolytic system related to the mammalian calcium-dependent system recently has been detected in drosophila heads,[35] but an attempt to find calpains or calpastatin in a plant species failed.[36] The amino acid sequences of both μ- and m-calpain, and of calpastatin are highly conserved among the species analyzed thus far, with approximately 90% sequence homology among the different species studied (chicken, rabbit, pig, and human). This high degree of homology and the almost ubiquitous presence of the calpain system suggest that this system has a fundamentally important role in vertebrate (and perhaps other animal) cells. The nature of this role, and how activity of the calpain system is regulated *in vivo*, however, remains an enigma. Because the calpains rapidly degrade many of the cytoskeletal proteins, it has been suggested that they function in turnover or remodeling of the cell cytoskeletal proteins such as occurs during cell division or platelet aggregation. Cytoskeletal remodeling may be envisioned as a broader based version of the original hypothesis that the calpains were involved in myofibrillar protein turnover.[4,6-8] The calpains, however, do not degrade native actin (they do degrade denatured actin and this has caused some confusion about their effects on actin) and degrade native α-actinin, another important cytoskeletal protein, very slowly if at all. Consequently, if the calpain system is involved in cytoskeletal protein turnover, it has a specialized role that does not involve all the cytoskeletal proteins. The calpains also rapidly degrade a number of hormone receptors, such as the progesterone, estrogen, and EGF receptors, and a number of kinases, such as phosphorylase kinase (the original report of its activity[3]) protein kinase C, myosin light chain kinase, and a calmodulin-dependent phosphatase. Murachi[31] has grouped the possible substrates of the calpains into three categories: (1) cytoskeletal proteins, especially those involved in cytoskeletal/membrane interaction; (2) hormone receptors: and (3) enzymes, especially kinases and phosphatases. At the present time, it is not easy to envision a common function involving all three of these categories of substrates. Because a protein is rapidly degraded *in vitro* does not, of course, indicate that the protein is a physiological substrate of the calpains *in vivo*. Consequently, the physiological substrates and function of the calpain system are very unclear at the present time, regardless of the number of hypotheses that have been proposed.

A. THE MEMBRANE/ACTIVATION THEORY FOR REGULATION OF CALPAIN ACTIVITY

Regulation of the proteolytic activity of the calpains *in vivo* also is a mystery. Of the four forms of the calpains that might exist in cells, only autolyzed μ-calpain has a Ca^{2+} requirement similar to the free Ca^{2+} concentration of 200 to 1000 n*M* that exists in living, normal cells.[33] The other three forms of the calpains all require Ca^{2+} concentrations much higher than 1 μ*M* for even half maximal activity: m-calpain requires 150 to 1000 μ*M* Ca^{2+} for half maximal activity; autolyzed m-calpain requires 10 to 300 μ*M* Ca^{2+}; and μ-calpain requires 5 to 50 μ*M* Ca^{2+} for half maximal activity. Although autolysis lowers the Ca^{2+} requirement for both μ- and m-calpain,[19,20,37] the Ca^{2+} concentration required for autolysis also is much higher that the *in vivo* Ca^{2+} concentration; 150 to 1000 μ*M* Ca^{2+} is required for a half maximal rate of autolysis of m-calpain and 125 to 200 μ*M* Ca^{2+} for μ-calpain.[37] A possible solution to this dilemma was offered by the discovery that phosphatidylinositol lowers the Ca^{2+} concentration required for a half maximal rate of autolysis of bovine aorta m-calpain from 680 μ*M* to 87 μ*M*.[38] This finding eventually led to formulation of what may be called the membrane/activation theory for regulation of calpain activity.[39,40] This theory proposes that the unautolyzed calpains, as they exist in the cell cytoplasm with 80-kDa and 28-kDa subunits, are inactive "proenzymes" and become active proteolytically only after they have been autolyzed to 78 (or 76 for μ-calpain)-kDa and 18-kDa subunits. Activation by proteolytic cleavage is analogous to activation of trypsinogen or prothrombin, where a specific proteolytic cleavage converts the inactive zymogen form of these proteases into their active forms. In the calpain example, however, the proteolytic cleavage is autocatalytic rather than from another protease. The membrane/activation theory proposes that autocatalytic activation of the calpains does not (cannot) occur until the calpain molecule attaches to a phospholipid (phosphatidylinositol)-binding site on the cell membrane. Binding to the membrane-phospholipid is proposed to lower the Ca^{2+} concentration required for autolysis sufficiently to permit such autolysis to occur at physiological concentrations of free Ca^{2+}. The autolyzed enzymes then either are released into the cytosol where, with their reduced Ca^{2+} requirement, they are now able to hydrolyze their substrate proteins, or they remain associated with the cell membrane and degrade cytoskeletal proteins involved in interaction of the cell cytoskeleton with the plasma membrane. Activity of the autolyzed calpains is then subject to a final regulation by calpastatin.

An attractive feature of the membrane/activation theory is that it offers a number of experimentally testable concepts. Rozanov and Mellgren[41] have recently shown that a number of phospholipids, including phosphatidylserine, phosphatidylethanolamine, and phosphatidylcholine, all lower the concentration required for autolysis of m-calpain, although phosphatidylinositol is most effective. Consequently, several types of membrane-binding sites could be involved in activating the calpains. A fundamental problem remains, however, because, even in the presence of phosphatidylinositol, the Ca^{2+} concentration required for autolysis is at least five to ten times higher than the free Ca^{2+} concentrations that exist in cells. It may be suggested that transient pulses of high Ca^{2+} concentrations occur in cells or that Ca^{2+} concentrations near Ca^{2+} channels might momentarily reach 1 to 10 μM. Studies of cytosolic Ca^{2+} transients using fura 2-injected cells, however, have failed to find intracellular Ca^{2+} concentrations higher than 900 nM, even near Ca^{2+} channels.[42,43] Although these measurements may not have been sufficiently sensitive to detect very rapid Ca^{2+} pulses of high concentrations in very localized areas of cells, the experimental evidence, thus far, does not indicate the existence of pulses of high Ca^{2+} concentration in cells (e.g., proteins such as calmodulin and troponin C, which are known to have Ca^{2+}-activated regulatory roles, have Ca^{2+} binding constants of 10^7 to 10^6 M^{-1}).

Because of these differences between intracellular free Ca^{2+} concentration and the Ca^{2+} concentration required for calpain activity or autolysis, we initiated a careful study into the role of autolysis in activity of μ- and m-calpain from bovine skeletal muscle.[37] These studies showed that μ-calpain required 40 μM Ca^{2+} for half maximal proteolytic activity, but required 200 μM Ca^{2+} for a half maximal rate of autolysis. Phosphatidylinositol lowered the Ca^{2+} concentration required for a half maximal rate of autolysis of μ-calpain from 200 to 140 μM. Hence, unautolyzed μ-calpain clearly is an active protease and not a proenzyme; indeed the high concentration of Ca^{2+} required for autolysis of μ-calpain, even in the presence of phosphatidylinositol, raises questions about why μ-calpain should need to be autolyzed. On the o ther hand, m-calpain required 700 μM Ca^{2+} for half maximal proteolytic activity, 720 μM Ca^{2+} for a half maximal rate of autolysis in the absence of phosphatidylinositol, and 380 μM Ca^{2+} for a half maximal rate of autolysis in the presence of phosphatidylinositol. Recent kinetic studies done in our laboratory show that unautolyzed m-calpain also is an active protease, although it autolyzes rapidly at any Ca^{2+} concentration that activates it. Consequently, the calpains are not inactive proenzymes, and there is no need to propose that autolysis is needed to activate them. Autolysis of the calpains occurs so rapidly and so specifically, however, that it is sagacious to suspect that it has some physiological significance. For example, autolysis could be involved in regulation of calpain activity by other, as yet unknown controls, or it could be the first step in degradation and turnover of the calpain molecules.

One regulation of the calpains that might be affected by autolysis is its interaction with calpastatin. It has been shown that less Ca^{2+} (530 μM) is required for calpastatin to bind to bovine heart m-calpain than is required for half maximal activity of the same m-calpain (920 μM)[23] Therefore, m-calpain could never be active if it exists in the same cellular locations as calpastatin, even if Ca^{2+} concentrations became high enough to activate it, because calpastatin would bind to and inactivate it before it could function proteolytically.

Could autolysis be a means to alter the Ca^{2+} regulation of the calpain/calpastatin interaction? We have recently studied the effects of Ca^{2+} on binding of all four forms of the calpains to calpastatin, using proteins purified from bovine skeletal muscle. Half maximal binding of autolyzed m-calpain to calpastatin occurs at 24 μM Ca^{2+}, but half maximal proteolytic activity of autolyzed μ-calpain requires 94 μM Ca^{2+}. In agreement with our earlier studies on bovine cardiac muscle m-calpain and calpastatin,[23] half maximal binding of unautolyzed m-calpain to calpastatin requires 400 μM Ca^{2+}, but half maximal activity of unautolyzed m-calpain requires 760 μM Ca^{2+}. Half maximal binding of unautolyzed μ-calpain to calpastatin requires 42 μM Ca^{2+} and half maximal activity of unautolyzed μ-calpain requires 34 μM Ca^{2+}. Half maximal binding of autolyzed μ-calpain to calpastatin requires only 42 nM Ca^{2+}, and half maximal activity of autolyzed μ-calpain requires 600 nM Ca^{2+}. Hence, autolysis does not alter the Ca^{2+}-induced control of calpain activity by calpastatin. Indeed, of the four forms of the calpains, only unautolyzed μ-calpain could ever be active in living cells unless the calpains were located in areas of the cell that did not contain calpastatin, or the calpastatin/calpain interaction was subject to some kind of regulation, such as phosphorylation, that would alter the Ca^{2+} requirement for calpain activity or for the calpastatin/calpain interaction.

It has been suggested that neither μ-calpain[44] nor m-calpain[41] is subject to inhibition by calpastatin if they are associated with erythrocyte membranes. If membrane-association is the only means for the calpains to escape calpastatin inhibition, however, the calpains, even in their autolyzed forms, could never be active in the cell cytoplasm unless calpastatin was absent. Therefore, to understand how activity of the calpain system is regulated, it is important to know in some detail the relative locations of the calpains and calpastatin in cells, and whether the calpains are preferentially located near the plasma membrane as would be predicted by the membrane/activation theory.

B. LOCALIZATION OF THE CALPAINS AND CALPASTATIN

After the initial subcellular fractionation studies showing that m-calpain was not a lysosomal enzyme but was either free in the cell cytoplasm or was associated with myofibrils or the plasma membrane in muscle cells,[8] a large number of studies have described immunolocalization of the calpains and calpastatin. All these studies agree that both μ- and m-calpain and calpastatin are found exclusively inside cells; hence, the calpains do not function extracellularly where the Ca^{2+} concentration is high enough to activate them. Some μ- and m-calpain and calpastatin were found extracellularly in muscle tissue from starved or denervated rats,[45] but creatine kinase assays showed that this extracellular location was due to "leakage" of these proteins, probably through damaged areas of the muscle cell plasma membrane rather than to secretion of these proteins from the muscle cells. Despite the substantial number of immunolocalization studies of μ-calpain, m-calpain, or calpastatin, only three studies have described immunolocalization of all three proteins in the same tissue.[45-47] Also, only two studies have done immunolocalization at the electron microscope level. In general, the immunolocalization studies have found that μ-calpain, m-calpain and calpastatin all are widely distributed throughout the cytoplasm in a variety of cells. Some early studies suggested that m-calpain was localized at the Z-disk of striated muscle cells,[48] but more recent quantitative immunogold studies indicate that both μ- and m-calpain and calpastatin are located in all regions of the myofibril and the remainder of the muscle cell,[45] but that both calpains and calpastatin are approximatley two times more prevalent at the Z-disk than in the A- or I-band areas of the myofibril. This higher concentration in the Z-disk may have given the impression, in fluorescence localization, that the calpains or calpastatin were located exclusively at the Z-disk.[48] Several studies using immunofluorescence localization or immunoperoxidase at the light microscope level have reported some preferential localization of m-calpain near the plasma membrane of human skeletal muscle[48] or of myoblasts treated with A23187, a Ca^{2+} ionophore,[49] or just before fusion.[50] Preferential localization of m-calpain near the plasma membrane has also been reported in PtK_1 cells in metaphase or interphase.[51] In all these instances, preferential membrane localization was observed with m- but not μ-calpain. Other investigators, however, find no association of either μ- or m-calpain with the cytoskeleton in PK 15 cells, and no relocation of either calpain in response to A23187.[52]

We recently initiated a careful study, using both immunogold localization at the electron microscope level and immunofluorescence at the light microscope level, to localize μ- and m-calpain and calpastatin in the same tissue.[33,45] Our studies thus far have been done only on rat soleus muscle. Using quantitative immunogold localization, we found, as indicated earlier, that the apparent concentration of both μ- and m-calpain and calpastatin is two times greater at the Z-disk than at the I- or A-band areas of myofibrils. Both calpains and calpastatin are widely distributed throughout the interior of muscle cells from starved or denervated animals. Hence, the stress caused by starvation or denervation does not alter the intracellular distribution of the calpains or calpastatin. Occasionally, some preferential localization of m-calpain was observed near the plasma membrane (Figure 6), but even in such instances, most of the m-calpain (over 80% of total m-calpain in the cells) was distributed throughout the interor of the cell.

We also used fluorescence microscopy in an attempt to assess preferential membrane localization of the calpains. Our subjective evaluation after examination of several hundred samples taken from several dozen rats is that m-calpain and μ-calpain are preferentially localized near the plasma membrane in approximately 30 to 40% and 10 to 20%, respectively, of the cells examined, but that calpastatin is only rarely located preferentially near the plasma membrane[33] (Figures 7 to 9). It is important to note that, because of the way fluorescence micrographs are sometimes photographed, only the higher fluorescence intensity near the plasma membrane may be detected in photographs, whereas in fact, the remainder of the cell also is fluorescent, although at a lower intensity (see Figures 7 and 8). Consequently, even in those cells that exhibit preferential membrane localization of μ- or m-calpain, most of the total calpain is distributed throughout the cell because of the much larger volume of the interior of the cell compared with the volume near the plasma membrane. We estimate conservatively that, even in those cells that exhibit preferential membrane localization of the calpains (which is only 30 to 40% of all cells for m-calpain), over 80% of total cellular calpain is distributed throughout the interior of the cell.

C. SUMMARY

The experimental results obtained during the last several years and summarized in the two preceding sections indicate that the membrane/activation theory cannot explain at least three important properties of the calpain system:

1. All the calpains except autolyzed μ-calpain require for proteolytic activity Ca^{2+} concentrations significantly higher than those that exist *in vivo*. Autolysis, even in the presence of a phospholipid such as phosphatidylinositol, requires Ca^{2+} concentrations 5 to 100 times greater than exist in normal cells. There is no evidence that pulses of Ca^{2+} concentration high enough to activate even the small fraction of the calpains that might be exposed to such pulses ever occur in living cells. There must be some as yet unknown mechanism to lower the Ca^{2+} requirement of the calpains.
2. Because the Ca^{2+} requirement for the calpain/calpastatin interaction is less than the Ca^{2+} requirement for proteolytic activity of that calpain, all the calpains except μ-calpain would bind to and be immediately inactivated by calpastatin, even if they did ever encounter Ca^{2+} concentrations high enough to activate them. Extensive immunolocalization studies show that the calpains and calpastatin colocalize in cells, so there are no large intracellular areas where the calpains are not subject to inhibition

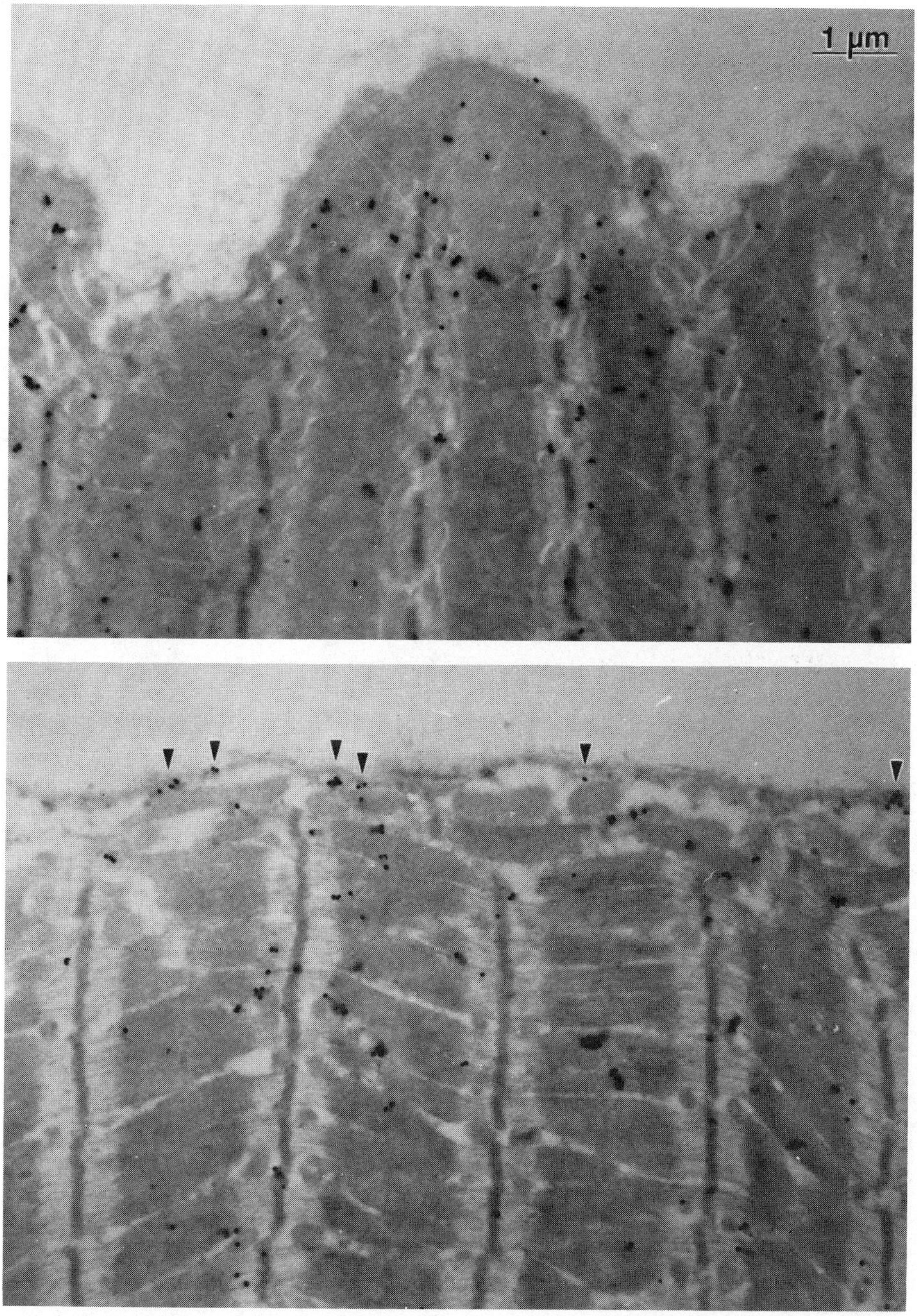

FIGURE 6. Electron micrographs of rat soleus muscle that had been embedded in LR-white resin and then incubated with monoclonal antibodies specific for the 80-kDa subunit of μ-calpain (upper) or of m-calpain (lower). Goat anti-mouse secondary antibodies coupled with 15 n*M* colloidal gold particles were used to detect the sites of antibody binding. Colloidal gold particles are evenly distributed throughout the muscle cells in the samples incubated with either anti-μ-calpain or anti-m-calpain antibodies, indicating that both these proteins are widespread in muscle cells. It is difficult to discern the higher concentration of colloidal gold particles at the Z-disk in single micrographs, but counts on many micrographs indicated that the density of colloidal gold particles was two times higher at the Z-disk than in the A-or I-band areas of myofibrils or in nuclei or mitochondria. In some instances, patches of colloidal gold particles (arrows) were observed at the plasma membrane in cells that had been incubated with anti-m-calpain but not in cells incubated with anti-μ-calpain. These patches of membrane-associated colloidal gold particles were not observed in all cells, but indicate that occasionally m-calpain is clustered near the plasma membrane in skeletal muscle cells.

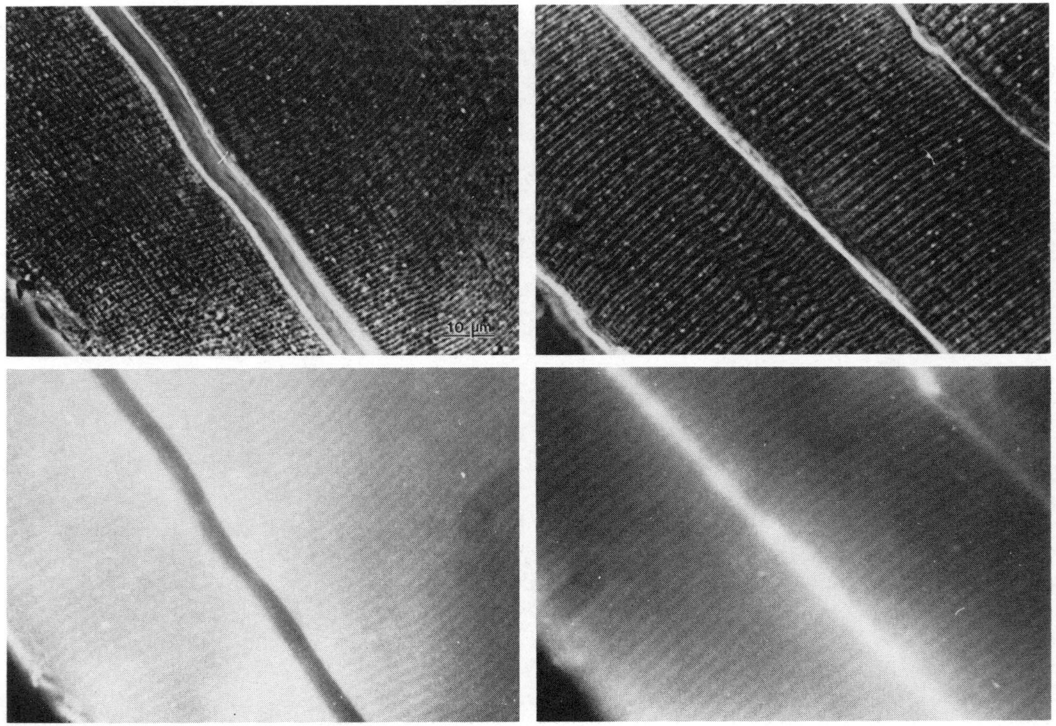

FIGURE 7. Phase contrast and fluorescence micrographs showing distribution of μ-calpain in rat soleus muscle. Muscle sections were incubated with a monoclonal antibody specific for the 80-kDa subunit of μ-calpain followed by reaction with a goat anti-mouse secondary antibody coupled to fluorescein isothiocyanate. Upper micrographs are phase contrast images of two sections and the lower micrographs are fluorescence images of the same two sections. In most instances, μ-calpain is widely distributed through the cell with greater fluorescence intensity in the I-band than in the A-band regions (lower left). In approximately 10 to 20% of the cells, however, fluorescence intensity was greater at the periphery than in the interior of the cell (lower right). Note that, even in those cells having high fluorescence intensity at the periphery of the cell, the remainder of the cell also fluoresces and displays the characteristic periodicity of high fluorescence in the I-bands.

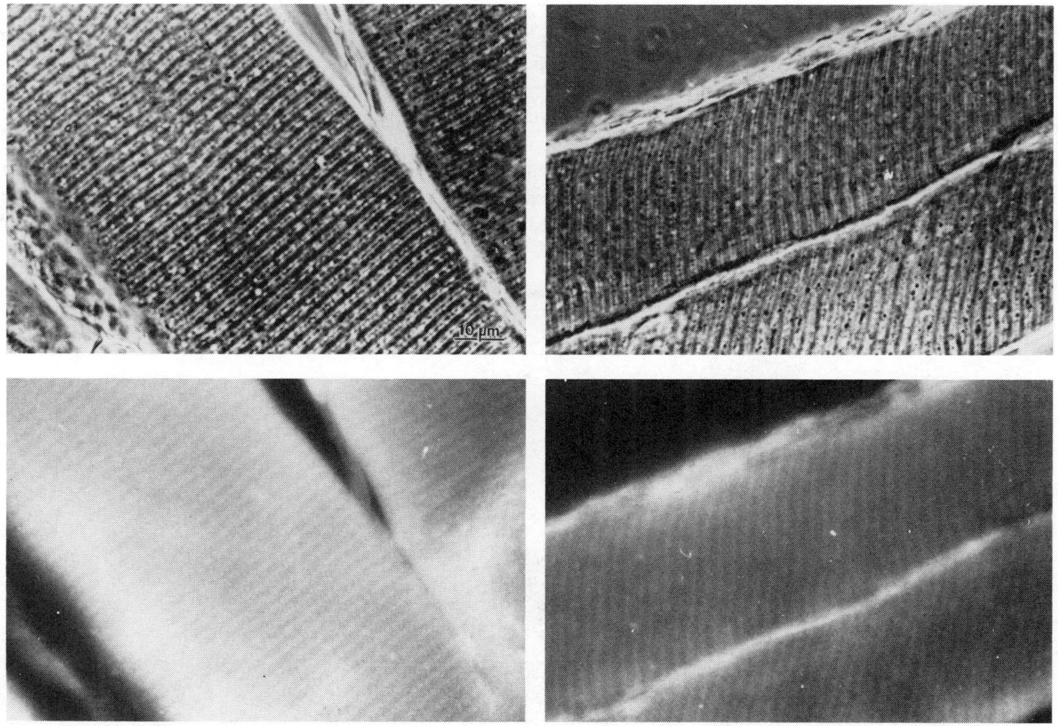

FIGURE 8.. Phase contrast and fluorescence micrographs showing distribution of m-calpain in rat soleus muscle. Muscle sections were incubated with a monoclonal antibody specific for the 80-kDa subunit of m-calpain, followed by reaction with a secondary antibody as described in the legend to Figure 7. Upper micrographs are phase contrast images of two sections, and the lower micrographs are fluorescence images of the same two sections. The distribution of m-calpain in rat soleus muscle is identical to the distribution of μ-calpain, at least as it can be determined by fluorescence microscopy; periodicity due to greater fluorescence intensity in the I-band than in the A-band regions is observed in sections incubated with anti-m-calpain antibodies (lower left and right), just as in sections incubated with anti-μ-calpain antibodies. This fluorescence intensity in the I-band may be due to the high concentration of both calpains in the Z-disk, as determined by colloidal gold localization. In approximately 30 to 40% of the cells examined, some preferential localization of m-calpain was observed near the plasma membrane of muscle cells (lower right), but even in these cells having preferential membrane localization of m-calpain, the remainder of the cell also contains m-calpain. Membrane localization of either μ- or m-calpain seemed to occur in some cells and not in others, rather than occurring at part of the membrane in one cell and then not in the remainder of the membrane in the same cell.

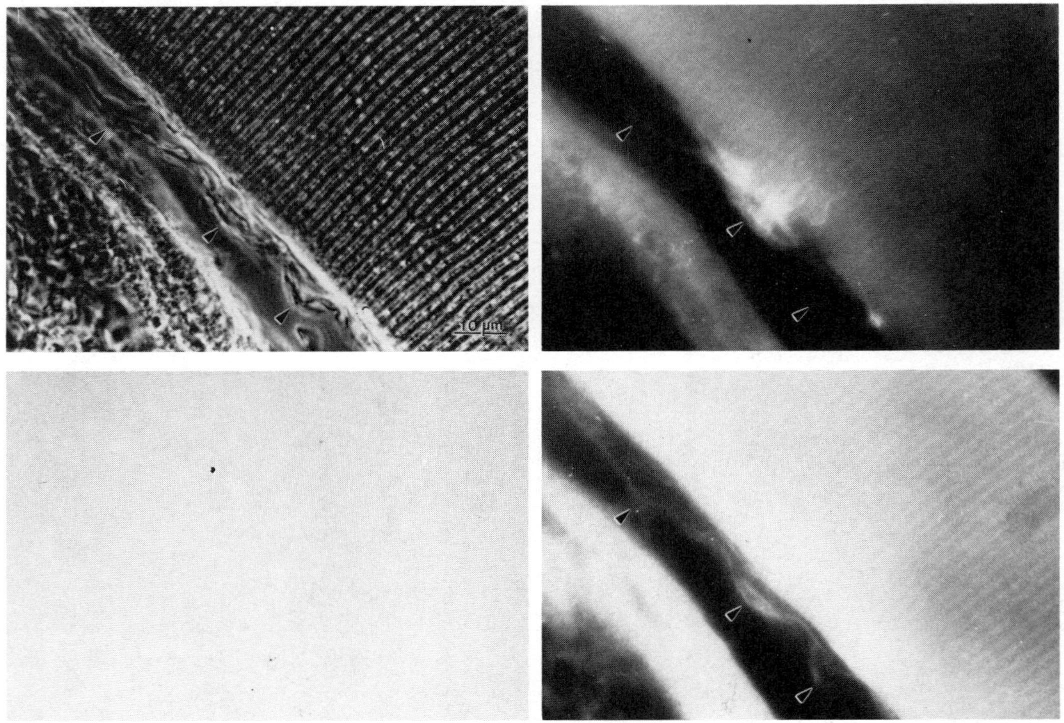

FIGURE 9. Phase contrast and fluorescence micrographs showing distribution of calpastatin and m-calpain in rat soleus muscle. Muscle sections were incubated with a polyclonal antibody specific for calpastatin, followed by incubation with a goat anti-rabbit secondary antibody coupled to tetramethylrhodamine isothiocyanate. After washing, the section was incubated with a monoclonal antibody specific for the 80-kDa subunit of m-calpain, followed by incubation with a goat anti-mouse secondary antibody coupled to fluorescein isothiocyanate. Upper left is a phase contrast image of a muscle section, the upper right is a fluorescence image of the same sample incubated with anti-m-calpain, and the lower right is a fluorescence image of the same sample incubated with anti-calpastatin. This cell is one of the 30 to 40% of rat soleus muscle cells that exhibits some preferential peripheral localization of m-calpain (upper right), but calpastatin is widely distributed throughout the interior of this same cell with no indication of any peripheral localization. Note that calpastatin also seems to be periodically distributed in muscle cells, with a greater fluorescence intensity in the I-band area than in the A-band. Both calpastatin and m-calpain are located in cells that exist near the larger striated muscle cell (arrows).

by calpastatin. Either the caplain/calpastatin interaction is regulated in some as yet unknown way, or a mechanism for lowering the Ca^{2+} requirement for proteolytic activity of the calpains exists.

3. Immunolocalization results consistently show that the calpains and calpastatin are widely distributed throughout the cell cytoplasm and that only a small proportion, at most, of the total cellular calpain or calpastatin content is preferentially located near the plasma membrane. Why would the cell store so much cytosolic calpain if it is doomed to inactivity?

A number of results now suggest that activity of μ- and m-calpain is regulated by two different mechanisms, although these two proteinases may eventually degrade the same substrates.

1. Autolysis of μ-calpain requires significantly higher Ca^{2+} concentrations that proteolytic activity of μ-calpain, whereas the Ca^{2+} concentration required for autolysis of m-calpain is very similar to the Ca^{2+} concentration required for its activity.[37]
2. Phosphatidylinositol lowers the specific activity of autolyzed μ-calpain to 10% of its original activity but has no effect on the specific activity of autolyzed m-calpain.[37]
3. Monoclonal antibodies specific for the 28-kDa subunit that is common to both μ- and m-calpain lower the Ca^{2+} concentration required for autolysis of m-calpain by three to four-fold but do not affect the Ca^{2+} concentration required for autolysis of μ-calpain.[53]
4. The effect of Ca^{2+} on binding of the calpains to calpastatin differs significantly between μ- and m-calpain; m-calpain requires less Ca^{2+} for binding to calpastatin than for proteolytic activity, whereas μ-calpain requires less Ca^{2+} for proteolytic activity than for binding to calpastatin.

Clearly, many important questions about the function and regulation of the calpain system remain unanswered. Recent results with monoclonal antibodies provide some indication that the 28-kDa subunit, which has no catalytic site and whose function remains unknown, does in fact affect the Ca^{2+} requirement[53] and specific activity[54] of the catalytic 80-kDa subunit of m- but probably not μ-calpain. Because the 28-kDa subunits of μ- and m-calpain are identical, these findings indicate that the interaction of the 28-kDa subunit with 80-kDa subunit of m-calpain differs from its interaction with the 80-kDa subunit of μ-calpain.

Ca^{2+} has a central role in activity of the calpain system. Slightly different Ca^{2+} concentrations are required for proteolytic activity of the calpains, for autolysis of the calpains, and for the calpain/calpastatin interaction, suggesting that different sets of Ca^{2+} binding sites are involved in each of these three activities. It may be that Ca^{2+} is also involved in the as yet unknown mechanism that must exist to lower the Ca^{2+} concentration required for activity of the calpains. Microinjecting (and hence overloading) Ca^{2+} into PtK_1 cells[51] or incubating erythrocytes with A23187 in the presence of a membrane-mobility agent A_2C and Ca^{2+} (which also would overload Ca^{2+} into the erythrocyte)[55] activates m-calpain and μ-calpain activity, respectively, in these two experimental systems. Why such Ca^{2+} overload would negate the calpastatin control and activate the calpains is currently a mystery.

ACKNOWLEDGMENTS

We thank Janet Christner for her excellent help and maintaining a cheerful disposition while preparing this manuscript, Drs. William R. Dayton, R. M. Robson, and M. H. Stromer for their candid comments about one of the authors, (DEG) recollections of past events, and Ann Peterson and Kathy McGee for their skillful assistance with some of the experimental results described in this chapter. We also gratefully acknowledge Professor Takashi Murachi's helpful comments concerning this manuscript. The research that was described in this chapter and that was done in the authors' laboratory was supported by grants from the National Institutes of Health (HL-20984), the USDA Competitive Grants Program (85-CRCR-1-1818 and 87-CRCR-1-2283), and the Arizona Agriculture Experiment Station, Project No. 28, a contributing project to USDA Regional Project NC-131. J. Cong was Postdoctoral Fellow of the American Heart Association, Arizona Affiliate.

REFERENCES

1. **Busch, W. A., Stromer, M. H., Goll, D. E., and Suzuki, A.,** Ca^{2+}-specific removal of Z-lines from rabbit skeletal muscle, *J. Cell Biol.,* 52, 367, 1972.
2. **Goll, D. E., Arakawa, N., Stromer, M. H., Busch, W. A., and Robson, R. M.,** Chemistry of muscle proteins as a food, in *Physiology and Biochemistry of Muscle As a Food,* Vol. 2, Briskey, E. J., Cassens, R. G., and Marsh, B. B., Eds., University of Wisconsin Press, Madison, WI, 1970, 755.
3. **Huston, R. B. and Krebs, E. G.,** Activation of skeletal muscle phosphorylase kinase by Ca^{2+}. II. Identification of the kinase activating factor as a proteolytic enzyme, *Biochemistry,* 7, 2116, 1968.
4. **Dayton, W. R., Goll, D. E., Stromer, M. H., Reville, W. J., Zeece, M. G., and Robson, R. M.,** Some properties of a Ca^{2+}-activated protease that may be involved in myofibrillar protein turnover, in *Cold Spring Harbor Conf. Cell Proliferation, Vol. 2, Proteases and Biological Control,* Reich, E., Rifkin, D. B., and Shaw, E., Eds., Cold Spring Harbor Laboratory, Cold Spring Harbor, New York, 1975, 551.
5. **Guroff, G.,** A neutral, calcium-activated proteinase from the soluble fraction of rat brain, *J. Biol. Chem.,* 239, 149, 1964.
6. **Dayton, W. R., Goll, D. E., Zeece, M. G., Robson, R. M., and Reville, W. J.,** A Ca^{2+}-activated protease possibly involved in myofibrillar protein turnover. Purification from porcine muscle, *Biochemistry,* 15, 2150, 1976.
7. **Dayton, W. R., Reville, W. J., Goll, D. E., and Stromer, M. H.,** A Ca^{2+}-activated protease possibly involved in myofibrillar protein turnover. Partial characterization of the purified enzyme, *Biochemistry,* 15, 2159, 1976.
8. **Reville, W. J., Goll, D. E., Stromer, M. H., Robson, R. M., and Dayton, W. R.,** A Ca^{2+}-activated protease possibly involved in myofibrillar protein turnover. Subcellular localization of the protease in porcine skeletal muscle, *J. Cell Biol.,* 70, 1976.
9. **Mellgren, R. L.,** .Canine cardiac calcium-dependent proteases: resolution of two forms with different requirements for calcium, *FEBS Lett.,* 109, 129, 1980.
10. **Dayton, W. R., Schollmeyer, J. V., Lepley, R. A., and Cortes, L. R.,** A calcium-activated protease possibly involved in myofibrillar protein turnover. Isolation of a low-calcium-requiring form of the protease, *Biochim. Biophys. Acta,* 659, 48, 1981.
11. **Szpacenko, A., Kay, J., Goll, D. E., and Otsuka, Y.,** A different form of the Ca^{2+}-dependent proteinase activated by micromolar levels of Ca^{2+}, *Proc. Symp. Proteinases and Their Inhibitors: Structure, Function, and Applied Aspects,* Turk, V. and Vitale, Lj., Eds., Pergamon Press, Oxford, 1981, 151.
12. **Ohno, S. Emori, Y., Imajoh, S., Kawasaki, H., Kisaragi, M., and Suzuki, K.,** Evolutionary origin of a calcium-dependent protease by fusion of genes for a third protease and a calcium-binding protein?, *Nature,* 312, 566, 1984.
13. **Sakihama, T., Kakidani, H., Zenita, K., Yumoto, N., Kikuchi, T., Sasaki, T., Kannagi, R., Nakaniski, S., Ohmori, M., Takai, K., Titani, K., and Murachi, T.,** A putative Ca^{2+}-binding protein: structure of the light subunit of porcine calpain elucidated by molecular cloning and protein sequence analysis, *Proc. Natl. Acad. Sci. U.S.A.,* 82, 6075, 1985.
14. **Aoki, K., Imajoh, S., Ohno, S., Emori, Y., Koike, M., Kosaki, G., and Suzuki, K.,** Complete amino acid sequence of the large subunit of the low-Ca^{2+}-requiring form of human Ca^{2+}-activated neutral protease (μ-CANP) deduced from its cDNA sequence, *FEBS Lett.,* 205, 313, 1986.
15. **Emori, Y., Kawasaki, H., Imajoh, S., Kawashima, S., and Suzuki, K.,** Isolation and sequence analysis of cDNA clones for the small subunit of rabbit calcium-dependent protease, *J. Biol. Chem.,* 261, 9472, 1986.
16. **Imajoh, S., Aoki, K., Ohno, S., Emori, Y., Kawasaki, H., Sugihara, H., and Suzuki, K.,** Molecular cloning of the cDNA for the large subunit of the high Ca^{2+}-requiring form of human Ca^{2+}-activated neutral protease, *Biochemistry,* 27, 8122, 1988.

17. **Suzuki, K., Tsuji, S., Ishiura, S., Kimura, Y., Kubota, S., and Imahori, K.,** Autolysis of calcium-activated neutral protease of chicken skeletal muscle, *J. Biochem.*, 90, 1787, 1981.
18. **Hathaway, D. R., Werth, D. K., and Haeberle, J. R.,** Limited autolysis reduces the Ca^{2+}-requirement of a smooth muscle Ca^{2+}-activated protease, *J. Biol. Chem.*, 257, 9072, 1982.
19. **Dayton, W. R.,** Comparison of low- and high-calcium-requiring forms of the calcium-activated protease with their autocatalytic breakdown products, *Biochim. Biophys. Acta* 709, 166, 1982.
20. **Nagainis, P. A., Sathe, S. K., Goll, D. E., and Edmunds, T.,** Autolysis of high-Ca^{2+} and low-Ca^{2+} forms of the Ca^{2+}-dependent proteinase from bovine skeletal muscle, *Fed. Proc.*, 42, 1780, 1983.
21. **Okitani, A., Goll, D. E., Stromer, M. H., and Robson, R. M.,** Intracellular inhibitor of a Ca^{2+}-activated protease involved in myofibrillar protein turnover, *Fed. Proc.*, 35, 1746, 1976.
22. **Otsuka, Y. and Goll, D. E.,** Purification of the Ca^{2+}-activated-protease-inhibitor from bovine cardiac muscle, *Fed. Proc.*, 39, 2044, 1980.
23. **Otsuka, Y. and Goll, D. E.,** Purification of the Ca^{2+}-dependent proteinase inhibitor from bovine cardiac muscle and its interaction with the millimolar Ca^{2+}-dependent proteinase, *J. Biol. Chem.*, 262, 5839, 1987.
24. **Takano, E. and Murachi, T.,** Purification and some properties of human erythrocyte calpastatin, *J. Biochem.*, 92, 2021, 1982.
25. **Takano, E., Maki, M., Hatanaka, M., Mori, H., Zenita, K., Sakihama, T., Kannagi, R., Marti, T., Titani, K., and Murachi, T.,** Evidence for the repetitive domain structure of pig calpastatin as demonstrated by cloning of complementary DNA, *FEBS Lett.*, 208, 199, 1986.
26. **Takano, E., Maki, M., Mori, H., Hatanaka, M., Marti, T., Titani, K., Kannagi, R., Ooi, T., and Murachi, T.,** Pig heart calpastatin: identification of repetitive domain structures and anomalous behavior in polyacrylamide gel electrophoresis, *Biochemistry*, 27, 1964, 1988.
27. **Emori, Y., Kawasaki, H., Imajoh, S., Imahori, K., and Suzuki, K.,** Endogenous inhibitor for calcium-dependent cysteine protease contains four internal repeats that could be responsible for its multiple reactive sites, *Proc. Natl. Acad. Sci. U.S.A.*, 84, 3590, 1986.
28. **Emori, Y., Kawasaki, H., Imajoh, S., Minami, Y., and Suzuki, K.,** All four repeating domains of the endogenous inhibitor for calcium-dependent protease independently retain inhibitory activity, *J. Biol. Chem.*, 263, 2364, 1988.
29. **Imajoh, S., Kawasaki, H., Emori, Y., and Suzuki, K.,** Calcium-activated neutral protease inhibitor from rabbit erythrocytes lacks the N-terminal region of the liver inhibitor but retains three inhibitory units, *Biochem. Biophys. Res. Comm.*, 146, 630, 1987.
30. **Takano, E., Kitahara, A., Sasaki, T., Kannagi, R., and Murachi, T.,** Two different molecular species of pig calpastatin. Structural and functional relationship between 107 kDa and 68 kDa molecules, *Biochem. J.*, 235, 97, 1986.
31. **Murachi, T.,** Intracellular regulatory system involving calpain and calpastatin, *Biochem. Int.*, 18, 263, 1989.
32. **Goll, D. E., Kleese, W. C., Sloan, D. A., Shannon, J. D., and Edmunds, T.,** Properties of the Ca^{2+}-dependent proteinases and their protein inhibitor, *Cienc. Biol.*, 11, 75, 1986.
33. **Goll, D. E., Kleese, W. C., Kumamoto, T., Cong, J., and Szpacenko, A.,** In search of the function and regulation of the Ca^{2+}-dependent proteinases (calpains), in *Intracellular Proteolysis; Mechanisms and Regulation* Katunuma, N. and Kominami, E., Eds., Japanese Scientific Societies Press, Tokyo, 1989, 82.
34. **Murachi, T.,** Intracellular Ca^{2+} protease and its inhibitor protein: calpain and calpastatin, in *Calcium and Cell Function,* Vol. 4, Cheung, W. Y. Ed., Academic Press, New York, 1983, 377.
35. **Pinter, M. and Friedrich, P.,** The calcium-dependent proteolytic system calpain-calpastatin in *Drosophila melanogaster, Biochem. J.*, 253, 467, 1988.
36. **Wolfe, F. H., Szpacenko, A., McGee, K., and Goll, D. E.,** Failure to find Ca^{2+}-activated proteinase (calpain) activity in a plant species, *Elodea densa, Life Sci.*, 45, 2093, 1989.
37. **Cong, J., Goll, D. E., Peterson, A. M., and Kapprell, H. P.,** The role of autolysis in activity of the Ca^{2+}-dependent proteinases (μ-calpain and m-calpain), *J. Biol. Chem.*, 264, 10096, 1989.
38. **Coolican, S. A. and Hathaway, D. R.,** Effect of L-α-phosphatidylinositol on a vascular smooth muscle Ca^{2+}-dependent protease, *J. Biol. Chem.*, 261, 11627, 1986.
39. **Mellgren, R. L.,** Calcium-dependent protease: an enzyme system active at cellular membranes?, *FASEB J.*, 1, 110, 1987.
40. **Suzuki, K., Imajoh, S., Emori, Y., Kawasaki, H., Minami, Y., and Ohno, S.,** Calcium-activated neutral protease and its endogenous inhibitor, *FEBS Lett.*, 220, 271, 1987.
41. **Rozanov, C. B. and Mellgren, R. L.,** Studies on the regulation of bovine myocardial calpain II by interaction with biological membranes, *J. Cell Biol.*, 107, 391, 1988.
42. **Lipscombe, D., Madison, D. V., Poenie, M., Reuter, H., Tsien, R. Y., and Tsien, R. W.,** Spatial distribution of calcium channels and cytosolic calcium transients in growth cones and cell bodies of sympathetic neurons, *Proc. Natl. Acad. Sci. U.S.A.*, 85, 2398, 1988.
43. **Lee, H.-C., Smith, N., Mohabir, R., and Clusin, W. T.,** Cytosolic calcium transients from the beating mammalian heart, *Proc. Natl. Acad. Sci. U.S.A.*, 84, 7793, 1987.

44. **Hatanaka, M., Yoshimura, N., Murakami, T., Kannagi, R., and Murachi, T.,** Evidence for membrane-associated calpain I in human erythrocytes. Detection by an immunoelectrophoretic blotting method using monospecific antibody, *Biochemistry,* 23, 3272, 1984.
45. **Kumamoto, T., Kleese, W. C., Cong, J., Goll, D. E., Pierce, P. R., and Allen, R. E.,** Localization of the Ca^{2+}-dependent proteinases and their inhibitor in normal, starved, and denervated rat skeletal muscle, submitted.
46. **Kitahara, A., Takano, E., Ohtsuki, H., Kirihata, Y., Yamagata, Y., Kannagi, R., and Murachi, T.,** Reversed distribution of calpains and calpastatin in human pituitary gland and selective localization of calpastatin in adrenocorticotropin-producing cells as demonstrated by immunohistochemistry, *J. Clin. Endocrin. Metab.,* 63, 343, 1986.
47. **Kleese, W. C., Goll, D. E., Edmunds, T., and Shannon, J. D.,** Immunofluorescent localization of the Ca^{2+}-dependent proteinase and its inhibitor in tissues of *Crotalus atrox, J. Exp. Zool.,* 241, 277, 1987.
48. **Dayton, W. R. and Schollmeyer, J. V.,** Immunocytochemical localization of a calcium-activated protease in skeletal muscle cells, *Exp. Cell Res.,* 136, 423, 1981.
49. **Schollmeyer, J. V.,** Role of Ca^{2+} and Ca^{2+}-activated protease in myoblast fusion, *Exp. Cell Res.,* 162, 411, 1986.
50. **Schollmeyer, J. V.,** Possible role of calpain I and calpain II in differentiating muscle, *Exp. Cell Res.,* 163, 413, 1986.
51. **Schollmeyer, J. V.,** Calpain II involvement in mitosis, *Science,* 240, 911, 1988.
52. **Yoshimura, N., Hatanaka, M., Kitahara, A., Kawaguchi, N., and Murachi, T.,** Intracellular localization of two distinct Ca^{2+}-proteases (calpain I and calpain II) as demonstrated by using discriminative antibodies, *J. Biol. Chem.,* 259, 9847, 1984.
53. **Cong, J., Szpacenko, A., Goll, D. E., Pierce, P. R., and Allen, R. E.,** Monoclonal antibodies to the 28-kDa subunit affect the Ca^{2+} requirement of m- but not μ-calpain, submitted.
54. **Mellgren, R. L. and Lane, R. D.,** Myocardial calpain II is inhibited by monoclonal antibodies specific for the small noncatalytic subunit, *Biochim. Biophys. Acta,* 954, 154, 1988.
55. **Glaser, T. and Kosower, N.,** Fusion of rat erythrocytes by membrane-mobility agent A_2C depends on membrane proteolysis by a cytoplasmic calpain, *Eur. J. Biochem.,* 159, 387, 1986.

Chapter 2

THE STRUCTURE OF CALPAINS AND THE CALPAIN GENE

Koichi Suzuki

TABLE OF CONTENTS

I. STRUCTURES OF THE LARGE AND SMALL SUBUNITS OF CALPAINS

Ordinary mammalian cells have two calpain isozymes, calpain I and calpain II, or μCANP (calcium activated neutral protease) and mCANP, respectively, with different calcium sensitivities.[1-4] Both are heterodimers composed of large (80 kilodalton; 80K) and small (30 kilodalton; 30K) subunits. The 80K subunit is unique to each isozyme and responsible for protease activity, while the 30K subunit is a regulatory subunit and is common to both isozymes.[1-4] The structure of the 80K subunit was first elucidated in 1984 for chicken CANP* by analyzing the sequence of its cDNA isolated with oligonucleotide probes synthesized according to partial amino acid sequences.[5] This cDNA was next used as a probe to isolate two sets of rabbit cDNA clones for the μ and m 80K subunits,[6] which were in turn used to obtain the human CANP μ[7] and m 80K[8] subunits. The complete sequences of the human CANP μ and m 80K subunits and partial sequences of the rabbit CANP μ and m 80K subunits have been reported. These sequences are compared in Figure 1 and the amino acid compositions and some molecular properties of the 80K and the 30K subunits are summarized in Tables 1 and 2, respectively. The predicted structures of the chicken 80K subunit, and the human μ and m 80K subunits are composed of 705, 714, and 700 amino acid residues, respectively, and the molecular weights are about 80,000. The N- and C-terminal sequences of the chicken 80K subunit reveal that the initiation methionine is removed during posttranslational modification and the acetylated Met-2 is the N-terminus of the mature chicken protein.[5] No processing is found in the C-terminal region. A similar modification also occurs with the 80K subunits of rabbit and human μ and mCANPs, since blocked N-terminal tryptic peptides having sequences identical to the N-terminal regions of human μ and m 80K subunits but without the initiation methionine have been isolated from the rabbit μ and m 80K subunits, respectively[9] The results also demonstrate that the near N-terminal sequences of the rabbit and human μ and m 80K subunits are almost identical. The structures of the 80K subunits are generally highly homologous, irrespective of the origins and types of isozymes, as summarized in Table 3. Sequence comparison between the human and rabbit μ and m 80K subunits clearly indicates that the same isozymes from different species are about 95% identical and much more homologous than different isozymes of the same species, which show about 60% identity. The chicken 80K subunit lies between the mammalian μ and m subunits, judging from the sequence homology, although it presumably corresponds to the mammalian m type.[10]

The 80K subunit can be divided into four domains as illustrated in Figure 2 on the basis of sequence homology to other known proteins.[1,3,5] The second domain (No. 81-330)** is homologous to typical cysteine proteinases like papain and cathepsins B, L, and H. Sequence homologies are significantly high around residues Cys-108 and His-265, corresponding to the active site cysteine and histidine residues of various cysteine proteinases. In fact, Cys-108 of the chicken 80K subunit has been identified as its active site by selective modification with cysteine proteinase inhibitors such as iodoacetic acid, E64, and leupeptin.[11] Clear evidence indicating that His-265 is involved in catalysis has not yet been obtained. Although CANP is clearly distinct from other cysteine proteinases in its requirement of calcium for activity, molecular weight (about five times larger), and subunit structure (dimer in contrast to monomer), its protease activity can be ascribed to the ''cysteine proteinase'' domain, domain II. CANP has no sequence similarity to metalloproteinase in spite of its dependence on calcium.

* Chicken was first presumed to contain a single isozyme of CANP. Nevertheless, two types of CANP were later found as in mammals.[10] One type, which is the main molecular species in muscle, presumably corresponds to mammalian mCANP, and the other type, which has not yet been characterized, will be a μ type enzyme.

** Chicken CANP numbering.

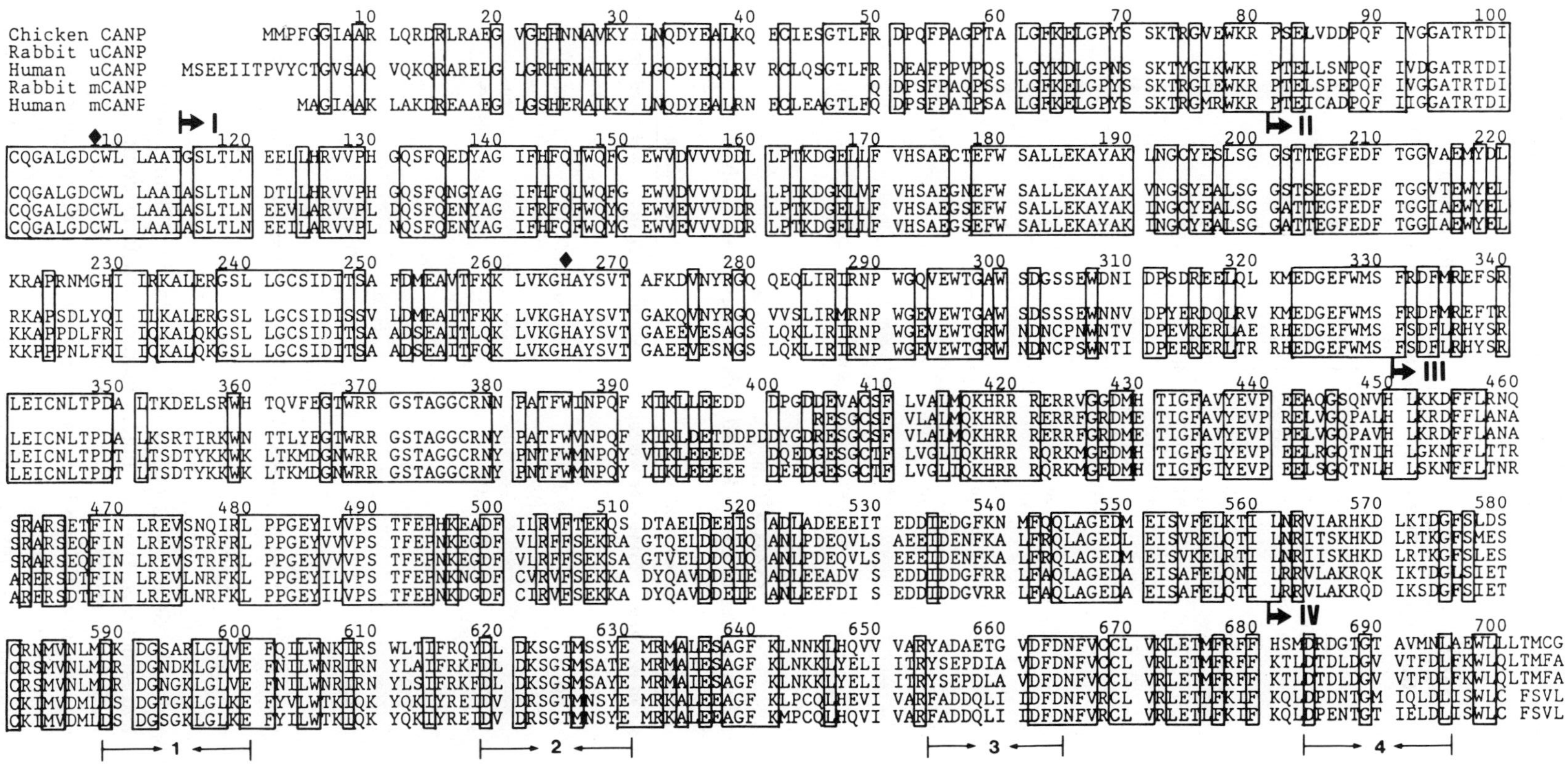

FIGURE 1. Diagram showing the orientation of oxidoreductase enzymes in the plasma membrane.

TABLE 1
Amino Acid Composition of the Primary Translation Products of the CANP Subunits

Amino acid	80K Subunit			30K Subunit		
	Chicken[5]	Human μ[2]	Human m[8]	Rabbit[14]	Human[13]	Porcine[15]
Asp	53	47	47	15	15	15
Asn	26	28	27	14	14	13
Thr	36	36	36	11	12	11
Ser	41	49	40	15	17	17
Glu	63	55	68	16	15	15
Gln	28	26	26	9	9	9
Pro	21	25	24	7	7	7
Gly	54	51	51	48	51	51
Ala	46	39	46	19	17	17
Cys	13	11	15	3	3	2
Val	40	46	26	12	11	12
Met	21	16	12	10	10	10
Ile	31	32	46	13	15	15
Leu	66	73	68	23	22	22
Tyr	14	22	20	7	7	7
Phe	43	44	39	14	14	14
Lys	33	37	41	10	9	10
His	15	9	10	4	4	4
Arg	45	52	42	13	13	12
Trp	16	16	16	3	3	3
Total	705	714	700	266	268	266

TABLE 2
Molecular Properties of the 80K and 30K Subunits of CANP

	Subunit	Primary translation product: Number of residues	Primary translation product: Molecular weight	N-Terminus of: Mature protein	N-Terminus of: Activated protein
80K	Subunit				
	Chicken	705	80,350	N-Ac-Met-2	Ala-18
	Human μ	714	81,888	—	—
	Rabbit μ	—	—	N-Ac-Ser-2[a]	Leu-28[a]
	Human m	700	80,019	—	—
	Rabbit m	—	—	N-Ac-Ala-2[a]	Ser-20[a]
30K	Subunit				
	Human	268	28,315	—	—
	Porcine	266	28,068	N-Ac-Met-1	—
	Rabbit	266	28,254	N-Ac-Met-1	Ala-92
	Chicken	—	—	—	Val-88[b]

[a] Residue numbers are determined assuming that the corresponding N-terminal sequences of the human and rabbit subunits are identical.
[b] Assuming that the human and chicken sequences are identical as above.

The C-terminal domain (No. 561-705) is a calmodulin-like calcium-binding domain with four consecutive EF hand structures, typical calcium-binding structures comprising helix-loop-helix structures of about 30 residues with calcium binding to the loop region.[12] This structural motif of four consecutive EF hand structures distributed along the peptide chain of about 150 residues with distinct spacings is a typical characteristic of calcium-binding proteins of the calmodulin family including calmodulin, troponin C, myosin light chain,

TABLE 3
Sequence Homology of the 80K Subunits of CANP in Domains

Combination of 80K subunit	Homology (%) in domain I	II	III	IV	Total
Hm/Rm	87[a]	93	95	94	94
Hμ/Rμ	—	—	96[a]	97	96
Hm/Hμ	51	73	62	48	62
Rm/Rμ	—	—	62[a]	51	57
C/Hm	70	75	64	53	66
C/Hμ	51	78	71	65	70

Note: H, R and C indicate human, rabbit and chicken, respectively. μ and m denote μ and mCANPs, respectively.

[a] Known partial sequence was compared.

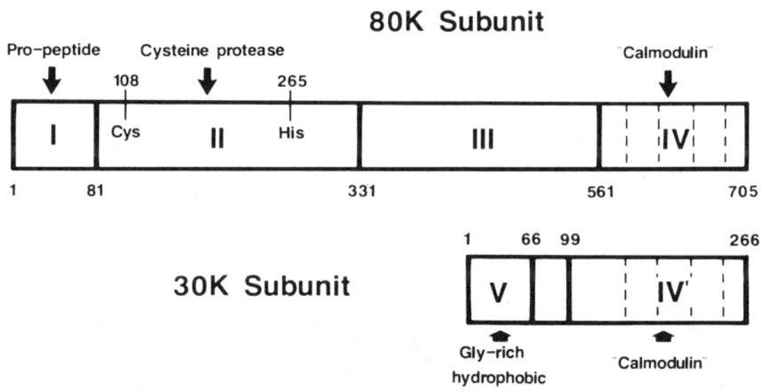

FIGURE 2. Domain structure of CANP. Residue numbers shown for the 80K and 30K subunits are for chicken and rabbit enzymes, respectively. Vertical broken lines denote the positions of loop regions of EF hand structures.

etc.[3,12] The other two domains, domains I and III, have no clear sequence homology with known proteins.

The structures of the human,[13] rabbit,[14] and porcine[15] 30K subunits have been determined as shown in Figure 3. They are composed of 266 to 268 residues with molecular weights of about 28,000 (Table 2). The three known sequences show more than 90% sequence identity (Table 3), and their mature proteins start with the N-acetylated initiation methionine,[15,16] though the identification of the blocking N-acetyl group is still tentative. As with the 80K subunit, C-terminal modification of the 30K subunit does not occur.

The most interesting structural feature of the 30K subunit is an extremely high glycine content in the N-terminal 60 to 70 residues. The longest glycine stretch in the human 30K subunit is 20 residues and a block of 11 residues also exists. The glycine content of the No. 10 to 56 region is more than 80%. Because residues other than glycine are mostly hydrophobic, this region (No. 1-66, domain V) is termed a glycine-rich hydrophobic domain. Domain V of the human 30K subunit contains 40 glycine (60%) and 17 hydrophobic (26%) residues, accounting for 86% of the total residues.

Another interesting structural feature of the 30K subunit is the existence of a 2nd calmodulin-like calcium-binding domain (domain IV', No. 99 to 266) in the C-terminal region. This domain shares marked sequence homology (50%) with the corresponding domains (domain IV) in the μ and m 80K subunits, as compared in Figure 4. Further, the four EF hand structures in these sequences can be aligned exactly at the same positions.

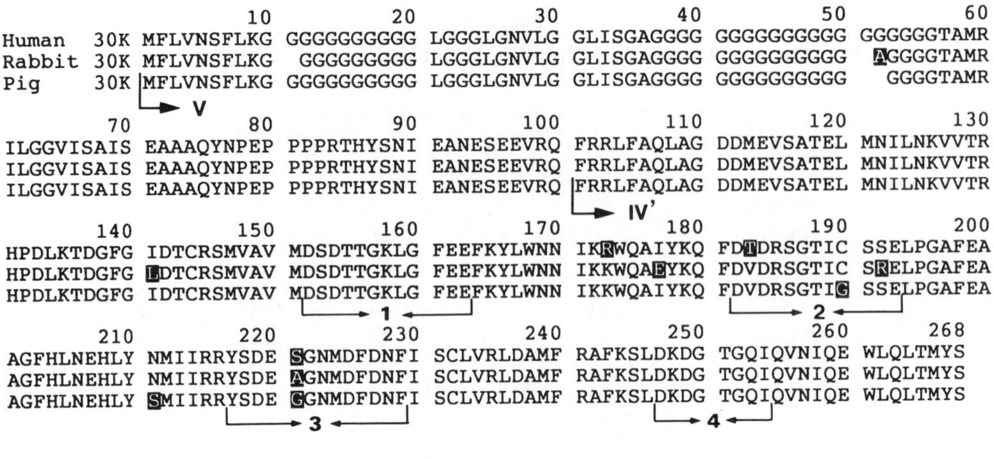

FIGURE 3. Amino acid sequences of the 30K subunits of human, rabbit and porcine CANPs. Residues different among the three sequences are marked. Loop regions of EF hand structures (1 to 4) and starting points of domains V and IV′ are also shown.

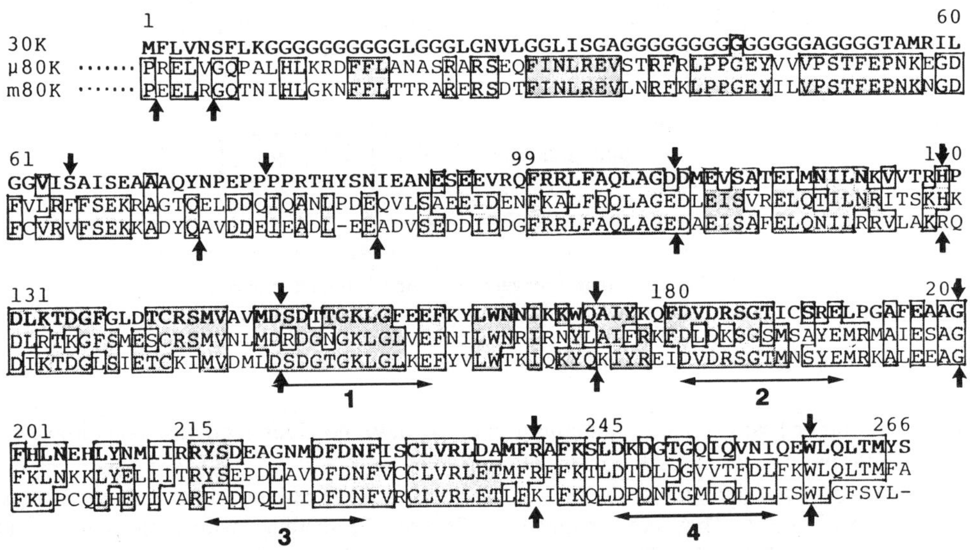

FIGURE 4. Comparison of amino acid sequences of the 30K subunit with the C-terminal region of the 80K subunit. Sequences are for the rabbit enzymes. Upward and downward arrows indicate intron positions for the 80K and 30K subunits, respectively.

This sequence homology of the 30K subunit with the C-terminal region of the μ and m 80K subunits is observed from the C-terminus to residue No. 99, where the homology disappears abruptly. The N-terminal sequence of the 30K subunit (No. 1 to 98) is entirely different from the corresponding sequences in the μ and m 80K subunits, although the μ and m 80K subunits have high sequence homology beyond this point to the N-terminus, as discussed above. This indicates that the 30K subunit is composed of at least two domains. It is worth noting that two regions with characteristic amino acid compositions exist between domains V and IV′, a proline-rich region (No. 78-83), and a glutamic acid and hydrophilic amino acid-rich region (No. 91-97).

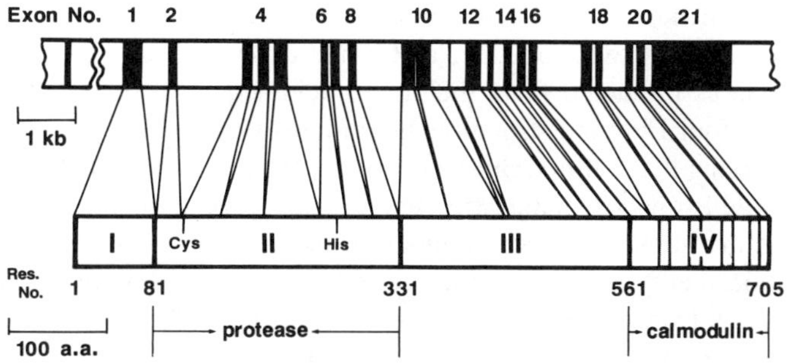

FIGURE 5. Gene structure of the chicken CANP 80K subunit and its correspondence to the traslational product. Filled and open regions are exons and introns, respectively.

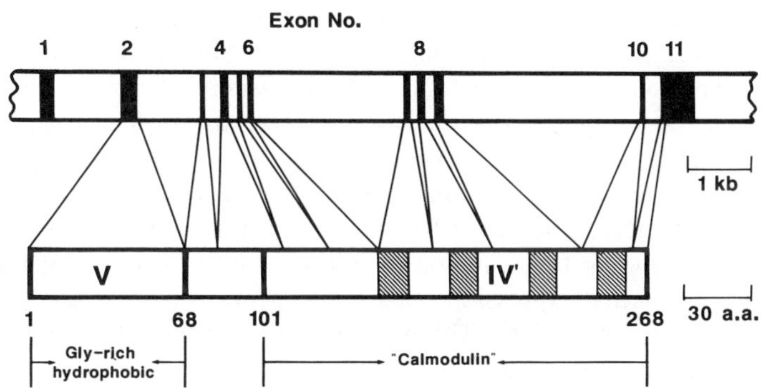

FIGURE 6. Gene structure of the human 30K subunit and its correspondence to the translational product. Symbols are the same as in Figure 5.

II. GENE STRUCTURES

Both genes for the chicken 80K[17] and rabbit 30K[18] subunits are unique and their structures have been clarified. The gene for the chicken 80K subunit is about 10 kb and composed of 21 exons of various sizes ranging from 12 bp to 1.2 kbp as shown in Figure 5. Intron-exon junctions exist at or very close to the domain junctions defined from the sequence homology. This confirms the predicted domain structure and supports the notion that the 80K subunit is derived by fusion of the genes for the four domains. In the protease domain, the regions around the active site Cys-108 and His-265, which are highly conserved among cysteine proteases, are split by introns. The gene organization does not show any clear correlation to the functional unit in this domain. A similar situation is also found in the gene structures of cysteine proteases thus far analyzed.[19] In the calmodulin domain, each of the four EF hand structures is encoded by one exon, showing strong correlation between the gene structure and the functional unit.

The gene for the rabbit 30K subunit is about 11 kb long, comprising 11 exons (Figure 6). The first exon encodes the 5′-noncoding sequence. Again, correspondence of the domain junctions to the intron-break points is observed. The second exon codes for the glycine-rich hydrophobic domain, and each of the four consecutive EF hand structures in the C-terminal calmodulin domain (domain IV′) is encoded by one exon. Further, the seven intron positions in domain IV′ completely coincide at the nucleotide level with those in domain IV of chicken CANP (Figure 4),[18] whereas those in the near N-terminal region of the 30K subunit are entirely different from those in the 80K subunit. The gene structure implies that the 30K subunit arose from at least two domains, and domains IV and IV′ have evolved from the same ancestral gene as predicted from the amino acid sequence homology. Moreover, the fact that each of the four EF hand structures in domains IV and IV′ corresponds to one exon supports the hypothesis that calmodulin has evolved from a mini-gene encoding a calcium-binding sequence with one EF hand structure by two gene duplication steps.[20] The gene organization of calmodulin and other calmodulin-like proteins is not fully consistent with the gene duplication theory.[21] The original gene structures are lost presumably because of rearrangment of the intron positions after gene duplication. The structure of the prototype calmodulin gene is preserved in CANP rather than in calmodulin.

Precise analyses of the 5′-upstream region have identified multiple translation initiation sites both for the 30K and 80K subunits, six and eleven positions, respectively.[18,22] These upstream regions lack typical promoter elements such as CAAT and TATA boxes and show very high GC content. These structural features are common to genes for various housekeeping enzymes. The promoter region of the human 80K subunit exists between −220 and −80 and can be divided into upstream (−202 to −130, P1) and downstream (−130 to −80, P2) promoter elements.[22] P2 contains potential binding sites for Sp1 and AP-1. Upstream of the promoter region are four tandemly reiterated regulatory regions which negatively regulate the promoter activity. Furthermore, the presence of cellular factors mediating these positive (promoter) and negative regulatory elements was suggested. Ubiquitous, but cell-type specific, expression of the CANP genes may be controlled by these cis- and trans-acting factors. The genes for the human 30K and 80K subunits share 50% sequence homology in the upstream region (−360 to −60), including Sp1 and AP-1 recognition sites, suggesting that expression of these two genes is co-regulated at the transcriptional level by the same mechanism. The mRNA levels for the 30K and 80K subunits in various rabbit and human tissues analyzed by Northern-blotting vary significantly but in a parallel manner. This partly supports the idea of co-regulation of the two subunits at the transcriptional level.

III. FUNCTIONAL IMPLICATIONS OF DOMAIN STRUCTURES

Both 80K and 30K subunits are multidomain proteins (Figure 2). Among six domains, functions of a protease domain (II) and two calmodulin domains (IV and IV′) are clear. Since the 30K subunits of μ and mCANPs are identical, the difference in their calcium sensitivities should arise from differences in the 80K subunits. This has been confirmed by the exchangeability of the 30K subunits between μ- and mCANPs without affecting their calcium sensitivity.[23] Sequence comparison of the μ and m 80K subunits shows that domain II has significantly higher homology (about 75%) for various combinations (Table 3). The homologies for domains I and IV are much lower (about 50%). Hence μ- and mCANPs have essentially the same protease domain, and their differences exist in domains I and IV; the latter will define their calcium sensitivities.

From its homology with cysteine proteinases, isolated domain II should itself have protease activity in the absence of calcium. However, since CANP absolutely requires calcium for activity, the protease activity intrinsic to domain II is repressed or inhibited by interaction with other domains. The function of the two calmodulin domains, IV and IV′, is to activate the repressed or inhibited protease activity through conformational changes induced by the binding of calcium. This type of activation is usually observed with calmodulin-dependent enzymes, e.g., phosphodiesterase, calcineurin, etc. CANP is similar to protein kinase C, cGMP-dependent protein kinase, and other enzymes where the built-in active enzymes express their activities only in the presence of cofactors. The built-in active enzymes do not require any cofactors when they are isolated from regulatory domains. That a similar situation holds also for CANP, i.e., that domain II is an active protease which does not require calcium, is not yet established, although calcium-independent protease activity was detected in a proteolytic digest of CANP.

Recently it has been proposed from various lines of evidence that CANP, which has been regarded as an active enzyme, is actually an inactive proenzyme and is converted to the active enzyme upon removal of the N-terminal region by autolysis.[2-4] In the cases of chicken, rabbit μ and m 80K subunits, 17, 27, and 19 N-terminal residues are removed.[19] This situation is similar to the activation of cathepsins B, L, and H, which are synthesized as inactive pre-proenzymes and require the removal of propeptides for expression of activity. Thus, domain I may be a part of the protease domain and termed a propeptide domain, although all of domain I is not removed during activation, in contrast to propeptides of cathepsins.

The function of domain III, with relatively high sequence conservation among CANPs, is not clear at present, but it might be a calmodulin-binding protein and control the intrinsic protease of domain II by a flip-flop mechanism.[3] Domain III interacts with domain II and inhibits its protease activity in the absence of calcium. When calcium binds to domain IV, domain III interacts with domain IV as a calmodulin-binding protein, leaving domain II free and active. Domain V is essential for the interaction with phospholipid and reduces the calcium concentration required for autolysis.[24] This domain is completely removed together with the the N-terminal region of the 80K subunit during autocatalytic activation.[3] CANP without domain V can be prepared by limited digestion with chymotrypsin.[24] The enzymatic properties of this enzyme preparation are almost identical to those for native CANP except for calcium sensitivity in the presence of phospholipid. Thus domain V has virtually no effect on protease activity measured *in vitro* and is not essential for interaction with the cell membrane and the 80K subunit.

Not all of the four EF hand structures in domains IV and IV′ are active in calcium binding. Domain IV′ of chicken CANP binds 2 mol whereas domains IV′ of rabbit μ- and mCANPs bind 2 and 3 mol of calcium, respectively. The apparent average dissociation constants of domain IV of rabbit μ, and chicken CANPs prepared by expression of the corresponding cDNAs in *E. coli*, are 35, 79, and 61 μ*M*, respectively, values proportional to the calcium concentration required for activity.[25] This is a direct indication that domain IV, which shows the lowest sequence similarity bctwccn μ- and mCANPs, determines the calcium sensitivity. The active calcium-binding sites or rabbit domain IV′ have been identified to be the 1st and 4th EF hand structures by site-directed mutagenesis.[26] Sequence homology of domain IV with IV′ suggests that the same two EF hand structures are the calcium-binding sites of domain IV of rabbit μCANP and that the second EF hand structure will be the additional calcium-binding sites for domain IV of rabbit mCANP. It is difficult to predict from the sequence why the second EF hand structure of rabbit mCANP domain IV binds calcium, while the corresponding sites in rabbit μCANP domain IV and IV′ do not. Domains IV and IV′ are essential not only as calcium-binding sites but also for interaction between the two subunits and with the cell membrane. Like native CANP, CANP without domain V is a dimer and can still bind to the membrane in the presence of calcium, implying that domain V is not the primary site for association with either the cell membrane or the 80K subunit. The 30K subunit binds to the 80K subunit only through domain IV′. Removal of more than three residues from either domain IV or IV′ causes dissociation of the subunits.[27] CANP binds to the cell membrane in the presence of calcium, indicating that domains IV and/or IV′ themselves, or some other hydrophobic regions exposed to the surface by the binding of calcium, are indispensable for interaction of CANP with the cell membrane.

At present CANP is the only enzyme in which calmodulin binds covalently to its target enzyme (domain II), though some calmodulin-dependent enzymes have calmodulin as subunits. CANP also has a subunit type calmodulin as domain IV′. Calmodulin usually exists dissociated from its target enzymes. The activation of target enzymes requires a complex formation with calmodulin being activated by calcium binding. Thus the activation of calmodulin-dependent enzymes by free calmodulin proceeds in two steps. In contrast, the control of the target enzyme by covalently bound calmodulin is a single step, and the regulation is more directly and accurately controlled by the presence of calcium. A more rapid response to changes in calcium level would be possible. This strict regulation of CANP activity seems quite reasonable to control the intracellular protease activity which, if uncontrolled, would be extremely toxic to the cell. However, questions as to why 2 mol of calmodulin are required for the regulation of protease activity, what their functional shares are, and how the two calmodulin domains work in a concerted manner, still remain to be answered.

REFERENCES

1. **Suzuki, K.,** Calcium activated neutral protease: domain structure and activity regulation, *Trends Biochem. Sci.*, 12, 103, 1987.
2. **Suzuki, K., Imajoh, S., Emori, Y., Kawasaki, H., Minami, Y., and Ohno, S.,** Calcium-activated neutral protease and its endogenous inhibitor, *FEBS Lett.*, 220, 271, 1987.
3. **Suzuki, K., Ohno, S., Emori, Y., Imajoh, S., and Kawasaki, H.,** Calcium-activated neutral protease (CANP) and its biological and medical implications, *Prog. Clin. Biochem. Med.*, 5, 44, 1987.
4. **Mellgren, R. L.,** Calcium-dependent proteases: an enzyme system activated at the cell membrane?, *FASEB J.*, 2, 110, 1988.
5. **Ohno, S., Emori, Y., Imajoh, S., Kawasaki, H., Kisaragi, M., and Suzuki, K.,** Evolutionary origin of calcium-dependent protease by fusion of genes for a thiol protease and a calcium-binding protein?, *Nature*, 312, 566, 1984.
6. **Emori, Y., Kawasaki, H., Sugihara, H., Imajoh, S., Kawashima, S., and Suzuki, K.,** Isolation and sequence analyses of cDNA clones for the large subunits of two isozymes of rabbit calcium-dependent protease, *J. Biol. Chem.*, 261, 9465, 1986.
7. **Aoki, K., Imajoh, S., Ohno, S., Emori, Y., Koike, M., Kosaki, G., and Suzuki, K.,** Complete amino acid sequence of the large subunit of the low-Ca^{2+}-requiring from of human Ca^{2+}-activated neutral protease (μCANP) deduced from its cDNA sequence, *FEBS Lett.*, 205, 313, 1986.
8. **Imajoh, S., Aoki, K., Ohno, S., Emori, Y., Kawasaki, H., Sugihara, H., and Suzuki, K.,** Molecular cloning of the cDNA for the large subunit of the high-Ca^{2+}-requiring form of human Ca^{2+}-activated neutral protease, *Biochemistry*, 27, 8122, 1988.
9. **Kawasaki, H., Imajoh, S., and Suzuki, K.,** Separation of peptides on the basis of the difference in positive charge: simultaneous isolation of C-terminal and blocked N-terminal peptides from tryptic digests, *J. Biochem.*, 102, 393, 1987.
10. **Murakami, T., Ueda, M., Hamakubo, T., and Murachi, T.,** Identification of both calpains I and II in nucleated chicken erythrocytes, *J. Biochem.*, 103, 168, 1988.
11. **Suzuki, K., Hayashi, H., Hayashi, T., and Iwai, K.,** Amino acid sequence around the active site cysteine residue of calcium-activated neutral protease (CANP), *FEBS Lett.*, 152, 67, 1983.
12. **Kretsinger, R. H.,** Calcium-binding proteins, *Ann. Rev. Biochem.*, 45, 239, 1976.
13. **Ohno, S., Emori, Y., and Suzuki, K.,** Nucleotide sequence of a cDNA coding for the small subunit of human calcium-dependent protease, *Nucleic Acids Res.*, 14, 5559, 1986.
14. **Emori, Y., Kawasaki, H., Imajoh, S., Kawashima, S., and Suzuki, K.,** Isolation and sequence analysis of cDNA clone for the small subunit of rabbit calcium-dependent protease, *J. Biol. Chem.*, 261, 9472, 1986.
15. **Sakihama, T., Kakidani, H., Zenita, K., Yumoto, N., Kikuchi, T., Sasaki, T., Kannagi, R., Nakanishi, S., Ohmori, M., Takio, K., Titani, K., and Murachi, T.,** A putative Ca^{2+}-binding protein: structure of the light subunit of porcine calpain elucidated by molecular cloning and protein sequence analysis, *Proc. Natl. Acad. Sci. U.S.A.*, 82, 6075, 1985.
16. **Kawasaki, H., Imajoh, S., Kawashima, S., Hayashi, H., and Suzuki, K.,** The small subunits of calcium-dependent proteases with different calcium sensitivities are identical, *J. Biochem.*, 99, 1525, 1986.
17. **Emori, Y., Ohno, S., Tobita, M., and Suzuki, K.,** Gene structure of calcium-dependent protease retains the ancestral organization of calcium-binding protein gene, *FEBS Lett.*, 194, 249, 1986.
18. **Miyake, S., Emori, Y., and Suzuki, K.,** Gene organization of the small subunit of human calcium-activated neutral protease, *Nucleic Acids Res.*, 14, 8805, 1986.

19. **Suzuki, K., Imajoh, S., Kawasaki, H., Minami, Y., and Ohno, S.,** Regulation of activity of calcium-activated neutral protease, *Adv. Enzyme Regul.*, 27, 153, 1988.
20. **Iida, Y.,** Molecular evolution of protein. Internal homology in the amino acid sequence of calmodulin, *J. Mol. Biol.*, 159, 167, 1982.
21. **Smith, V. L., Doyle, K. E., Maune, J. F., Munjaal, R. P., and Beckingham, K.,** Structure and sequence of the *Drosophila meranogaster* calmodulin gene, *J. Mol. Biol.*, 196, 471, 1987.
22. **Hata, A., Ohno, S., Akita, Y., and Suzuki, K.,** Tandemly reiterated negative enhancer-like elements regulate transcription of a human gene for the large subunit of calcium-dependent protease, *J. Biol. Chem.*, 264, 6404, 1989.
23. **Kikuchi, T., Yumoto, N., Sasaki, T., and Murachi, T.,** Reconstitution of calpain I and II from their subunits: interchangeability of the light subunits, *Arch. Biochem. Biophys.*, 234, 639, 1984.
24. **Imajoh, S., Kawasaki, H., and Suzuki, K.,** The amino-terminal hydrophobic region of the small subunit of calcium-activated neutral protease (CANP) is essential for its activation by phosphatidylinositol, *J. Biochem.*, 99, 1281. 1986.
25. **Minami, Y., Emori, Y., Kawasaki, H., and Suzuki, K.,** EF hand structure-domain of calcium-activated neutral protease (CANP) can bind Ca^{2+} ions, *J. Biochem.*, 101, 889, 1987.
26. **Minami, Y., Emori, Y, Imajoh-Ohmi, S., Kawasaki, H., and Suzuki, K.,** Carboxyl-terminal truncation and site-directed mutagenesis of the EF hand structure-domain of the small subunit of rabbit calcium-dependent protease, *J. Biochem.*, 104, 927, 1988.
27. **Imajoh, S., Kawasaki, H., and Suzuki, K.,** The COOH-terminal EF hand structure of calcium-activated neutral protease (CANP) is important for the association of subunits and resulting proteolytic activity, *J. Biochem.*, 101, 447, 1987.

Chapter 3

STRUCTURE-FUNCTION RELATIONSHIP OF CALPASTATINS

Masatoshi Maki, Masakazu Hatanaka, Emiko Takano, and Takashi Murachi

TABLE OF CONTENTS

I. INTRODUCTION

The clear demonstration of the occurrence of a specific endogenous inhibitor for calpain appeared almost concurrently from two independent laboratories,[1,2] although such an inhibitor had been predicted much earlier,[3] and also a preliminary report had been presented.[4] Soon after its discovery, the inhibitor was found to be as widely distributed in nature as the enzyme calpain.[5] Calpastatins have been isolated from various organs of mammals and their biochemical properties have been characterized by many investigators (see References 6 to 8 for reviews). When calpastatin was first identified chromatographically, it was reported to behave as a protein having molecular mass of approximately 280 kDa[9] or 300 kDa[2]. After the purification of calpastatin from human erythrocytes to homogeneity, the polyacrylamide gel electrophoresis of the product in the presence of sodium dodenyl sulfate (SDS) revealed a molecular mass of 70 kDa.[10] When the electrophoresed calpastatin protein was eluted from the gel and chromatographed again on a Sephacryl S-300 column, the peak elution position of the material corresponded to 280 kDa, indicating self-association of calpastatin protomers. Such diversity arising from protomer-oligomer transformation has also been noted with calpastatins from bovine cardiac muscle,[11,12] human platelets,[13] rabbit liver,[14] and rat liver,[15] rabbit skeletal muscle,[16] porcine erythrocytes,[17] and porcine cardiac muscle.[17]

The values of reported molecular mass of isolated calpastatin ranged from 24 kDa to 172 kDa in SDS-polyacrylamide gel electrophoresis (SDS-PAGE).[8] Although Western-blotting analyses using specific anticalpastatin antibody also revealed calpastatins with different molecular sizes (50 to 110 kDa) in various organs of human and pig. Major forms were 110 kDa proteins for all tissues except erythrocytes, which showed 70 kDa proteins as major forms.[18] Surprisingly, high molecular weight (107 kDa) and low molecular weight (68 kDa) forms of calpastatins purified from pig heart muscle and pig erythrocytes, respectively, showed close similarity in amino acid compositions and in isoelectric points (pI 4.7 to 4.8).[17] They also showed immunological cross-reactivity. The number of calpain-binding sites per molecule of calpastatin was proportional to the size of the calpastatin molecule (107-kDa form, 8 sites; 68-kDa form, 5 sites). The findings led Takano et al. to propose that the calpastatin molecule may comprise several functional regions composed of homologous amino acid sequences.[17] Recently recombinant DNA-techniques have facilitated analyses of primary structures of calpastatins, and the hypothetical multidomain structure model has been substantiated.[8-11] Structure-function relationships of calpastatins will be described in this chapter.

II. PRIMARY STRUCTURES

Sequences of complementary DNAs (cDNAs) encoding calpastatins of pig,[19,20] rabbit,[21] and human[22] have been reported by two independent groups. They employed similar approaches to obtain the cDNA clones. Calpastatins were first purified from pig heart muscle, pig erythrocytes, and rabbit liver. Partial amino acid sequences of the fragmented peptides were then determined by the Edman degradation method. Synthetic oligonucleotides corresponding to the determined amino acid sequences were used for screening pig and rabbit cDNA libraries. Human calpastatin cDNA clones were isolated by screening a commercially available human cDNA library using a pig calpastatin cDNA fragment as a probe.

Figure 1 shows the deduced amino acid sequences of human, pig and rabbit calpastatins. They consist of 673, 713, and 718 amino acid residues, respectively. They have four repeats of mutually homologous sequences at the interval of about 140 amino acid residues. For convenience, a unique N-terminal leader region is referred to as domain L, and the internally repetitive regions of approximately 140 residues are called domains 1, 2, 3, and 4 (Figure 2). Human calpastatin has two prominent deletions of 22 and 13 residues in domains L and

```
(H) MNPTETKAV-----------------------KTEPEKKSQSTKLSVVHEKKSQEGKPKEHTEQKSLPKQASDTGS  53
(P) MNPTETKAIPVSKQLEGPHSPNKKRHKKQAVKTEPEKKSQSTKPSVVHEKKTQEVKPKEHPEPKSLPTHSADAGS  75
(R) MNPAEAKAVPISKEMEGPHPHSKKRHRRQDAKTEPEK-SQSTKPPVDHEKKAQEGKPKEHTKPKSTHKHASDGEG  74

(H) KDAHNKKAVSRSAEQQPSEKSAEPKTKPQDTVSAGGESVA-GVAATSGKPGDKKKEKKSLAPALPVESKPDKPSG 127
(P) KRAHKEKAVSRSNEQPTSEKSTKPKAKPQDPTPSDGKLSVTGVSAASGKPAETKKDDKSLTSSVPAESKSSKPSG 150
(R) KHGRNEKTASRSKEPVTPAKRTEPETKPQDTKPAGGKSVAAGTTAAPGKAGDPKKEKKSLPAAALAEPKPDEPSG 149

                                                    ┌─CCS─┐
    ~~~~~~~~
(H) KSGMDAALDDLIDTLGGPEETEEENTTYTGPEVSDPMSSTYIEELGKREVTIPPKYRELLAK------------- 189
(P) KSDMDAALDDLIDTLGGPEETEEDNTTYTGPEVLDPMSSTYIEELGKREVTLPPKYRELLDKKEGIPVPPPDTSK 225
(R) KSGMDAALDDLIDTLGEPSETQEDSTAYTGPEISDPMSSTYIEELGKREVTIPPKYRELLEKKTGVAGPPPDSVT 224

       ______                                                                ~~~
(H) PIGPDDAIDALSSDFTCGSPTAAGKKTEKE---EST-EVLKAQSAGTVRSAA-PPQEKKRKVEKDTMSDQALEAL 259
(P) PLGPDDAIDALSLDLTCSSPTADGKKTEKE---KSTGEVLKAQSVGVIKSAAAPPHEKKRRVEEDTMSDQALEAL 297
(R) PLGPDDAIDALSSDFTCSSPVASGKEAGKEAAKSAG-EVLEAESAKVMRAAA-PPQEKKRKVEEDAMSDQALEAL 297

                                          ┌─CCS─┐                                -
 ~~~
(H) SASLGTRQAEPELDLRSIKEVDEAKAKEEKLEKCGEDDETIPSEYRLKPATDKDGKPLLPEPEEKPKPRSESELI 334
(P) SASLGSRKSEPELDLSSIKEIDEAKAKEEKLKKCGEDDETVPPEYRLKPAMDKDGKPLLPEAEEKPKPLSESELI 372
(R) SASLGTRMAEPELDLSSIKEVAEAKRKEEKVEKCGEDDETVPAEYRLKPATDKDGKPLLPEPAEKPKPRSESELI 372

  _____                                                             ~~~~~~~~
(H) DELSEDFDRSECKEKPSKPTEKTEESKAAAPAPVSEAVSRTSMCSIQSAPPEPATL-KGTVPDDAVEALADSLGK 408
(P) DELSEDFDQSKRKEKQSKPTEKTKESQATAPTPVGEAVSRTSLCCVQSAPPKPAT---GMVPDDAVEALAGSLGK 444
(R) DELSKDFSQAKSNEKQPKPTGKTEESKAAVPAPVAEAVPRTSMCSIQPVPPKPASLQKSTVPDDAVEALAGSLGR 447

                                 ┌─CCS─┐                                 _____
(H) KEADPEDGKPVMDKVKEKAKEEDREKLGEKEETIPPDYRLEEVKDKDGKPLLPKESKEQLPPMSEDFLLDALSED 483
(P) KEADPEDGKPVEDKVKEKAKEEDREKLGEKEETIPPDYRLEEVKDKDGKTLPHKDPKEPVLPLSEDFVLDALSQD 519
(R) KEADPEEGKPVADKIKEKSKEEEREKLGEKEETIPPDYRLEEAKDKDGKPLLPSEPTAQLPALSEDLLLDALSED 522

                                                         ~~~~~~~~
(H) FSGPQNASSLKFEDAKLAAAISEVVSQTPASTTQAGAPPRDTSQSDKDLDDALDKLSDSLGQRQPDPDENKPMGD 558
(P) FAGPPAASSL-FEDAKLSAAVSEVVSQTSAPTTHSAGPPPDTVSDDKKLDDALDQLSDSLGQRQPDPDENKPIED 593
(R) FSGPSSASSLKFDDAMLSAAVSEVVSQSPASITRATAPPPDTRPSNKELDDALDKLSDSLGQRQPDPDENKPMED 597

                 ┌─CCS─┐                                  _____
(H) KVKEKAKAEHRDKLGERDDTIPPEYRHLLDDNGQDKPVKPPTKKSEDSKKPADDQDPIDALSGDLDSCPSTTETS 633
(P) KVKEKAEAEHRDKLGERDDTIPPEYRHLLDKDEEGKSTKPPTKKPEAPKKPEAAQDPIDALSGDFDRCPSTTETS 668
(R) KVKERAKKEHKDKLGERDDTIPPEYRHLLDQGEQDKPEKPPTKKSKEIKKPAGDQDPIDALSGDLDSCPPAAETS 672

(H) QNTAKDKCKKA-ASSSKAPKNGGKAKDSAKTTEETSKPKDD      673
(P) ENTTKDKDKKT-ASKSKAPKNGGKAKDSTKAKEETSKQKSDGKSTS 713
(R) QATEKDKSKTTTASSSKAAKHGDKAKDSAQTTEETSKPKANEKNAS 718
```

FIGURE 1. Comparison of amino acid sequence of the three calpastatins. Amino acid sequences of human (H), pig (P), and rabbit (R) calpastatins are aligned in order from top to bottom. Gaps were introduced to obtain a best-fit alignment. Residues identical in the three calpastatins are shaded. The regions indicated with CCS are functionally important and have highly conserved sequences which are referred to as "central consensus sequence".[34] Conserved regions in amino-terminal and carboxyl-terminal sides in each domain are marked with waved and thin lines above the human calpastatin sequences, respectively, as shown in Figure 3.

1, respectively, compared with the sequences of pig and rabbit calpastatins. Amino acid sequence homology between any pair of the three species is about 70%. When the whole regions of four repetitive domains are compared with one another, taking pig calpastatin, for instance, the homology ranges from 23% (domain 1 vs. domain 2) to 36% (domain 2 vs. domain 3).[20] However, conserved residues among the four domains are not evenly distributed but are clustered in three regions in each domain (Figures 2, 3).

No significant sequence homology has been found between calpastatins and any other known proteins. A large number of cysteine proteinase inhibitors including kininogens[23] are classified as the "cystatin superfamily" which has a consensus sequence Gln-Val-Val-Ala-

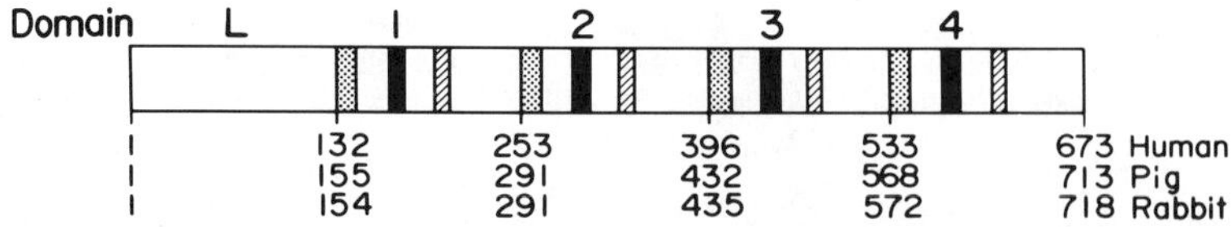

FIGURE 2. Schematic structures of calpastatins. The calpastatins contain five domains: amino-terminal nonhomologous domain (leader domain, domain L) and four mutually homologous domains (domains 1, 2, 3, and 4) of approximately 140 amino acid residues. Each internally repetitive domain has three clustered regions of conserved sequences. Dotted, closed, and hatched areas indicate amino-terminal, central, and carboxyl-terminal conserved regions, respectively. Numbers below the dotted areas indicate the first residues of the amino-terminal conserved regions in each repetitive domain of human, pig and rabbit calpastatins.

Gly.[24] Such a sequence is not present either in a unique N-terminal domain L or internally repetitive domains 1 to 4. Although domain 2 of low molecular weight kininogen is known to possess inhibitory activity against calpain,[25] no sequence homology is found between any portions of calpastatin and kininogen. Therefore, it is concluded that calpastatins constitute a unique family of cysteine proteinase inhibitors quite independent from the cystatin superfamily. A portion of the central consensus sequence of calpastatin was found to be similar to the sequence Ser-Ile-Pro-Pro in the reactive sites of some serine proteinase inhibitors and to the sequence Glu-Gly-Ile-Pro-Pro-Asp of ubiquitin (Table 1). Since the similarities between calpastatin and these peptides are observed only in restricted short segments, there may be no real biological significance in the observed similarities except that the corresponding segments might be exposed on the surface of protein molecules.

Amino termini of calpastatins isolated from pig heart and pig erythrocytes were blocked, and could not be determined by Edman degradation.[20] The isolated unblocked proteolytic peptide fragment which was located nearest the translation initiating methionine started from Gln[53] for pig heart calpastatin.[20] However, the amino terminus of calpastatin purified from rabbit liver was not blocked, and could be determined as glutamic acid, which corresponded to the 80th residue from the initiation methionine.[21] Imajoh et al.[26] analyzed the structure of rabbit erythrocyte calpastatin by peptide mapping. Although the amino-terminus was blocked, they estimated that the erythrocyte calpastatin started from Ser[290] and lacked domain L and domain 1.

III. MOLECULAR WEIGHT

The molecular weight of the pig calpastatin calculated from the deduced primary sequence (M_r 77,122) is significantly lower than the value of pig heart calpastatin (107 kDa) estimated by SDS-PAGE.[20] The discrepancy between the molecular weight estimated by SDS-PAGE (110,000) and that based on the amino acid sequence (68,000) was also reported in the case of rabbit liver calpastatin by the group of Suzuki.[21] A possibility of posttranslational modification such as glycosylation was first proposed to explain the molecular weight discrepancy.[21,27] However, from the results of the following experiments we ruled out such a possibility.[20] As shown in Figure 4, the expression of full and truncated forms of the calpastating gene in *E. coli* and the detection of translation products either by enzyme inhibition assay or by the antibody revealed that the proteins migrated considerably slower on SDS-PAGE.[20,28] The values of the molecular weight of the translated products estimated by SDS-PAGE were about 50% greater than the values calculated from the translated amino acid sequence (Table 2).

Regardless of the different sizes of the translation products expressed in *E. coli*, they must have similar amino acid compositions to one another due to the occurrence of repetitive

```
                       ~~~~~~~~~~                                         ┌──CCS──┐                                                  ────────
Domain 1 (H) 132 DAALDDLIDTLGGPEETEEENTTYTGPEVSDPMSSTYIEELGKREVTIPPKYRELLAK------------------PIGPDDAIDALSSDF 204
         (P) 155 DAALDDLIDTLGGPEETEEDNTTYTGPEVLDPMSSTYIEELGKREVTLPPKYRELLDK-KEGIPVPPPDTSK-----PLGPDDAIDALSLDL 240
         (R) 154 DAALDDLIDTLGEPSETQEDSTAYTGPEISDPMSSTYIEELGKREVTIPPKYRELLEK-KTGVAGPPPDSVT-----PLGPDDAIDALSSDF 239

Domain 2 (H) 253 DQALEALSASLGTRQAEPELDLRSIKEVDEAKAKEEKLEKCGEDDETIPSEYRLKPATDKDGKPLLPEPEEKPK---PRSESELIDELSEDF 341
         (P) 291 DQALEALSASLGSRKSEPELDLSSIKEIDEAKAKEEKLKKCGEDDETVPPEYRLKPAMDKDGKPLLPEAEEKPK---PLSESELIDELSEDF 379
         (R) 291 DQALEALSASLGTRMAEPELDLSSIKEVAEAKRKEEKVEKCGEDDETVPAEYRLKPATDKDGKPLLPEPAEKPK---PRSESELIDELSKDF 379

Domain 3 (H) 396 DDAVEALADSLGKKEADPEDGKPVM-DKVKEKAKEEDREKLGEKEETIPPDYRLEEVKDKDGKPLLPKESKEQLP--PMSEDFLLDALSEDF 484
         (P) 432 DDAVEALAGSLGKKEADPEDGKPVE-DKVKEKAKEEDREKLGEKEETIPPDYRLEEVKDKDGKTLPHKDPKEPVL--PLSEDFVLDALSQDF 520
         (R) 435 DDAVEALAGSLGRKEADPEEGKPVA-DKIKEKSKEEEREKLGEKEETIPPDYRLEEAKDKDGKPLLPSEPTAQLP--ALSEDLLLDALSEDF 523

Domain 4 (H) 533 DDALDKLSDSLGQRQPDPDENKPMG-DKVKEKAKAEHRDKLGERDDTIPPEYRHLLDDNGQDKPVKPPTKKSEDSKKPADDQDPIDALSGDL 623
         (P) 568 DDALDQLSDSLGQRQPDPDENKPIE-DKVKEKAEAEHRDKLGERDDTIPPEYRHLLDKDEEGKSTKPPTKKPEAPKKPEAAQDPIDALSGDF 658
         (R) 572 DDALDKLSDSLGQRQPDPDENKPME-DKVKERAKKEHKDKLGERDDTIPPEYRHLLDQGEQDKPEKPPTKKSKEIKKPAGDQDPIDALSGDL 662
```

FIGURE 3. Sequence alignment of conserved regions in internally repetitive domains of human (H), pig (P), and rabbit (R) calpastatins. Conserved regions in amino-terminal, central, and carboxyl-terminal areas are indicated by waved line, CCS and thin line above the sequence of human calpastatin domain 1. Conserved residues in all four domains of the three calpastatins are shaded.

TABLE 1
Quantitative Comparison of Similar Amino Acid Sequences

Calpastatin		Peptide compared			
Residue No.[a]	Sequence	Name	Residue No.[a]	Sequence	Av. CC
		Central consensus sequence region			
475	EETIPPDYRL	MAI-DE-4(Bowman-Birk family)	22	TKSIPPQCHC	0.63
336	ETVPPEYRLKP	Bradykinin-potentiating peptide B	1	QGLPPRPKIPP	0.62
611	DDTIPPEYRHL	Ubiquitin	33	KEGIPPDQQRL	0.61
474	KEETIPPDYR	Alpha-1-antitrypsin	356	IPMSIPPEVK	0.59
		Other region			
303	SRKSEPELDLS	Phosphatase inhibitor-1	43	SDQSSPEVDED	0.65
654	LSGDFDRCPSTTET	Ovomucoid	1	VEVDCSRFPNTTNE	0.62
538	AVSEVVSQTSA	Alpha-1-antichymotrypsin	70	TLTEILKASSS	0.62
586	DENKPIEDKVKEKA	Rat liver TPI (Stefin family)	14	TETQEIADKVKSQL	0.61

[a] The residue number indicated that of the amino-terminal residue of the segment in its parent protein.

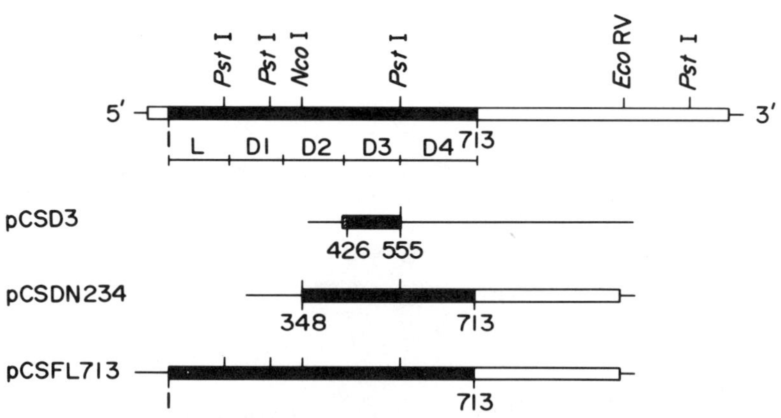

A

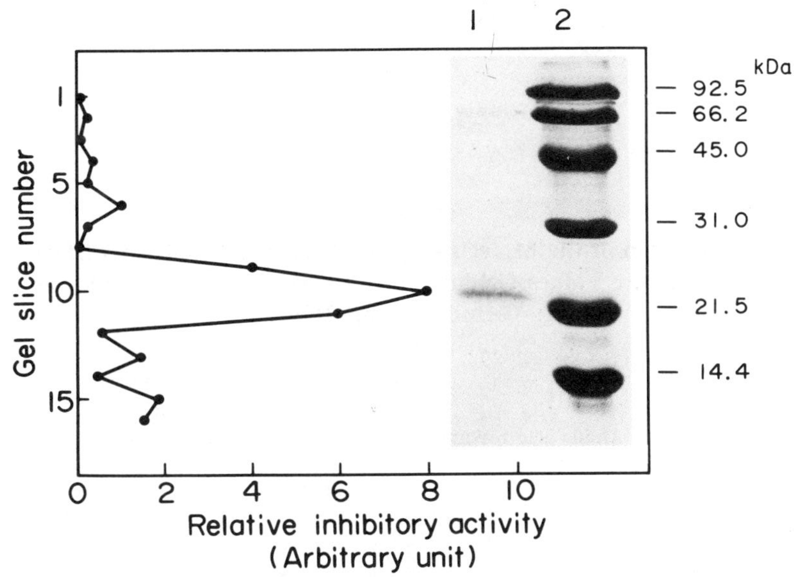

B

FIGURE 4. (A.) Structures of cDNAs for producing pig heart calpastatin in *E. coli*. The top bar indicates the cDNA structure of pig heart calpastatin. The protein coding region (■) and untranslated regions (□) are indicated. Translation product of pCSD3 has 10 extra noncalpastatin residues (hatched box). (B.) Inhibitory activity of pig calpastatin domain 3 protein expressed from clone pCSD3. After electrophoresis of the partially purified calpastatin segment on a polyacrylamide gel containing SDS, the inhibitor was extracted from the gel slices and assayed for calpain inhibition. Lane 1, inhibitor; lane 2, molecular marker proteins. (From Maki, M., Takano, E., Mori, H., Kannagi, R., Murachi, T., *Biochem. Biophys. Res. Commun.*, 143, 300, 1987. With permission.) (C.) Western blot analysis of purified authentic pig heart calpastatin and the extracts from *E. coli* cells. (Lane 1) purified authentic pig heart calpastatin; (other lanes) extracts from *E. coli* cells harboring clones pCSFL713 (lanes 2 and 5), pCSDN234 (lanes 3 and 6), and pUC119 (lane 4, negative control). For lanes 2 to 4, the extracts were prepared by directly dissolving the harvested and washed cells into the sample buffer for SDS-PAGE and boiling. For lanes 5 and 6, the extracts were prepared by treating the cells with lysozyme followed by freeze-thawing and centrifugation. Proteolytic degradation of the translation products were observed by this treatment. (From Takano, E., Maki, M., Mori, H., Hatanaka, M., Marti, T., Titani, K., Kunnagi, R., Ooi, T., and Murachi, T., *Biochemistry,* 27, 1964, 1988. With permission.)

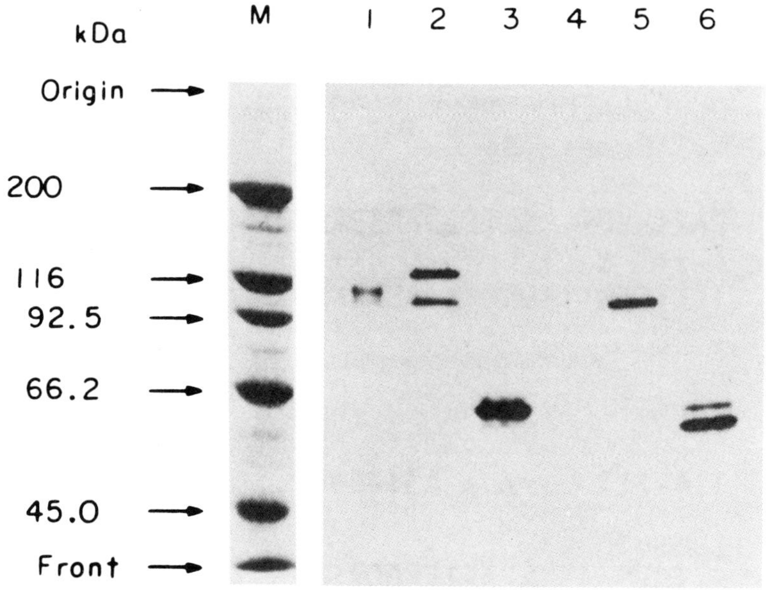

FIGURE 4C.

TABLE 2
Comparison of the Molecular Mass Values Estimated by SDS-PAGE to the Values Calculated from Amino Acid Residues as Deduced from cDNA

		Molecular mass values (kDa)		
Plasmid	Amino acid residues[a]	(A) SDS-PAGE	(B) Calculated[b]	A/B
pCSFL713	1—713	120	77.1	1.56
pCSDN234	348—713	64	39.7	1.61
pCSD3	426—555	22	15.0	1.47

[a] Residue numbers of calpastatin shown in Figure 4.
[b] Translation initiation methionine is included for the calculation of molecular mass.

sequences. Since posttranslational modification responsible for an increase in apparent molecular weight is unlikely to occur in *E. coli* cells, the retardation of the electrophoretic migration should be attributable to the nature of the calpastatin polypeptide. We concluded that the calpastatin molecule, or the repetitive domain(s) thereof, show an anomalous electrophoretic behavior in SDS-PAGE, giving an overestimation of the true molecular weight. The physicochemical properties of calpastatin responsible for such an anomalous behavior are yet to be defined, but they must reflect rather unique amino acid compositions of calpastatins. Calpastatins have very few aromatic amino acids (1.5 to 2.8 mol%), and they are rich in proline (9.2 to 11.1 mol%) and charged amino acids (Glu, 11.5 to 12.1 mol%; Asp, 9.4 to 11.4 mol%; Lys, 10.6 to 14.7 mol%). The numerical values shown above in parentheses were taken from the sequence-deduced amino acid compositions of the translation products of pCSFL713, pCSDN234, and pCSD3 (Figure 4 and Table 2), and from the amino acid analysis of purified pig heart calpastatin.[17] High contents of negatively charged amino acids may restrict the binding of SDS to protein molecules. Anomalous electrophoretic behaviors have also been reported for several other proteins rich in proline or acidic residues.[29-31]

Overestimation of the calpastatin molecular weight by SDS-PAGE should elicit reconsideration on the previously reported stoichiometry of calpain inhibition. Previously Takano et al.[17] reported that pig heart and erythrocyte calpastatins could bind approximately 8 and 5 calpain molecules per molecule of calpastatin, respectively. The calculation was based on the assumption that the molecular mass values obtained by SDS-PAGE, 107 kDa for heart calpastatin and 68 kDa for erythrocyte calpastatin, should directly represent the respective molecular weights. Since overestimation by SDS-PAGE became evident, the calculation of the calpain-binding capacities of calpastatin species was corrected accordingly. For such correction, we utilized and averaged ratio of the overestimation, 1.53, obtained from the data shown in Table 2. It followed that one molecule of pig heart and erythrocyte calpastatins could bind approximately 5 and 3 calpain molecules, respectively. Imajoh et al.[26] also reported that 1 molecule of rabbit liver and erythrocyte calpastatin bound approximately 4 and 3 calpain molecules, respectively, based on their molecular masses calculated from their amino acid sequences. These corrected numbers may correspond to the numbers of repetitive domains, 4 in heart calpastatin and 3 in erythrocyte calpastatin.

IV. STRUCTURE-FUNCTION RELATIONSHIP

The good agreement between the numbers of the repetitive domains in calpastatin sequences and the numbers of calpain inhibition sites per calpastatin molecule implies that the inhibitory function of calpastatin is attributed to each repetitive domain. The evidence that supported the hypothesis was reported first by Maki et al.[28,32] and later by Emori et al.[33] Complementary DNA portions coding for each domain of pig calpastatin were subcloned into *E. coli* plasmids to express the respective portions of the proteinase inhibitor gene in bacteria. Cell extracts of *E. coli* harboring recombinant plasmids were assayed for calpain inhibition. All four internally repetitive domains showed inhibitory activities, essentially similar to one another, against calpains I and II (Figure 5). No inhibition was observed in the case of the N-terminal nonhomologous domain (domain L).

The structure-function relationship of the unit inhibitory domain was further analyzed by expression of serial mutated cDNAs of pig calpastatin domain 3 in *E. coli*.[34] Conserved regions on both the N- and C-terminal sides were predicted to form α-helix structures. Deletion of either α-helix structure caused a drastic loss of inhibitory activity against calpain I and, to a lesser degree, against calpain II (Figure 6). Inhibitory activities were below the detectable level in mutants deleted further toward the central consensus sequence. Substitution of two amino acids immediately preceding the central consensus sequence of the wild-type domain 3 protein caused a drastic loss of inhibitory activity against both calpains.

The location of each amino acid residue at the surface or in the interior of the protein was predicted and expressed as the relative location measure (Figure 7).[24] The relative location measure profiles of domains 1 to 4 are similar to one another, suggesting similarities in tertiary structures. Conserved regions in the N- and C-terminal halves of each internally repetitive domain, marked with open boxes under letters N and C, respectively, were predicted to be located in the interior of calpastatin, as shown by positive values in the relative location measure in Figure 7. In contrast, the region containing the central consensus sequences was predicted to be rather exposed on the surface and to form turns (negative values and troughs in the profile in Figure 7).

The mutation experiments and the prediction of amino acid residue location (on the surface or in the interior of the protein) suggest that the central conserved region of each repetitive domain is an area for direct interaction with calpain, and the rest of the domain is a region stabilizing the functionally important tertiary structure of the domain. The degree

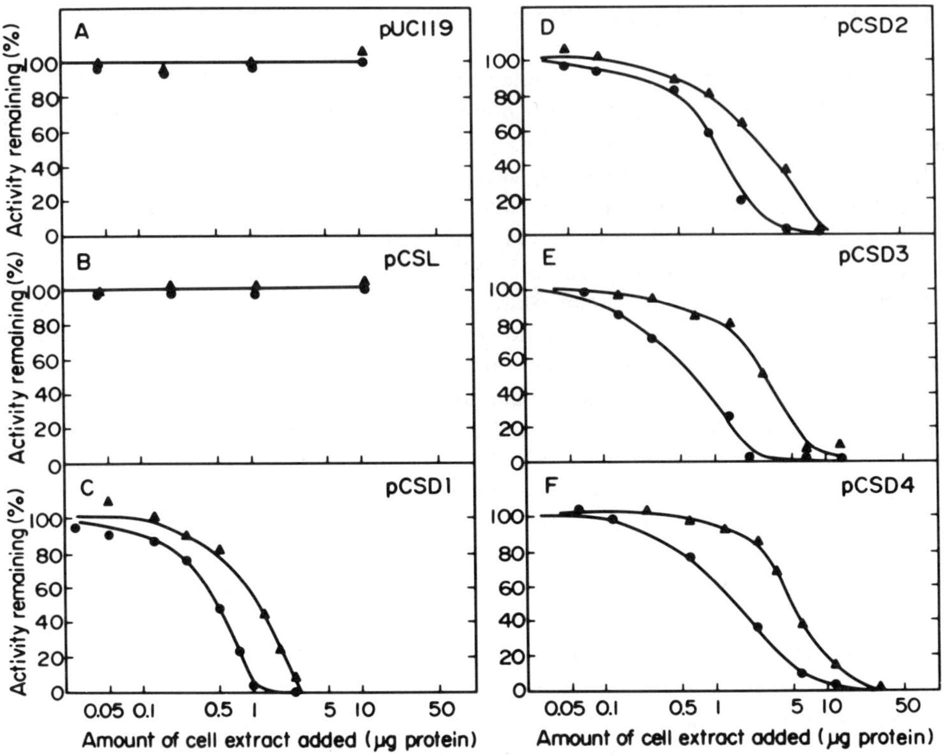

FIGURE 5. Inhibition profiles of calpains I and II. Proteolytic activities of calpain I (▲) and calpain II (●) were assayed in the presence of various amounts of crude cell extracts of *E. coli* harboring recombinant plasmids which expressed each domain of pig calpastatin: (A) pUC119, negative control; (B) pCSL, 1—140, domain L; (C) pCSD1, 137—272, domain 1; (D) pCSD1, 273—416, domain 2 (E) pCSD3, 426—555, domain 3; (F) pCSD4, 555—713, domain 4 (Numbers of the first and the last residue of the expressed segment of calpastatin are indicated). (From Maki, M., Takano, E., Mori, H., Sato, A., Murachi, T., and Hatanaka, M., *FEBS Lett.*, 223, 174, 1987. With permission.)

of stabilizing effects of N- or C-terminal regions may vary among domains. Emori et al. reported that by deletion of N- or C-terminal regions of rabbit calpastatin domains, the activity of domain 3 was most affected, whereas domain 1 retained a significant activity.[33] Recently synthetic oligopeptides covering the central consensus sequence of rabbit and human calpastatin domain 1, have been shown to possess inhibitory activities.[35,62,63]

V. INHIBITION MECHANISM

The mechanism of the inhibition of calpain by calpastatin has not yet been clarified. The inhibition is not due to the sequestering of Ca^{2+} by the inhibitor. The lines of supporting evidence include no effect of higher Ca^{2+} concentration on the degree of inhibition, no effect of calpastatin on the Ca^{2+} concentration required for half-maximal enzyme activity, and inability of the inhibitor to bind Ca^{2+}.[2,36-39] However, Ca^{2+} is known to be required for calpastatin to bind calpain, and the inhibitor can be dissociated from calpain by chelating Ca^{2+}.[2,40,41] Tsuji et al.[42] reported that the role of Ca^{2+} in calpain activation is to change the conformation of the enzyme into a form that is able to bind the substrate. A similar hypothesis should be applicable to the calpain-calpastatin interaction.

Cottin et al.[43] and Shiba et al.[13] reported that the mode of inhibition of calpains by calpastatin was noncompetitive by Dixon plot analysis. As pointed out by Laskowski and Sealock,[44] the graphical kinetic analyses are not appropriate for tight-binding inhibitors like

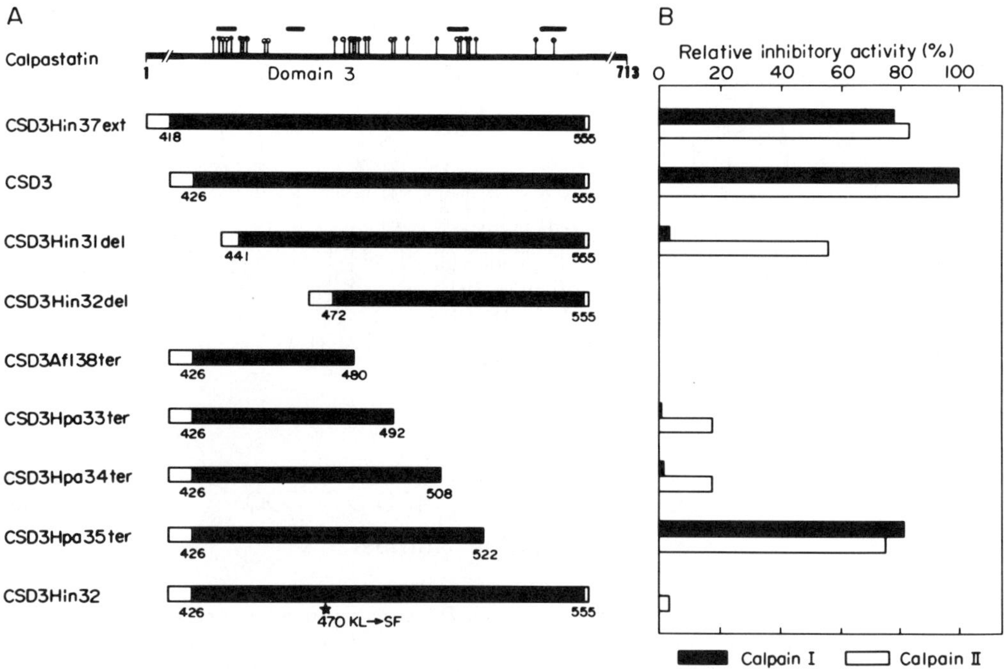

FIGURE 6. Relative inhibitory activities of mutated calpastatin domain 3 proteins. (A) Domain 3 proteins with different molecular structures were synthesized in *E. coli* by expressing calpastatin cDNA sequences fused with the amino-terminal region of the *lacZ'* gene. Closed bars and open bars represent calpastatin and noncalpastatin sequences, respectively. Numbers under the bars indicate the first and the last amino acid residue numbers of each calpastatin protein. Closed and open circles marked above the top line indicate identical residues and preferred substitutions among pig calpastatin repetitive domains, respectively. Regions of α-helices predicted are marked with short horizontal bars on top. (B) Inhibitory activities of mutated calpastatin domain 3 proteins were assayed using crude extracts of *E. coli* cells harboring recombinant plasmids as test samples. The inhibitory activities of the mutants are expressed as relative activities which were obtained by calculating the ratio of specific activity (calpastatin units/mg protein) of each mutant to that of CSD3 on calpain I (closed bars) or calpain II (open bars). Activities were below the detectable level in the cases of CSD3Hin32del (calpains I and II), CSD3Afl38ter (calpains I and II), and CSD3Hin32 (calpain I). (From Maki, M., Takano, E., Osawa, T., Ooi, T., Murachi, T., and Hatanaka, M., *J. Biol. Chem.*, 263, 10254, 1988. With permission.)

the calpastatins. Correct K_i values are not given by the conventional kinetic analyses, and the plots indicative of noncompetitive inhibition should not be taken as defining the mechanism of inhibition.[8] To partially circumvent this problem, we used gene-engineered calpastatins with low binding affinities for kinetic analyses (Figure 8).[34] The double-reciprocal plots indicated that the calpastatins inhibited calpain competitively, suggesting direct interaction either with the active center of calpain or in close proximity to it. The calculated inhibition constants of CSD3Hpa33ter, for example, gave K_i values of 3.8×10^{-8} *M* for calpain I and 4.6×10^{-9} *M* for calpain II (Table 3). Although inhibition constants for calpastatins purified from tissues could not be determined, due to the tight-bindings, the inhibition constants should be considerably lower, by at least a magnitude of 10 to 100, than the above values.

Inactivation of calpain activity by carboxymethylation of a cysteine residue in its catalytic site has been shown not to prevent the binding of calpastatin to the enzyme.[43,45] Cottin et al.[43] claimed that this was good evidence for the noncompetitive mechanism of calpain inhibition by calpastatin. However, the carboxymethyl group would not necessarily inhibit the binding of inhibitor with the enzyme if the two molecules interact at multiple sites,

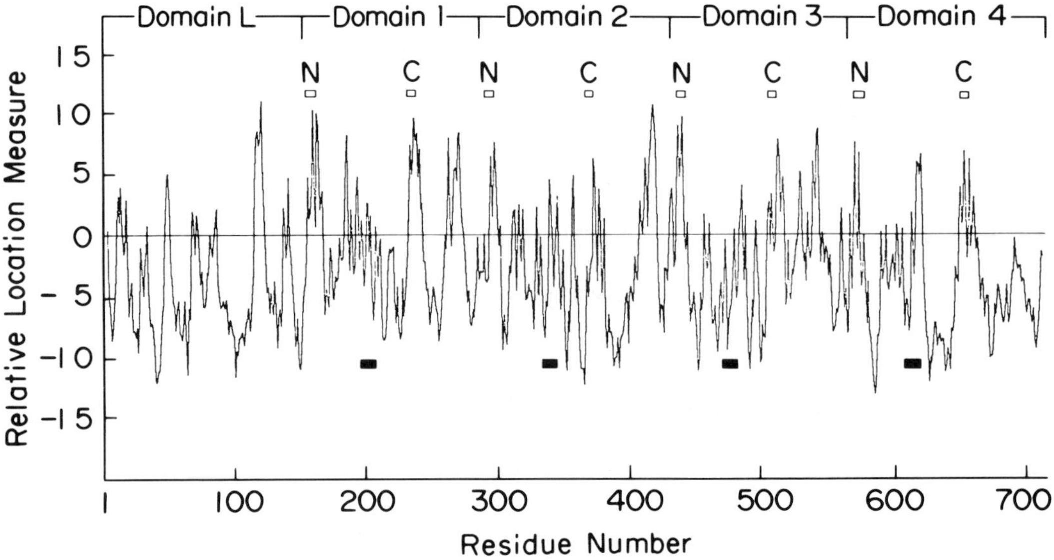

FIGURE 7. Computed profile of relative location measures for 713 amino acid residues of pig calpastatin. Location of an amino acid residue on the surface or in the interior portion of the calpastatin molecule was predicted and expressed as the relative location measure. Positive and negative values correlate with location preference of being in the interior (usually hydrophobic) and on the surface (usually hydrophilic) portions, respectively. Conserved regions at the amino-terminal and carboxyl-terminal sides of each repetitive domain are marked with open boxes under the letters N and C, respectively (for N: domain 1, 155—166; domain 2, 291—302; domain 3, 432—443; domain 4, 568—579; for C: domain 1, 233-239; domain 2, 372—378; domain 3, 513—519; domain 4, 651—657). The "central consensus sequence" regions are marked with closed boxes (domain 1, 196—207; domain 2, 332—343; domain 3, 472—483; domain 4, 608—619). (From Maki, M., Takano, E., Osawa, T., Ooi, T., Murachi, T., and Hatanaka, M., *J. Biol. Chem.*, 263, 10254, 1988. With permission.)

including sites close to the catalytic site. Carboxymethylation of active site Cys25 of papain is also known to be noneffective for the abolition of papain-cystatin interaction,[46] although papain is inhibited by cystatin competitively.[47] Recently, an X-ray crystal structure of chicken egg white cystatin has been reported.[48] Two hairpin loops and the N-terminal portion form a wedge-shaped edge which is quite complementary to the active cleft of papain. Docking of cystatin and papain by computer graphics suggests that the interaction sites are in the neighborhood of the active site Cys25 of papain but too far away to be attacked by the active site residue. According to the proposed model, the mechanism of binding of cystatin inhibitors to papain-like proteinases is different from the mechanism observed for the inhibition of serine proteinases by most of the small protein inhibitors which bind like substrates. It is premature to decide whether or not such a multiple-site hypothesis can also be applied to calpain-calpastatin interaction. Nevertheless, it should be noted that the 12-residue consensus sequence in the central region of calpastatin (Figures 2, 3,) does not contain any calpain-preferred P_2-P_1 site, Leu (or Val)-X-, where X stands for a rather bulky residue such as methione, phenylalanin, tyrosine, lysine, and arginine.[49,50] This would mean that the central consensus sequence region may not serve as a substrate-like site which interacts with the catalytic site of the enzyme. This thought agrees with the observation of Imajoh and Suzuki,[45] that calpastatin released from the complex with calpain by chelating calcium ions retains full inhibitory activity, and the calpain-calpastatin interaction is reversible. The competitive and reversible inhibitory nature of calpastatin suggests that the inhibition mechanism of this inhibitor is closer to that of cystatin than to that of serine proteinase inhibitors.[47,48,51]

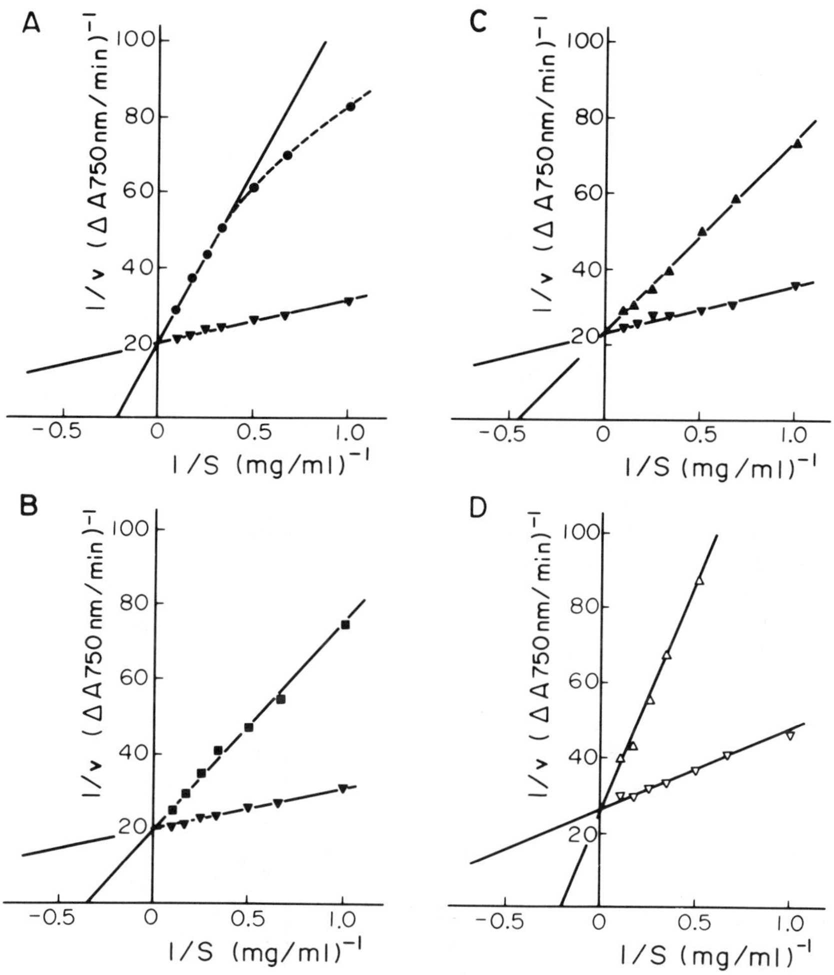

FIGURE 8. Double-reciprocal plots of calpain activity in the presence and in the absence of calpastatin domain 3 proteins. The concentration of casein in the reaction mixture was varied from 1 to 10 mg/ml. A, B, C, calpain I; D, calpain II. The reaction was performed in the absence (▼ or ▽) of the inhibitor and in the presence of 20.4 n*M* CSD3 (●); 236 n*M* CSD3Hin31del (■); 100 n*M* CSD3Hpa33ter (▲); or 30.3 n*M* CSD3Hpa33ter (△). (From Maki, M., Takano, E., Osawa, T., Ooi, T., Murachi, T., and Hatanaka, M., *J. Biol. Chem.*, 263, 10254, 1988. With permission.)

TABLE 3
Inhibition Constants of Pig Calpastatin Domain 3 Proteins Produced in *E. coli*.

	Calpain I		Calpain II	
Calpastatin	**IC50**[a]	**K_i**[b]	**IC50**[a]	**K_i**[b]
	n*M*		n*M*	
CSD3	18	3.0	4.7	ND[c]
CSD3Hin31del	230	72	6.8	ND
CSD3Hpa33ter	105	38	14	4.6

[a] The concentration of calpastatin which was required to cause 50% inhibition of calpain: calpain I, 15 n*M*, calpain II, 10 n*M*.
[b] Calculated from the results shown in Figure 8.
[c] ND, not determined.

VI. CONCLUSIONS

In the last few years gene technology has made a great contribution toward elucidation of the molecular structures of calpastatins and their structure-function relationships. The amino acid sequences deduced from cDNAs show no homology with any other proteins reported to date. This fact allows one to conclude that calpastatins constitute a unique family of proteinase inhibitors quite independent from the cystatin superfamily. Calpastatin possesses four functionally active internally repetitive domains, each composed of approximately 140 amino acid residues. Detailed analysis of calpain-calpastatin interactions, including inhibition mechanisms, will be facilitated by using single-domain inhibitors either produced in bacteria or synthesized as synthetic peptides.

The multiple and internally repetitive domain structure of a proteinase inhibitor is not unique to calpastatin. Kininogens[52] and ovomucoids[53] are other examples. Although multiplication of a unit inhibitor domain in calpastatin during evolution must have contributed to an increase of inhibitory capacities per one molecule of protein (fourfold, if disregarding any steric effect), the biological significance of this multidomain structure remains to be determined. DeMartino et al. reported that extensive digestion of purified bovine heart calpastatin with trypsin, chymotrypsin or *Staphylococcus aureus* V8 protease generated lower molecular weight peptides, with little or no effect on inhibitory activity.[54] Recently we isolated a human calpastatin genomic DNA clone. Partial DNA sequence analysis has revealed that domain 1 is split into four exons (Figure 9).[55] The three conserved regions are located on separate exons, 1A, 1B, and 1C. Exon 1D is linked to the 5′ end of the next exon which corresponds to exon 2A. This fact would suggest that the entire coding region of human calpastatin is split into at least 16 exons for the internally repetitive domains 1 to 4, and a few exons for the N-terminal nonhomologous domain L. Interestingly, erythrocyte calpastatin lacks domains L and 1, and the N-terminal residue is blocked.[26] Although molecular structures have not yet been clarified, calpastatins smaller than full-sized forms have also been reported.[56,57] It is not known whether such smaller molecules are formed merely by proteolytic degradation by cellular proteases including calpain,[41,58-60] or by difference in transcriptional controls, or in pre-mRNA processing. The question will be answered by molecular biological studies using a cloned genomic DNA as a probe.

ACKNOWLEDGMENTS

We wish to thank Dr. T. R. Shearer for his helpful discussions and Dr. T. Ooi for the quantitative sequence comparison of calpastatin with other inhibitors. The experiments and preparation of this manuscript were partially supported by Grants for Scientific and Cancer Research from the Ministry of Education, Science and Culture, Japan.

```
        Exon 1A
tttccagTCAGGCATGGATGCTGCTTTGGATGACTTAATAGATACTTTAGGAGGACCTGAAGAAACTGAAGAAGAAAATACAACGTATACTGGACCAGAAGTTTCAgtatgtg
        S  G  M  D  A  A  L  D  D  L  I  D  T  L  G  G  P  E  E  T  E  E  E  N  T  T  Y  T  G  P  E  V  S
       129          N-term. conserved seq.                                                          161

        Exon 1B
gttccagGATCCAATGAGTTCCACCTACATAGAGGAATTGGGTAAAAGAGAAGTCACAATTCCTCCAAAATATAGGGAACTATTGGCTgtaagtt
        D  P  M  S  S  T  Y  I  E  E  L  G  K  R  E  V  T  I  P  P  K  Y  R  E  L  L  A
       162                                     Central consensus sequence          188

        Exon 1C
tctttagAAACCCATAGGGCCAGATGATGCTATAGACGCCTTGTCATCTGACTTCACCTGTGGGTCGCCTACAGCTGCTGGAAAGAAAACTGAAAAAGAGgtattgt
        K  P  I  G  P  D  D  A  I  D  A  L  S  S  D  F  T  C  G  S  P  T  A  A  G  K  K  T  E  K  E
       189                      C-term. conserved seq.                                    219

        Exon 1D
ttttcagGAATCTACAGAAGTTTTAAAAGCTCAGTCAGCAGGGACAGTCAGAAGTGCTGCTCCACCCCAAGAGAAGAAAAGAAAGGTGGAGAAGgtatagt
        E  S  T  E  V  L  K  A  Q  S  A  G  T  V  R  S  A  A  P  P  Q  E  K  K  R  K  V  E  K
       220                                                                         248

        Exon 2A
ttaacagGATACAATGAGTGATCAAGCACTCGAGGCTCTGTCGGCTTCACTGGGCACCCGGCAAGCAGAACCTGAGCTCGACCTCCGCTCAATTAAGGAAGTCGATGAGgtactga
        D  T  M  S  D  Q  A  L  E  A  L  S  A  S  L  G  T  R  Q  A  E  P  E  L  D  L  R  S  I  K  E  V  D  E
       249                N-term. conserved seq.                                                       282
```

FIGURE 9. Partial nucleotide sequence of human calpastatin genomic DNA. One repeat unit of calpastatin, domain 1, is split into four exons. Conserved amino acid sequences among internally repetitive domains (see Figure 3) are underlined.

REFERENCES

1. **Waxman, L. and Krebs, E. G.,** Identification of two proteinase inhibitors from bovine cardiac muscle, *J. Biol. Chem.*, 253, 5888, 1978.
2. **Nishiura, I., Tanaka, K., Yamato, S., and Murachi, T.,** The occurrence of an inhibitor of Ca^{2+}-dependent neutral protease in rat liver, *J. Biochem.*, 84, 1657, 1978.
3. **Drummond, G. I. and Duncan, L.,** The action of calcium ion on cardiac phosphorylase b kinase, *J. Biol. Chem.*, 241, 3097, 1966.
4. **Okitani, A., Goll, D. E., Stromer, M. H., and Robson, R. M.,** Intracellular inhibitor of a Ca^{2+}-activated protease involved in myofibrillar protein turnover, *Fed. Proc.*, 35, 1746, 1976.
5. **Murachi, T., Hatanaka, M., Yasumoto, Y., Nakayama, N., and Tanaka, K.,** A quantitative distribution study on calpain and calpastatin in rat tissues and cells, *Biochem. Int.*, 2, 651, 1981.
6. **Murachi,T., Tanaka, K., Hatanaka, M., and Murakami, T.,** Intracellular Ca^{2+}-dependent protease (calpain) and its high-molecular-weight endogenous inhibitor (calpastatin), *Adv. Enzyme Regul.*, 19, 407, 1981.
7. **Murachi, T.,** Calpain and calpastatin, *Trends Biochem. Sci.*, 8, 167, 169.
8. **Parkes, C.,** Calpastatins, in *Proteinase Inhibitors,* Barrett, A. J. and Salvesen, G., Eds., Elsevier, Amsterdam, 1986, chap. 19.
9. **Waxman, L. and Krebs, E. G.,** Identification of two protease inhibitors from bovine cardiac muscle, *J. Biol. Chem.*, 253, 5888, 1978.
10. **Takano, E. and Murachi, T.,** Purification and some properties of human erythrocyte calpastatin, *J. Biochem.*, 92, 2021, 1982.
11. **Waxman, L.,** Characterization of two unique proteinase inhibitors and a calcium-activated protease from bovine cardiac muscle, in *Protein Turnover and Lysosome Function,* Segal, H. L. and Doyle, D. J., Eds., Academic Press, New York, 1978, 363.
12. **Otsuka, Y. and Goll, D. E.,** Purification of the Ca^{2+}-dependent proteinase inhibitor from bovine cardiac muscle and its interaction with the millimolar Ca^{2+}-dependent proteinase, *J. Biol. Chem.*, 262, 5839, 1987.
13. **Shiba, E., Tsujinaka, T., Kambayashi, J., and Kosaki, G.,** Purification and characterization of Ca^{2+}-activated neutral protease inhibitor from human platelets, *Thromb. Res.*, 32, 207, 1983.
14. **Melloni, E., Salamino, F., Sparatore, B., Michetti, M., Pontremoli, S., and Horecker, B. L.,** Regulation of the Ca^{2+}-dependent neutral proteinases from rabbit liver by an endogenous inhibitor, *Arch. Biochem. Biophys.*, 232, 513, 1984.
15. **DeMartino, G. N. and Croall, D. E.,** Purification and characterization of a protein inhibitor of calcium-dependent proteases from rat liver, *Arch. Biochem. Biophys.*, 232, 713, 1984.
16. **Nakamura, M., Inomata, M., Hayashi, M., Imahori, K., and Kawashima, S.,** Purification and characterization of 213,000-dalton inhibitor of calcium-activated neutral protease from rabbit skeletal muscle and its relation to 50,000-dalton inhibitor, *J. Biochem.*, 98, 757, 1985.
17. **Takano, E., Kitahara, A., Sasaki, T., Kannagi, R., and Murachi, T.,** Two different molecular species of pig calpastatin: structural and functional relationship between 107 kDa and 68 kDa molecules, *Biochem. J.*, 235, 97, 1986.
18. **Takano, E., Yumoto, N., Kannagi, R., and Murachi, T.,** Molecular diversity of calpastatin in mammalian organs, *Biochem. Biophys. Res. Commun.*, 122, 912, 1984.
19. **Takano, E., Maki, M., Hatanaka, M., Mori, H., Zenita, K., Sakihama, T., Kannagi, R., Marti, T., Titani, K., and Murachi, T.,** Evidence for the repetitive domain structure of pig calpastatin as demonstrated by cloning of complementary DNA, *FEBS Lett.*, 208, 199, 1986.
20. **Takano, E., Maki, M., Mori, H., Hatanaka, M., Marti, T., Titani, K., Kannagi, R., Ooi, T., and Murachi, T.,** Pig heart calpastatin: identification of repetitive domain structures and anomalous behavior in polyacrylamide gel electrophoresis, *Biochemistry*, 27, 1964, 1988.
21. **Emori, Y., Kawasaki, H., Imajoh, S., Imahori, K., and Suzuki, K.,** Endogenous inhibitor for calcium-dependent cysteine protease contains four internal repeats that could be responsible for its multiple reactive sites, *Proc. Natl. Acad. Sci. U.S.A.*, 84, 3590, 1987.
22. **Asada, K., Ishino, Y., Shimada, M., Shimojo, T., Endo, M., Kimizuka, F., Kato, I, Maki, M., Hatanaka, M., and Murachi, T.,** A cloning of human calpastatin: structural comparison between human, pig, and rabbit calpastatins., *J. Enzyme Inhibition*, 3, 49, 1989.
23. **Ohkubo, I., Kurachi, K., Takasawa, T., Shiokawa, H., and Sasaki, M.,** Isolation of a human cDNA for α_2-thiol proteinase inhibitor and its identity with low molecular weight kininogen, *Biochemistry*, 23, 5691, 1984.
24. **Müller-Esterl, W., Fritz, H., Kellermann, J., Lottspeich, F., Machleidt, W., and Turk, V.,** Geneology of mammalian cysteine proteinase inhibitors: common evolutionary origin of stefins, cystatins and kininogens, *FEBS Lett.*, 191, 221, 1985.

25. **Salvesen, G., Parks, C., Abrahamson, M., Grubb, A., and Barrett, A. J.,** Human low-M_r kininogen contains three copies of a cystatin sequence that are divergent in structure and in inhibitory activity for cysteine proetinases, *Biochem. J.*, 234, 429, 1986.
26. **Imajoh, S., Kawasaki, H., Emori, Y., and Suzuki, K.,** Calcium-activated neutral protease inhibitor from rabbit erythrocytes lacks the N-terminal region of the liver inhibitor but retains three inhibitory units, *Biochem. Biphys. Res. Commun.*, 146, 630, 1987.
27. **Imajoh, S., Kawasaki, H., Emori, Y., Ishiura, S., Minami, Y., Sugita, H., Imahori, K., and Suzuki, K.,** A fragment of an endogenous inhibitor produced in *Escherichia coli* for calcium-activated neutral protease (CANP) retains an inhibitory activity, *FEBS Lett.* 215, 274, 1987.
28. **Maki, M., Takano, E., Mori, H., Kannagi, R., Murachi, T., and Hatanaka, M.,** Repetitive region of calpastatin is a functional unit of the proteinase inhibitor, *Biochem. Biophys. Res. Commun.*, 143, 300, 1987.
29. **Kaufmann, E., Geisler, N., and Weber, K.,** SDS-PAGE strongly overestimates the molecular masses of the neurofilament proteins, *FEBS Lett.*, 170, 81, 1984.
30. **Kleinschmidt, J. A., Dingwall, C., Maier, G., and Frake, W. W.,** Molecular characterization of a karyophilic, histone-binding protein: cDNA cloning, amino acid sequence and expression of nuclear protein N1/N2 of *Xenopus laevis, EMBO J.*, 5, 3547, 1986.
31. **Iino, T., Takeuchi, K., Nam, S. H., Siomi, H., Sabe, H., Kobayashi, N., and Hatanaka, M.,** Structural analysis of p28 adult T-cell leukaemia-associated antigen, *J. Gen. Virol.*, 67, 1373, 1986.
32. **Maki, M., Takano, E., Mori, H., Sato, A., Murachi, T., and Hatanaka, M.,** All four internally repetitive domains of pig calpastatin possess inhibitory activities against calpains I and II, *FEBS Lett.*, 223, 174, 1987.
33. **Emori, Y., Kawasaki, H., Imajoh, S., Minami, Y., and Suzuki, K.,** All four repeating domains of the endogenous inhibitor for calcium-dependent protease independently retain inhibitory activity: expression of the cDNA fragments in *Escherichia coli, J. Biol. Chem.*, 263, 2364, 1988.
34. **Maki, M., Takano, E., Osawa, T., Ooi, T., Murachi, T., and Hatanaka, M.,** Analysis of structure-function relationship of pig calpastatin by expression of mutated cDNAs in *Escherichia coli, J. Biol. Chem.*, 263, 10254, 1988.
35. **Kawasaki, H., Emori, Y., Imajoh-Ohmi, S., Minami, Y., and Suzuki, K.,** Indentification and characterization of inhibitory domains of the endogenous inhibitor for calcium-dependent protease, *J. Biochem.*, 106, 274, 1989.
36. **Murakami, T., Hatanaka, M., and Murachi, T.,** The cytosol of human erythrocytes contains a highly Ca^{2+}-sensitive thiol proteinase (calpain I) and its specific inhibitor (calpastatin), *J. Biochem.*, 90, 1809, 1981.
37. **Takano, E. and Murachi, T.,** Purification and some properties of human erythrocyte calpastatin, *J. Biochem.*, 92, 2021, 1982.
38. **Ishiura, S., Tsuji, S., Murofushi, H., and Suzuki, K.,** Purification of an endogenous 68,000 dalton inhibitor of Ca^{2+}-activated neutral protease from chicken skeletal muscle, *Biochim. Biophys. Acta,* 701, 216, 1982.
39. **Cottin, P., Vidalenc, P. L., and Ducastaing, A.,** Ca^{2+}-dependent association between a Ca^{2+}-activated neutral proteinase (CaANP) and its specific inhibitor, *FEBS Lett.*, 136, 221, 1981.
40. **Melloni, E., Sparatore, B., Salamino, F., Michetti, M., and Pontremoli, S.,** Cytosolic calcium-dependent proteinase of human erythrocytes: formation of an enzyme-natural inhibitor complex induced by Ca^{2+} ions, *Biochem. Biophys. Res. Commun.*, 106, 731, 1982.
41. **Shigeta, K., Yumoto, N., and Murachi, T.,** Fragmentation of a 70,000 dalton calpastatin molecule upon its complex formation with calpain, *Biochem. Int.*, 9, 327, 1984.
42. **Tsuji, S., Ishiura, S., Takahashi-Nakamura, M., Katamoto, T., Suzuki, K., and Imahori, K.,** Studies on a Ca^{2+}-activated neutral proteinase of rabbit skeletal muscle. II. Characterization of sulfhydryl groups and a role of Ca^{2+} ions in this enzyme, *J. Biochem.*, 90, 1405, 1981.
43. **Cottin, P., Vidalenc, P. L., Merdaci, N., and Ducasting, A.,** Evidence for noncompetitive inhibition between two calcium-dependent activated neutral proteinases and their specific inhibitor. *Biochim. Biophys. Acta,* 743, 299, 1983.
44. **Laskowski, M., Jr., and Sealock, R. W.,** Protein proteinase inhibitors — molecular aspects, in *The Enzymes,* Vol. 3, Boyer, P. D., Ed., Academic Press, New York, 1971, 375.
45. **Imajoh, S. and Suzuki, K.,** Reversible interaction between Ca^{2+}-activated neutral protease (CANP) and its endogenous inhibitor, *FEBS Lett.*, 187, 47, 1985.
46. **Anastasi, A., Brown, M. A., Kembhavi, A. A., Nicklin, M. J. H., Sayers, C. A., Sunter, D. C., and Barrett, A. J.,** Cystatin, a protein inhibitor of cysteine proteinases: improved purification from egg white, characterization and detection in chicken serum, *Biochem. J.*, 211, 129, 1983.
47. **Nicklin, M. J. H. and Barrett, A. J.,** Inhibition of cysteine proteinases and dipeptidyl peptidase I by egg-white cystatin, *Biochem. J.*, 223, 245, 2984.

48. **Bode, W., Engh, R., Musil, D., Thiele, U., Huber, R., Karshikov, A., Brzin, J., Kos, J., and Turk, V.,** The 2.0 Å X-ray crystal structure of chicken egg white cystatin and its possible mode of interaction with cysteine proteinases, *EMBO J.*, 7, 2593, 1988.
49. **Sasaki, T., Kikuchi, T., Yumoto, N., Yoshimura, N., and Murachi, T.,** Comparative specificity and kinetic studies on porcine calpain I and calpain II with naturally occurring peptides and synthetic fluorogenic substrates, *J. Biol. Chem.*, 259, 12489, 1984.
50. **Sasaki, T., Kikuchi, T., Fukui, I., and Murachi, T.,** Inactivation of calpain I and calpain II by specificity-oriented tripeptidyl chloromethyl ketones, *J. Biochem.*, 99, 173, 1986.
51. **Read, R. J. and James, M. N. G.,** Introduction to the protein inhibitors: X-ray crystallography, in *Proteinase Inhibitors*, Barrett, A. J. and Salvesen, G., Eds., Elsevier, Amsterdam, 1986, chap. 7.
52. **Müller-Esterl, W., Iwanaga, S., and Nakanishi, S.,** Kininogens revisited, *Trends Biochem. Sci.*, 11, 336, 1986.
53. **Kato, I., Schrode, J., Kohr, W. J., and Laskowski, M., Jr.,** Chicken ovomucoid: determination of its amino acid sequence, determination of the trypsin reactive site, and preparation of all three of its domains, *Biochemistry*, 26, 193, 1987.
54. **DeMartino, G. N., Wachendorfer, R., McGuire, M. J., and Croall, D. E.,** Proteolysis of the protein inhibitor of calcium-dependent proteases produces lower molecular weight fragments that retain inhibitory activity, *Arch. Biochem. Biophys.*, 262, 189, 1988.
55. **Murachi, T., Adachi, Y., Takano, E., Maki, M., and Hatanaka, M.,** Recent progress in calpastatin research (Abstract), 4th Rinsho-ken Inter. Conf. New Horizon for Protease and Its Inhibitor Research, Tokyo, June 5 to 7, 1989.
56. **Adachi, Y., Takano, E., Murachi, T., and Hatanaka, M.,** Distribution and expression of calpastatin in human hematopoietic system cells, *Biol. Chem. Hoppe-Seyler*, 369 (Suppl. 223), 1988.
57. **Yamato, S., Tanaka, K., and Murachi, T.,** The appearance of a 34,000-dalton inhibitor of calpain (Ca^{2+}-dependent cysteine proteinase) in rat liver after the administration of phenylhydrazine, *Biochem. Biophys. Res. Commun.*, 115, 715, 1983.
58. **Mellgren, R. L. and Carr, T. C.,** The protein inhibitor of calcium-dependent proteases: purification from bovine heart and possible mechanism of regulation, *Arch. Biochem. Biophys.*, 225, 779, 1983.
59. **Mellgren, R. L., Mericle, M. T., and Lane, R. D.,** Proteolysis of the calcium-dependent protease inhibitor by myocardial calcium-dependent protease, *Arch. Biochem. Biophys.*, 246, 233, 1986.
60. **Nakamura, M., Inomata, M., Imajoh, S., Suzuki, K., and Kawashima, S.,** Fragmentation of a endogenous inhibitor upon complex formation with high-and low-Ca^{2+}-requiring forms of calcium-activated neutral proteases, *Biochemistry*, 28, 449, 1989.
61. **Kubota, Y., Nishikawa, K., Takahashi, S., and Ooi, T.,** Correspondence of homologies in amino acid sequence and tertiary structure of protein molecules, *Biochem. Biophys. Acta*, 701, 242, 1982.
62. **Maki, M., Bagči, H., Hamaguchi, K., Ueda, M., Murachi, T., and Hatanaka, M.,** Inhibition of calpain by a synthetic oligopeptide corresponding to an exon of the human calpastatin gene, *J. Biol. Chem.*, 264, 18866, 1989.
63. **Uemori, T., Shimojo, T., Asada, K., Asano, T., Kimizuka, F., Kato, I., Maki, M., Hatanaka, M., Murachi, T., Hanzawa, H., and Arata, Y.,** Characterization of a functional domain of human calpastatin, *Biochem. Biophys. Res. Commun.*, 166, 1485, 1990.

Chapter 4

CALPAIN SUBSTRATE SPECIFICITY

Kenji Takahashi

TABLE OF CONTENTS

I. INTRODUCTION

Calpains I and II are known to exist widely in animal tissues and appear to be involved in various aspects of physiological regulation at a cellular level through limited proteolysis of a variety of physiologically important proteins and peptides.[1,2] In order to shed light on the physiological functions of calpains, it is important to clarify their substrate specificities and define their native, physiological substrates. The study of the substrate specificity is also important to elucidate the molecular mechanism of calpain action. The present chapter will give an overview on what is known so far regarding the substrate specificity of the calpains toward proteins and peptides. The substrate specificity of calpains has been reviewed recently by Murachi et al.,[3] with special reference to its implication in neuropeptide metabolism.

II. SUBSTRATE SPECIFICITY OF CALPAIN TOWARD PEPTIDES

Previously, calpain was presumed to have a tendency to hydrolyze only certain proteins and large peptides and to scarcely hydrolyze smaller peptides or synthetic substrates. Guroff[4] first reported that calpain from rat brain cleaved none of the peptide substrates tested, most of which were dipeptides and their derivatives (Table 1), while it hydrolyzed oxidized insulin B chain at the Tyr^{16}-Leu^{17} bond and alkali-denatured casein. Waxman[5] reported that calpain from bovine cardiac muscle did not degrade any of the peptides tested including reduced and carboxymethylated insulin (Table 1). Ishiura et al.[6] also investigated the substrate specificity of calpain from chicken skeletal muscle, reporting that none of the synthetic substrates examined were susceptible to calpain (Table 1), whereas oxidized insulin B chain was hydrolyzed at a few specific sites including the Tyr^{16}-Leu^{17} bond. Based on these results, they concluded that calpain hydrolyzes only fairly long polypeptides without requiring specific amino acid sequences. Later, they reported a very slow hydrolysis of *N*-benzyloxycarbonyl-L-tyrosine-*p*-nitrophenyl ester as an exceptional case.[7]

More recently, however, various neuropeptides and other naturally occurring peptides have been examined as substrates for calpains, first by Hirao et al.,[8] Hirao and Takahashi,[9] and Sasaki et al.[10] These studies have demonstrated that calpain can hydrolyze several oligopeptides with a certain degree of peptide bond specificity, and thus provided some clue to the understanding of the substrate specificity of calpain.

A. SPECIFICITY OF ACTION ON OLIGOPEPTIDES AND SUBSITE SPECIFICITY

Table 2 summarizes the reported cleavage sites in various neuropeptides and other peptides by calpains from various sources including monkey cardiac muscle[8] and brain[9] calpains II, porcine erythrocyte calpain I[10] and porcine kidney calpain II,[10] chicken skeletal muscle calpain II,[11] human skeletal muscle calpains I and II,[12] and bovine spinal cord calpain II (soluble and cytoskeleton bound forms).[13] The cleavage sites are generally similar among the enzymes, and only one or a few peptide bonds, which are generally in the interior part of the peptides, were cleaved rather specifically. These results indicate that calpain is an endopeptidase with some degree of substrate specificity. The cleavage of the Tyr^{1}-Gly^{2} bond in Leu-enkephalin by calpain II from monkey cardiac muscle[8] is the only exception with respect to the endopeptidase nature of calpain. Since a partially purified enzyme was used in this case, the possibility cannot be completely excluded that the cleavage is due to some contaminating proteinase(s). On the other hand, calpain II purified from monkey brain to an apparent homogeneity had little activity toward both Leu- and Met-enkephalins.[9]

As can be seen from Table 2, the specificities of calpains are very similar, but not entirely identical, among animal species and tissues form which they were derived as well

TABLE 1
Peptides Unsusceptible to Calpain

Rat brain calpain II[4]

Z-Gly-Phe-NH_2	Bz-Arg-OMe	Gly-Gly	Gly-Trp
Z-Gly-Phe	Bz-Arg-OEt	Gly-Ala	Gly-His
Z-Glu-Tyr	Bz-Tyr-OEt	Gly-Ser	Ala-Gly
Z-Tyr-Leu	Bz-Gly-Phe	Gly-Val	Leu-Gly
Z-Tyr-Ser-OMe	Bz-Gly-Arg	Gly-Leu	Leu-Tyr
Z-Tyr-His	Ac-Tyr-NH_2	Gly-Met	Gly-Gly-Gly
Bz-Gly-NH_2	Ac-Tyr-Gly	Gly-Tyr	Leu-Gly-Gly
Bz-Arg-NH_2	Ac-Phe-Glu	Gly-Phe	

Bovine cardiac muscle calpain II[5]

Leu-Arg-Arg-Ala-Ala-Leu-Gly	His-Gly-Tyr-Tyr-Ser-Leu-Gly
Leu-Arg-Arg-Ala-Ala-Val-Leu-Gly	Ac-Gly-Gly-Tyr-Ser-Leu-Gly
Leu-Arg-Arg-Ala-Ser(P)-Leu-Gly	Leu-Arg-Arg-Ala-Cys-Leu-Gly
Leu-Arg-Lys-Ala-Ser-Leu-Gly	Reduced and carboxymethylated insulin
Arg-Gly-Tyr-Ser-Leu-Gly	

Chicken skeletal calpain II[6]

Z-Phe-Ala	Z-Phe-Leu-Ala-Ala	Bz-Arg-OEt
Z-Phe-Tyr	Z-Gly-Pro-Leu-Gly-Pro	Bz-Arg-*p*NA
Z-Glu-Tyr	Bz-Tyr-OEt	Suc-Ala-Ala-Ala-*p*NA
Z-Leu-Tyr-NH_2	Bz-Tyr-*p*NA	Leu-*p*NA
Z-His-Phe(NO_2)-Phe-OMe		

Note: Z, benzyloxycarbonyl; Bz, benzoyl; Ac, acetyl; Suc, succinyl; NH_2, amide; OMe, methyl ester; OEt, ethyl ester; *p*NA, *p*-nitroanilide.

as between calpains I and II and between the soluble and cytoskeleton-bound forms. The differences among animal species may also be due in part to the differences in experimental conditions used. In the case of porcine calpains,[10] calpain I caused, in most cases, more extensive hydrolysis of the substrate than an equal amount of porcine calpain II, and this was thought to be due to different stability of the two enzymes, reflecting different auto-digestibility. Neurotensin was hydrolyzed well by all the enzymes used, except for human skeletal muscle calpain I. This enzyme was reported to have failed to cleave neurotensin significantly.[12] The reason for this discrepancy is not certain at present.

In Table 3, the sites of peptide bond cleavage given in Table 2 are shown in four different ways to express the subsite specificity of calpains. For the P_1 site (Table 3, lane 3), amino acids with a bulky side chain, especially tyrosine (Y), and basic amino acids, especially arginine (R) and then lysine (K), seem to be preferred residues. The subsite specificity, however, appears to be more notable at the P_2 position (Table 3, lane 2). At this position, leucine (L) is the most predominant residue, followed by valine (V). This seems to indicate that calpain has a marked preference at the P_2 position for amino acids with bulky aliphatic side chains, especially leucine, then valine and possibly isoleucine. In addition, at the P_3 position (Table 3, lane 1), phenylalanine (F), tryptophan (W), leucine (L), valine (V), etc. appear to be the predominant residues. Thus amino acids with an aromatic or a bulky aliphatic side chain may be preferred at this position. At the P'_1 position (Table 3, lane 4), on the other hand, basic amino acids, especially arginine, but also lysine, and amino acids with a bulky aliphatic side chain, especially leucine, appear to be the preferred residues. The subsite specificity at the P_3 and P'_1 positions, however, is less clear than that at the P_1 and P_2

TABLE 2
Sites of Cleavage by Calpain in Various Peptides

```
Peptide / Calpain[a]          Site of cleavage[b]

Peptide bond cleavage

Leu-enkephalin  Tyr-Gly-Gly-Phe-Leu
  M II             ○
  M'II             ▲       ▲
Met-enkephalin  Tyr-Gly-Gly-Phe-Met
  M II             ○
Hexapeptide     Leu-Trp-Met-Arg-Phe-Ala
  P I                  ▲
  P II                 Δ
β-Neoendorphin  Tyr-Gly-Gly-Phe-Leu-Arg-Lys-Tyr-Pro
  M II                                 ▲
  M'II             ○               ▲   ▲   ▲
  B II                                 ▲
  B'II                                 ▲
α-Neoendorphin  Tyr-Gly-Gly-Phe-Leu-Arg-Lys-Tyr-Pro-Lys
  M II                                 ▲
  M'II             ○               ▲   ▲
  P I                                  ▲
  P II                                 Δ
  B II                                 ▲
  B'II                                 ▲
Angiotensin I   Asp-Arg-Val-Tyr-Ile-His-Pro-Phe-His-Leu
  M II                         ▲
  P I                          ▲
  P II                         Δ
  B II                         ▲               Δ
  B'II                         ▲               Δ
LH-RH           Glu-His-Trp-Ser-Tyr-Gly-Leu-Arg-Pro-Gly-NH2
  P I                              ▲   Δ
  P II                             Δ   Δ
  H I                              ▲
  H II                             ▲
BAM-12P         Tyr-Gly-Gly-Phe-Met-Arg-Arg-Val-Gly-Arg-Pro-Glu
  B II                                 x
  B'II                                 Δ
Dynorphin(1-13) Tyr-Gly-Gly-Phe-Leu-Arg-Arg-Ile-Arg-Pro-Lys-Leu-Lys
  M II                                 ▲
  M'II             ○               ○   ▲
  B II                                 ▲
  B'II                                 ▲
  H I                                  ▲
  H II                                 ▲
Neurotensin    <Glu-Leu-Tyr-Glu-Asn-Lys-Pro-Arg-Arg-Pro-Tyr-Ile-Leu
  M II                     ▲       ▲
  P I                      ▲                   Δ
  P II                     ▲                   Δ
  B II                     ▲   Δ                           Δ
  B'II                     ▲   ▲                           Δ
  H I                      x
  H II                     ▲
Somatostatin    Ala-Gly-Cys-Lys-Asn-Phe-Phe-Trp-Lys-Thr-Phe-Thr-Ser-Cys
  H I                                                  ▲
  H II                                                 ▲
Dynorphin       Tyr-Gly-Gly-Phe-Leu-Arg-Arg-Ile-Arg-Pro-Lys-Leu-Lys-Trp-Pro-Asn-Gln
  P I                                  ▲                           Δ
  P II                                 ▲                           ○
Glucagon        His-Ser-Glu-Gly-Thr-Phe-Thr-Ser-Asp-Tyr-Ser-Lys-Tyr-Leu-Asp
  P I
  P II

                -Ser-Arg-Arg-Ala-Glu-Asp-Phe-Val-Glu-Trp-Leu-Met-Asn-Thr
                Δ       Δ                   Δ           Δ       ▲
                Δ       ○                   Δ           ▲       ▲
```

TABLE 2 (continued)
Sites of Cleavage by Calpain in Various Peptides

```
Peptide / Calpain[a]        Site of cleavage[b]

Oxidized insulin B chain
                    Phe-Val-Asn-Gln-His-Leu-Cta-Gly-Ser-His-Leu-Val-Glu-Ala-Leu
  C II                                 ▴           ▴                   ▲
  P I                                                      Δ           Δ
  P II                                                     Δ           Δ
  B II                                                                 Δ
  B'II                                             Δ                   ▴
                    -Tyr-Leu-Val-Cta-Gly-Glu-Arg-Gly-Phe-Phe-Tyr-Thr-Pro-Lys-Ala
                        ▲                                       ▴
                        ▴                                       ▴
                        ▴                                       Δ
                        ▴                                       Δ
                        ▲                                       ▴

Protamine Y II      Pro-Arg-Arg-Arg-Thr-Arg-Arg-Ala-Ser-Arg-Pro-Val-Arg-Arg-Arg
  C II                                                                 ▴
                    -Arg-Pro-Arg-Arg-Val-Ser-Arg-Arg-Arg-Arg-Ala-Arg-Arg-Arg-Arg
                                                ▲
Protamine Z         Ala-Arg-Arg-Arg-Arg-Ser-Arg-Arg-Ala-Ser-Arg-Pro-Val-Arg-Arg-Arg
  C II                                                                     ▴
                    -Arg-Pro-Arg-Arg-Val-Ser-Arg-Arg-Arg-Arg-Ala-Arg-Arg-Arg-Arg
                                                ▲
γ-Lipotropin        Tyr-Gly-Gly-Phe-Met-Thr-Ser-Glu-Lys-Ser-Gln-Thr-Pro-Leu-Val-Thr
  (61-91)
  M II              -Leu-Phe-Lys-Asn-Ala-Ile-Ile-Lys-Asn-Ala-Tyr-Lys-Lys-Gly-Glu
                    ▲                               ▴               ▴

Amide bond cleavage

MSH-release inhibiting factor                               Pro-Leu-Gly-NH2
  P I                                                                  Δ
  P II                                                                 x
CCK-tetrapeptide                                        Trp-Met-Asp-Phe-NH2
  P I                                                                  ▴
  P II                                                                 ▴
Enkephalinamide                                     Tyr-D-Ala-Gly-Phe-Met-NH2
  P I                                                                  Δ
  P II                                                                 Δ
Eledoisin-related peptide                       Lys-Phe-Ile-Gly-Leu-Met-NH2
  P I                                                                  ▲
  P II                                                                 ▴
CCK-octapeptide                   Asn-Tyr(SO3H)-Met-Gly-Trp-Met-Asp-Phe-NH2
  H I                                                                  ▴
  H II                                                                 ▲
Substance P                     Arg-Pro-Lys-Pro-Gln-Gln-Phe-Phe-Gly-Leu-Met-NH2
  P I                                                                  ▲
  P II                                                                 ▲
  H I                                                                  x
  H II                                                                 ▲
Physalaemin                     Glu-Ala-Asp-Pro-Asn-Lys-Phe-Tyr-Gly-Leu-Met-NH2
  P I                                                                  ▲
  P II                                                                 ▲
Bombesin                Glu-Gln-Arg-Leu-Gly-Asn-Gln-Trp-Ala-Val-Gly-His-Leu-NH2
  P I                                                                  ▲
  P II                                                                 ▴
```

Note: <Glu, pyroglutamic acid: Cta, cysteic acid; LH-RH, leutenizing hormone-releasing hormone; BAM-12P, bovine adrenal medulla dodecapeptide; CCK, cholecytokinin.

[a] M II, monkey brain calpain II; M′II, monkey cardiac muscle calpain II; P I, porcine erythrocyte calpain I; P II, porcine kidney calpain II; B II, bovine spinal cord calpain II (soluble form); B′II, bovine spinal cord calpain II (cytoskeleton-bound form); H I, human skeletal muscle calpain I; H II, human skeletal muscle calpain II.

[b] The sites of cleavage are indicated as follows: ▲, high susceptibility site(extent of cleavage: 41 to 100 %); ▴, medium susceptibility site (extent of cleavage: 10 to 40 %); △, low susceptibility site (extent of cleavage: < 10 %); ○, very low susceptibility site (extent of cleavage: trace); x, uncleaved site (shown only when necessary). The discrimination among the various susceptibility sites are not strictly quantitative since the results reported in each paper were obtained under different experimental conditions.

TABLE 3
Subsite Specificity of Calpain Action on Peptides[a]

(1)					(2)					(3)					(4)				
P_3	P_2	P_1	↓P'_1	P'_2	P_3	**P_2**	P_1	↓P'_1	P'_2	P_3	P_2	**P_1**	↓P'_1	P'_2	P_3	P_2	P_1	↓**P'_1**	P'_2
F	L	R	K	Y	F	**L**	R	K	Y	A	L	**Y**	L	V	F	L	R	**R**	I
F	L	R	R	I	F	**L**	R	R	I	R	V	**Y**	I	H	V	S	R	**R**	R
F	F	Y	T	P	<E	**L**	Y	E	N	F	F	**Y**	T	P	V	R	R	**R**	R
(F)	M	R	R	V	A	**L**	Y	L	V	W	S	**Y**	G	L	G	F	L	**(R)**	K
W	L	M	N	T	W	**L**	M	N	T	<E	L	**Y**	E	N	F	M	R	**(R)**	V
W	S	Y	G	L		**L**	W	M	R			**(Y)**	G	G	K	P	R	**(R)**	P
W	K	T	F	T	K	**(L)**	K	W	P	R	P	**(Y)**	I	L	D	S	R	**(R)**	A
L	V	T	L	F	Y	**(L)**	D	S	R	F	L	**R**	K	Y	F	L	R	**K**	Y
L	V	E	A	L	R	**V**	Y	I	H	F	L	**R**	R	I	A	Y	K	**K**	G
(L)	Y	E	N	K	L	**V**	E	A	L	V	S	**R**	R	R	Y	E	N	**(K)**	P
(L)	R	K	Y	P	L	**V**	T	L	F	V	R	**R**	R	R	A	L	Y	**L**	V
(L)	C	G	S	H	I	**I**	K	N	A	F	M	**(R)**	R	V	L	V	T	**L**	F
V	S	R	R	R	A	**Y**	K	K	G	K	P	**(R)**	R	P	N	Q	H	**(L)**	C
V	R	R	R	R	L	**(Y)**	E	N	K	D	S	**(R)**	R	A	V	E	W	**(L)**	M
(V)	E	W	L	M	S	**(Y)**	G	L	R	I	I	**K**	N	A	G	S	H	**(L)**	V
I	I	K	N	A	F	**F**	Y	T	P	A	Y	**K**	K	G	S	Y	G	**(L)**	R
(Y)	E	N	K	P	G	**(F)**	L	R	K	L	R	**(K)**	Y	P	R	V	Y	**I**	H
(Y)	G	G	F	L	F	**(M)**	R	R	V	K	L	**(K)**	W	P	R	P	Y	**(I)**	L
(Y)	L	D	S	R	V	**R**	R	R	R	G	F	**(L)**	R	K	E	D	F	**(V)**	E
A	L	Y	L	V	L	**(R)**	K	Y	P	W	L	**M**	N	T	W	K	T	**F**	T
A	Y	K	K	G	W	**K**	T	F	T		L	**W**	M	R	Y	G	G	**(F)**	L
R	V	Y	I	H	W	**S**	Y	G	L	V	E	**(W)**	L	M	L	R	K	**(Y)**	P
(R)	P	Y	I	L	V	**S**	R	R	R	H	P	**(F)**	H	L	K	L	K	**(W)**	P
(K)	L	K	W	P	D	**(S)**	R	R	A	E	D	**(F)**	V	E		L	W	**M**	R
(K)	P	R	R	P	G	**(S)**	H	L	V	L	V	**E**	A	L	L	V	E	**A**	L
<E	L	Y	E	N	N	**(Q)**	H	L	C	L	Y	**(E)**	N	K	<E	L	Y	**E**	N
(E)	D	F	V	E	Y	**(E)**	N	K	P	W	K	**T**	F	T	I	K	K	**N**	A
(N)	Q	H	L	C	V	**(E)**	W	L	M	L	V	**T**	L	F	W	L	M	**N**	T
(D)	S	R	R	A	E	**(D)**	F	V	E	Y	E	**(N)**	K	P	L	Y	E	**(N)**	K
(H)	P	F	H	L	L	**(C)**	G	S	H	Y	L	**(D)**	S	R	F	F	Y	**T**	P
(S)	Y	G	L	R	Y	**(G)**	G	F	L	N	Q	**(H)**	L	C	H	P	F	**(H)**	L
(G)	F	L	R	K	K	**(P)**	R	R	P	G	S	**(H)**	L	V	L	C	G	**(S)**	H
(G)	S	H	L	V	R	**(P)**	Y	I	L	Y	G	**(G)**	F	L	Y	L	D	**(S)**	R
	L	W	M	R	H	**(P)**	F	H	L	L	C	**(G)**	S	H	W	S	Y	**G**	L
		Y	G	G			Y	G	G	S	Y	**(G)**	L	R			Y	**(G)**	G

[a] The data are based on the results given in Table 2, and the same P_3 to P'_2 site sequence occurring in different peptides is shown only once. Amino acids are shown in one-letter code. The same data are arranged in four different ways, (1) to (4), with special attention to the subsites P_3, P_2, P_1 and P'_1, respectively, as shown in bold face. The amino acid in parenthesis indicates that the P_1—P'_1 site in the peptide is a minor cleavage site. The vertical arrow indicates the cleavage site; C, cysteic acid; < E, pyroglutamic acid.

positions. Indeed, the cleavage of a hexapeptide Leu-Trp-Met-Arg-Phe-Ala at the Trp-Met bond (Table 2) indicates that the presence of a residue at the P_3 position is not necessarily required. As to the other sites beyond the P_3 and P'_1 positions, no significant subsite specificity can be noticed. Although a marked preference was thus found for certain groups of amino acids around the scissile bonds, especially at the P_1 and P_2 positions, calpain also cleaved several other peptide bonds rather nonspecifically (Tables 2, and 3), possibly suggesting the additional contributions of nearby residues other than those at the P_1 and P_2 positions. Therefore the deduced subsite specificity does not appear to be very rigid. However, the above results seem to indicate that X-Tyr and X-Arg, where X is Leu, Val, or Ile, may constitute the most preferred P_2-P_1 sites for calpain. In addition, it is worth noting that Pro never appears at the P_1 and P'_1 positions. This may explain why LH-RH and dynorphin were not cleaved just after Arg in the Leu-Arg-Pro and Ile-Arg-Pro sequences, respectively (Table 2). On the other hand, selective cleavages were often observed at Arg/

Lys-Arg/Lys bonds. These cleavages at paired basic residues have led to a suggestion,[8,9] that calpain may be involved in the processing of precursors of certain neuropeptides and peptide hormones.

In addition to peptide bond cleavage, calpain was also found[10,12,14] to hydrolyze the carboxyl-terminal amide bonds of certain peptides, such as substance P and its related peptides, as shown in Table 2. The deamidation by porcine calpain I was always more extensive than that by porcine calpain II, and this was thought to be due to different stability of the two enzymes as in the case of peptide bond cleavage.[14] Human skeletal muscle calpain I did not cleave the amide bond of substance P,[12] which is a good substrate for porcine calpains I and II and human calpain II. The reason for this difference is not clear.

Among these peptide amides, substance P and its related peptides with the carboxyl-terminal -Gly-Leu-Met-NH_2 structure were rapidly deamidated, whereas those with a different amino acid sequence preceding the amide group were less susceptible to calpain (Table 2). On the other hand, in glucagon the Trp-Leu-Met sequence is the major P_3 to P_1 site for the action of calpains (Table 2). Taken together, these results seem to indicate that Met is also a preferred residue at the P_1 position. Thus, the above results are consistent with the major subsite specificity observed for peptide bond cleavage by calpains. The presence of Gly at the P_3 position also suggests that the subsite specificity of calpain at the P_3 position is less rigid than those at the P_2 and P_1 positions. However, an amino acid with an aromatic or a bulky aliphatic side chain at the P_3 position may to some extent increase the susceptibility of the scissile bond to calpain. This may explain the selective cleavage of clupeins at only two sites, where the P_3 site is occupied by Val (Table 2).

B. SPECIFICITY OF ACTION ON SYNTHETIC SUBSTRATES

Ishiura et al.[7] reported that chicken skeletal muscle calpain II could hydrolyze *N*-benzyloxycarbonyl-L-tyrosine-*p*-nitrophenyl ester. This activity may be partly due to the *p*-nitrophenyl ester group in the substrate which is a very good leaving group. In addition, the benzyloxycarbonyl group at the P_2 position of the substrate may fit well with the S_2 site of the enzyme. Sasaki et al,[10] on the other hand, prepared various peptidyl 4-methylcoumaryl-7-amide (MCA) derivatives, based on the results obtained with oligopeptides, and investigated the rates of their hydrolysis by porcine calpains. The results are shown in Table 4. Although the rates of hydrolysis were not high, compared with hydrolysis by papain, synthetic MCA substrates containing Leu-Tyr, Leu-Met, and Leu-Lys in the P_2-P_1 positions were relatively susceptible, whereas the substitution of Leu at the P_2 position by Val or the blocking of the phenolic hydroxyl group of Tyr at the P_1 position with a 2,6-dichlorobenzyl group reduced the rate of hydrolysis by one order of magnitude. Suc-Leu-Met-MCA and Suc-Leu-Tyr-MCA were almost equally susceptible to both calpains, indicating that Met and Tyr are almost equally preferred at the P_1 position. Among the 15 MCA substrates tested for porcine calpains, 6 were not hydrolyzed. On the other hand, when 12 MCA derivatives, used as substrates for other proteinases, were tested as substrates for monkey brain calpain II, none were hydrolyzed, as shown in Table 4.[9] None of the MCA substrates unsusceptible to these calpains had Leu, Val, or other amino acids with a bulky hydrophobic side chain at the P_2 position, except for Boc-Val-Leu-Lys-MCA. Thus, the results are in general agreement with the subsite specificity of calpains deduced by using various oligopeptides. Boc-Val-Leu-Lys-MCA was hydrolyzed well by porcine calpains I and II, whereas it was not hydrolyzed by monkey brain calpain II, even though it meets the deduced subsite specificity of calpains. The reason for this apparent discrepancy is not certain at present. Porcine calpains did not hydrolyze the MCA substrates tested elsewhere than at the P_1 position. On the other hand, monkey calpain II hydrolyzed the Gly-Arg bond in Boc-Leu-Gly-Arg-MCA and the Thr-Arg bond in Boc-Leu-Thr-Arg-MCA instead of hydrolyzing the Arg-MCA bonds. These results, however, agree partially with the subsite specificity of calpains.

TABLE 4
Action of Porcine and Monkey Calpains and Papain on Various Peptidyl MCA-Derivatives as substrates.

		Specific activity × 10^2 (μmol min^{-1} mg^{-1})		
Substrate[a]	**Conc**	**Calpain I**	**Calpain II**	**Papain**
Porcine enzyme[b] and papain[10] (m*M*)				
Suc-Met-MCA	4.0	0.0	0.0	2.02
Suc-Leu-Met-MCA	4.0	6.84	5.05	306
Suc-Val-Met-MCA	4.0	0.43	0.31	554
Suc-Tyr(Cl_2-Bzl'-MCA	0.4	0.0	0.0	0.0
Suc-Leu-Tyr(Cl_2-Bzl'-MCA	0.4	0.42	0.33	0.58
Suc-Val-Tyr(Cl_2-Bzl'-MCA	0.4	0.07	0.04	1.58
Suc-Tyr-MCA	2.0	0.0	0.0	0.63
Suc-Leu-Tyr-MCA	2.0	7.63	4.00	38.3
Suc-Val-Tyr-MCA	2.0	0.81	0.37	30.7
Suc-Leu-Leu-Val-Tyr-MCA	0.2	1.81	1.79	161
Boc-Val-Leu-Lys-MCA	2.0	4.64	8.00	1220
Boc-Leu-Thr-Arg-MCA	4.0	0.84	0.36	860
Boc-Phe-Ser-Arg-MCA	4.0	0.0	0.0	51.7
Boc-Val-Pro-Arg-MCA	4.0	0.0	0.0	240
Boc-Leu-Gly-Arg-MCA	4.0	0.0	0.0	12.0
Monkey enzyme[c,9]				
Boc-Leu-Gly-Arg-MCA[d]			N.D.	
Boc-Leu-Gly-Arg-MCA[d]			N.D.	
Boc-Leu-Thr-Arg-MCA[e]			N.D.	
Boc-Gln-Arg-Arg-MCA			N.D.	
Boc-Phe-Ser-Arg-MCA			N.D.	
Boc-Val-Pro-Arg-MCA			N.D.	
Boc-Val-Leu-Lys-MCA			N.D.	
Boc-GLu-Lys-Lys-MCA			N.D.	
Suc-Ala-Ala-Pro-Phe-MCA			N.D.	
ZPhe-Arg-MCA			N.D.	
Bz-Arg-MCA			N.D.	
Leu-MCA			N.D.	
Arg-MCA				

Note: N.D., MCA-liberating activity not detected.

[a] Suc, succinyl; Boc, tert-butoxycarbonyl; Z, benzyloxycarbonyl; Bzl, benzyl; Bz, benzoyl; MCA, 4-methylcoumaryl-7-amide.
[b] Porcine erythrocyte calpain I and porcine kidney calpain II were used.
[c] Monkey brain calpain II was used.
[d] The Gly-Arg bond was cleaved.
[e] The Thr-Arg bond was cleaved.

Papain, tested for comparison,[10] hydrolyzed all the MCA substrates tested for porcine calpains at much greater rates than calpain. At the P_1 position, Arg and Met were preferred over Tyr, and at the P_2 position, Val was somewhat preferred over Leu. The subsite specificity of papain for the P_2 position is much less strict than with calpain. Therefore, the substrate specificity of calpain is considerably different from and much narrower than that of papain.

In addition, Sasaki et al.[15] also prepared several tripeptidyl chloromethyl ketones (Leu-Leu-Phe/Tyr/Lys-CH_2Cl), showing that they are strong inhibitors of porcine calpains I and II. Further, it had been found[16,17] that leupeptin, having the carboxyl terminal -Leu-arginal structure, inhibits calpain more strongly than antipain with its carboxyl terminal -Val-arginal structure. These results are consistent with the deduced subsite specificity of calpains. The

TABLE 5
Comparison of the Kinetic Parameters of Porcine Calpains I and II and Papain for Hydrolysis of Several Synthetic MCA Substrates and Deamidation of Substance P and its Related Neuropeptides[10,14]

Substrate[a]	Enzyme	K_m (mM)	k_{cat} (s^{-1})	$10^{-2} \times k_{cat}/K_m$ (M^{-1} s^{-1})
Suc-Leu-Met-MCA	I	1.21	0.062	0.510
	II	4.78	0.194	0.405
	P	0.343	0.926	27.0
Suc-Leu-Tyr-MCA	I	4.74	0.369	0.779
	II	2.21	0.092	0.417
	P	0.711	0.158	2.22
Suc-Leu-Leu-Val-Tyr-MCA	I	0.203	0.029	1.41
	II	0.466	0.065	1.39
	P	0.013	0.783	603
Boc-Val-Leu-Lys-MCA	I	5.92	0.491	0.830
	II	7.08	0.805	1.14
	P	13.0	208	160
Substance P	I	0.526	52.8	1004
	II	0.625	50.9	814
Physalaemin	I	4.54	172.6	379
	II	1.92	51.1	266
Eledoisin-related peptide	I	2.70	88.1	326
	II	1.21	84.7	271

Note: I, porcine erythrocyte calpain I; II, porcine kidney calpain II; and P, papain.

[a] Abbreviations used are the same as those in Table 4.

potent inhibition of calpains by E-64 (L-*trans*-epoxysuccinyl-leucyl-amido(4-guanidino)butane)[18] is also consistent with the deduced subsite specificity, since this compound has a carboxyl-terminal structure closely resembling Leu-Arg.

Table 5 compares the kinetic parameters of porcine calpains I and II and papain for hydrolysis of several synthetic MCA substrates as well as those for deamidation of substance P and related peptides.[10,14] The K_m values are not very different between the synthetic and neuropeptide substrates, whereas the k_{cat} values differ greatly between them. Thus the k_{cat}/K_m values are at least several hundredfold greater for the neuropeptides than for the synthetic substrates. The k_{cat}/K_m values for calpains I and II are not very different, whereas those for papain are several ten- to 100-fold greater than those for calpains. Similarly, chicken skeletal muscle calpain II was reported to hydrolyze *N*-benzyloxycarbonyl-L-tyrosine-*p*-nitrophenyl ester (K_m = 0.18 mM, k_{cat} = 0.19 s^{-1}) a hundred times more slowly than chymotrypsin (K_m = 0.03 mM, k_{cat} = 18.0 s^{-1})[7].

III. SUBSTRATE SPECIFICITY OF CALPAIN TOWARD PROTEINS

A. SPECIFICITY OF ACTION ON VARIOUS PROTEINS

There are increasing numbers of reports concerning the proteolysis by calpain of various endogenous proteins or their Ca^{2+}-dependent proteolysis presumably due to the action of calpain.[2,3] As shown in Table 6, they include myofibrillar proteins, cytoskeletal proteins, hormone receptors, protein kinases, and several others,[6,19-123] and in many of these cases, calpain appears to be involved in their limited proteolysis. For some of these proteins, such as neurofilament protein,[45] vimentin,[49] and hormone receptors,[70,72] K_m values were reported to be as low as 10^{-8} to 10^{-7} M. For example, Puca et al.[70] reported that calf uterus calpain

TABLE 6
Endogenous Protein Susceptible to Calpain[a]

Protein	Enzyme source or tissue
Myofibrillar proteins	
Z-Line proteins	Porcine,[19,20] rabbit,[21-23] and chicken[24] skeletal muscles
Troponin (T and I)	Porcine,[19] rabbit,[23,25] and chicken[6,26] skeletal muscles; monkey[27] and bovine[28] cardiac muscles
Tropomyosin	Porcine,[19] rabbit,[23] and chicken[6,26] skeletal muscles; monkey[27] and bovine[28] cardiac muscles
Myosin	Rabbit[23] and chicken[6,24] skeletal muscles; monkey cardiac muscle[27]
α-Actinin	Rabbit[21,22] and chicken[6] skeletal muscles
C-Protein	Porcine skeletal muscle[19]
Filamin	Bovine skeletal muscle[29]
Cytoskeletal proteins	
Neurofilament proteins (200K, 150—160K, 68—70K, etc.)	Bovine spinal cord,[30] rat sciatic nerve,[31-33] rat peripheral nerve,[34] rat spinal cord,[33,35-39] mouse optic nerve,[40] mouse retinal ganglion cell,[41] porcine,[42] calf,[43,44] and rat[45,46] brains; rat skeletal muscle[46]
Glial fibrillary acidic protein	Rat spinal cord,[37] mouse spinal cord[47]
180K Doublet protein	Rat brain[48]
Vimentin	Ehrich ascites tumor cell,[49,50] bovine lens,[51,52] porcine kidney[53]
Desmin	Ehrich ascites tumor cell[49,50]
Fodrin	Bovine skeletal muscle,[54] rat brain[55]
Tubulin	Calf brain,[43,44] rat spinal cord,[37] chicken skeletal muscle[6]
Microtubule-associated proteins	Porcine,[42] calf,[43,44] and rat[56-58] brains; rat retina[59]
Spectrin	Bovine cardiac muscle,[60] chicken skeletal muscle,[6] rabbit erythrocyte[61]
Band 4.1 protein	Human erythrocyte,[62-64] bovine cardiac muscle[60]
Actin	Bovine lens,[52] calf brain,[43] rat spinal cord[37]
Myosin	Calf brain[43]
α-Actinin	Turkey gizzard smooth muscle[65]
Actin-binding protein	Human platelet[66,67]
P235 protein	Human platelet[67]
Talin	Monkey kidney BS-C-1 cell[68]
Receptor proteins	
Estrogen receptor	Calf uterus,[69,70] chick oviduct[71]
Progesterone receptor	Chick oviduct[72]
Glutamate receptor	Rat brain[48,73]
α_1-Adrenergic receptor	Rat liver[74]
Epidermal growth factor receptor	A431 cell[75-77]
Platelet-derived growth factor receptor	Human fibroblast[78]
Thrombin receptor	Human platelet[79,80]
Enzyme proteins	
Glycogen phosphorylase kinase	Rabbit skeletal muscle,[81-83] bovine cardiac muscle,[83,84] bovine brain[83]
cAMP-dependent protein kinase	Rabbit skeletal muscle,[85] mouse lung[86]
Protein kinase C	Rat brain,[87,88] rat liver,[87,89] rat kidney,[87,90] rat lung, cerebellum, cardiac muscle, skeletal muscle and adipose tissue,[87] human neutrophil,[91,92] human platelet[93,94]
Calmodulin-dependent protein kinase	Human erythrocyte[95]
Myosin light chain kinase	Rabbit skeletal muscle[96]
Glycogen synthetase	Rat skeletal muscle,[97] rat liver[98]
Ca^{2+}-ATPase	Porcine erythrocyte membrane[99]
Calmodulin-dependent phosphatase	Human erythrocyte[100]
Phosphorylase phosphatase	Rabbit skeletal muscle[101]

TABLE 6 (continued)
Endogenous Protein Susceptible to Calpain[a]

Protein	Enzyme source or tissue
Ca^{2+}-dependent cyclic nucleotide phosphodiesterase	Rabbit skeletal muscle[96]
Phospholipase C	Several animal tissues[102]
Tryptophan hydroxylase	Rat brain[103]
Tyrosine hydroxylase	Bovine adrenal medulla[104]
Transglutaminase (platelet factor XIII)	Human erythrocyte[105]
Calpain	Autolysis : rabit skeletal muscle,[106] chicken gizzard smooth muscle[107]
Other proteins	
Myelin basic protein	Human brain,[108] human brain myelin,[109-111] rat brain myelin,[110] bovine spinal cord,[112] rat spinal cord myelin[38,39]
Myelin-associated glycoprotein	Human brain myelin,[110,111,113] rat brain myelin[110]
Proteolipid protein	Rat spinal cord myelin[38,39]
Wolfgram protein	Chicken skeletal muscle[6]
Platelet polypeptides	Human platelet[114]
Calmodulin-binding protein	Human platelet[115]
α-Crystalline	Bovine lens[116]
Chromogranin A	Insulinoma tissue[117]
von Willebrand factor	Human platelet[118]
Band 3 protein	Human erythrocyte,[62,63,119] bovine cardiac muscle,[60] rat erythrocyte[61]
Kyotorphin precursor	Rabbit skeletal muscle[120]
Histone	Rabbit skeletal muscle[121]
Globin	Human erythrocyte[122]
Calpastatin (suicide substrate)	Rabbit skeletal muscle[123]

[a] The entry is not exhaustive. The table includes Ca^{2+}-dependent proteolysis in crude tissues presumably due to the action of calpain.

has an extraordinary high affinity for native estrogen receptor (K_m 10^{-8} M) whereas it did not attack several other proteins (hemoglobin, plasma albumin, ovalbumin) tested both in the native state and after heat denaturation. Nelson and Traub[49] reported that calpain I purified from Ehrlich ascites tumor cells was vimentin- and desmin-specific and that it did not degrade any of the standard proteins tested (bovine serum albumin, ovalbumin, fibrinogen, azocasein, histones, and collagen) and various filament proteins (tubulin, actin, or any of the major myofibril proteins). Thus, it appears that calpains have some preference for specific types of proteins as substrates, possibly recognizing certain structural features yet to be defined. This may explain why the substrate specificity, especially toward endogenous proteins, is not always identical among calpains from various sources.

On the other hand, calpains generally seem to hydrolyze denatured proteins more rapidly and extensively than their native forms. For example, Pontremoli et al.[124] reported that rabbit liver calpain hydrolyzed the following proteins with relative activity shown in parentheses: denatured globin (100), native β-globin (16), casein (15), native human hemoglobin (2), bovine serum albumin (<1), rabbit liver fructose-1,6-bisphosphatase (<1), and rabbit muscle aldolase (<1). Thus, calpain is also thought to play an important role in the nonlysosomal degradation of denatured proteins *in vivo*.

B. LIMITED PROTEOLYSIS AND CLEAVAGE SITE SPECIFICITY

Although many proteins are known to be degraded by calpain by way of limited proteolysis, the nature of the peptide bonds cleaved by the enzyme has so far been reported in only a few cases of experiments *in vitro,* including hemoglobin (α and β chains),[122] troponin T,[25] α-crystalline (A and B chains),[116] vimentin,[53] histone,[121] α-fodrin,[54] myelin basic protein,[112] and protein kinase C.[90]

Melloni et al.[122] found that limited proteolysis occurred only in the NH_2-terminal regions of α- and β-globin chains of human hemoglobin by calpain I purified from human erythrocyte. The cleavages occurred at the Lys^{11}-Ala^{12} bond in the sequence of -Asn-Val-Lys-Ala-Ala- in the α-chain and at the Lys^{9}-Ser^{10} bond in the sequence of -Glu-Glu-Lys-Ser-Ala- in the β-chain. These cleavages are thus partially in accord with the deduced subsite specificity of calpains. The α- and β-chains contain one Leu-Arg sequence and one Val-Tyr sequence, respectively, in the interior regions. These should be potentially favorable P_2-P_1 sites as judged from the deduced subsite specificity of calpain, but the bonds preceded by these sequences were not cleaved. The failure of cleavage after the Val-Tyr sequence is presumably due in part to the presence of Pro after Tyr. The rates of hydrolysis of these two peptide bonds were reduced to one fifth with their corresponding native hemoglobin chains, whereas the native oligomeric hemoglobin itself was resistant to calpain. Further, the variant α-globin chain [α47(CE5) Asp → His] was hydrolyzed faster, while the variant β-globin chain [β7(A4) Glu → Gly], with the mutation near the site of cleavage, revealed a high degree of resistance to the proteolytic attack by calpain I. Some conformational factor in addition to the subsite specificity appears to be responsible for this limited proteolysis.

A similar limited cleavage in the NH_2-terminal region was also observed with troponin T (rabbit skeletal muscle).[25] In this case, the Thr^{45}-Ala^{46} bond in the sequence -Lys-Leu-Thr-Ala-Pro- was selectively cleaved by chicken skeletal muscle calpain II, partially in accord with the subsite specificity deduced with peptide substrates.

Yoshida et al.[116] demonstrated that calpain II purified from bovine lens produced a limited proteolysis of the A and B chains of bovine lens α-crystalline. The A chain was shown to be cleaved at the Arg^{163}-Glu^{164} bond in the sequence -Val-Ser-Arg-Glu-Glu- and the Ser^{162}-Arg^{163} bond in the sequence -Pro-Val-Ser-Arg-Glu-, the former being the major site of cleavage (Table 7). On the other hand, the B chain was shown to be cleaved at the Thr^{170}-Ala^{171} bond in the sequence -Ala-Val-Thr-Ala-Ala- and the Arg^{163}-Glu^{164} bond in the sequence -Ile-Thr-Arg-Glu-Glu-, the former being the major site of cleavage (Table 7). The amino acid sequences around the cleavage sites agree partly, but not perfectly, with the subsite specificity of calpains deduced from its action on peptide substrates. Although proteolysis was limited to the carboxyl-terminal regions of both chains, there is one Val-Arg sequence in the interior region of the A chain, and are one Ile-Arg and two Leu-Arg sequences in the interior region of the B chain which are potentially favorable P_2-P_1 sites. However, no cleavage occurred at the peptide bonds preceded by these sequences. Thus,only the carboxyl-terminal regions of these chains are thought to be accessible to calpain in their native conformations. This is consistent with the conformational characteristics of α-crystalline.[125]

Vimentin is also known to undergo a limited proteolysis by calpain.[50,52] Fischer et al.[53] investigated the cleavage sites of murine vimentin from Ehrlich ascites cells using calpain II purified from porcine kidney, and found that the major cleavage occurred at two peptide bonds. One is the Arg^{70}-Ser^{71} bond in the sequence -Arg-Leu-Arg-Ser-Ser- and the other is the Thr^{264}-Ala^{265} bond in the sequence -Asp-Leu-Thr-Ala-Ala- (Table 6). Thus the former cleavage is consistent with the subsite specificity deduced with peptide substrates. However, the latter cleavage agrees only partially with the deduced subsite specificity. Further, eight additional minor cleavage sites were identified. Among these, only one site agrees well with the deduced subsite specificity, whereas three sites agree partially, and the remaining three sites are not predicted cleavage sites. Moreover, in the amino-terminal 119-residue sequence, no cleavage occurred at the peptide bonds preceded by the Leu-Arg and two Val-Arg

TABLE 7
Sites of Cleavage by Calpain in Various Proteins

Protein[a]/ Calpain	Site of cleavage[b]
Hemoglobin (human) / Calpain I (human)[c] α-Chain	1 50 VLSPADKTNVKAAWGKVGAHAGEYGAEALERMFLSFPTTKTYFPHFDLSH 100 GSAQVKGHGKKVADALTNAVAHVDDMPNALSALSDLHAHKLRVDPVNFKL 141 LSHCLLVTLAAHLPAEFTPAVHASLDKFLASVSTVLTSKYR
β-Chain	1 50 VHLTPEEKSAVTALWGKVNVDEVGGEALGRLLVVYPWTQRFFESFGDLST 100 PDAVMGNPKVKAHGKKVLGAFSDGLAHLDNLKGTFATLSELHCDKLHVDP 146 ENFRLLGNVLVCVLAHHFGKEFTPPVQAAYQKVVAGVANALAHKYH
α-Crystalline (bovine) / Calpain II (bovine)[d] A Chain	1 50 *Ac*MDIAIQHPWFKRTLGPFYPSRLFDQFFGEGLFEYDLLPFLSSTISPYYRQ 100 SLFRTVLDSGISEVRSDRDKFVIFLDVKHFSPEDLTVKVQEDFVEIHGKH 150 NERQDDHGYISREFHRRYRLPSNVDQSALSCSLSADGMLTFSGPKIPSGV 173 DAGHSERAIPVSREEKPSSAPSS
B Chain	1 50 *Ac*MDIAIHHPWIRRPFFPFHSPSRLFDQFFGEHLLESDLFPASTSLSPFYLR 100 PPSFLRAPSWIDTGLSEMRLEKDRFSVNLNVKHFSPEELKVKVLGDVIEV 150 HGLHEERQDEHGFISREFHRKYRIPADVDPLAITSSLSSDGVLTVNGPRK 175 QASGPERTIPITREEKPAVTAAPKK
Vimentin (murine) / Calpain II (porcine)[e]	1 50 STRSVSSSSYRRMFGGSGTSSRPSSNRSYVTTSTRTYSLGSALRPSTSRS 100 LYSSSPGGAYVTRSSAVRLRSSVPGVRLLQDSVDFSLADAINTEFKNTRT 261 NEKVELQELNDRFANYIDK----------PDLTAALR-------------
α-Fodrin (human) / Calpain I (bovine)[f]	1 50 YHSLLELGEKRKGMLEKSCKKFMLFREANELQQWINEKEAALTSEEVGAD 100 LEQVEVLQKKFDDFQKDLKANESRLKDINKVAEDLESEGLMAEEVQAVQQ 131 QEVYGMMPRDETDSKTASPWKSARLMVHTVA
Myelin basic protein (bovine) / Calpain II (bovine)[g]	1 50 *Ac*AAQKRPSQRSKYLASASTMDHARHGFLPRHRDTGILDSLGRFFGSDRGAP 100 KRGSGKDGHHAARTTHYGSLPQKAQGHRPQDENPVVHFFKNIVTPRTPPP 150 SQGKGRGLSLSRFSWGAEGQKPGFGYGGRASDYKSAHKGLKGHDAQGTLS 169 KIFKLGGRDSRSGSPMARR
Histone (calf) / Calpain II (chicken)[h] H2A	1 50 SGRGKQGGKARAKAKTRSSRAGLQFPVGRVHRLLRKGNYAERVGAGAPVY 100 LAAVLEYLTAEILELAGNAARDNKKTRIIPRHLQLAIRNDEELNKLLGKV 129 TIAQGGVLPNIQAVLLPKKTESHHKAKGK
H2B	1 50 PEPAKSAPAPKKGSKKAVTKAQKKDGKKRKRSRKESYSVYVYKVLKQVHP 100 DTGISSKAMGIMNSFVNDIFERIAGEASRLAHYNKRSTITSREIQTAVRL 125 LLPGELAKHAVSEGTKAVTKYTSSK
H3	1 50 ARTKQTARKSTGGKAPRKQLATKAARKSAPATGGVKKPHRYRPGTVALRE 100 IRRYQKSTELLIRKLPFQRLVREIAQDFKTDLRFQSSAVMALQEACEAYL 135 VGLFEDTNLCAIHAKRVTIMPKDIQLARRIRGERA

[a] The amino acid sequence of each protein is shown using one-letter code. The amino acid sequence of rat vimentin has been determined only partially.[52] The amino acid sequence of only the repeat 11 domain is shown for human α-fodrin. *Ac*, acetyl.

[b] The sites of cleavage are shown as follows: ▲, Major cleavage site; ▴, minor cleavage site. The sequences Leu/Val/Ile-Arg/Tyr are underlined.

[c] Human erythrocyte enzyme.

[d] Bovine lens enzyme.

[e] Porcine kidney enzyme.

[f] Bovine cardiac muscle enzyme.

[g] Bovine spinal cord enzyme (soluble form).

[h] Chicken skeletal muscle enzyme.

sequences present. In the current models of vimentin structure,[126] the major cleavage site Thr^{264}-Ala^{265} lies at the amino-terminal end of the second coiled coil domain, while the other major cleavage site, Arg^{70}-Ser^{71}, precedes the first coiled coil domain thought to start around residue 100. Seven additional cleavage sites are also located in the amino-terminal head region. In addition, further cleavages occurred at two points in the carboxyl-terminal region after the second coiled coil domain of vimentin, of which the exact sites remain to be defined. Thus calpain acts primarily on the non-α-helical regions.

Fodrin (brain spectrin) is known to be rapidly and stoichiometrically cleaved by calpain I prepared from bovine heart at a hypersensitive site near the center of its α subunit.[127] Harris et al.[54] found this site of cleavage to be the Tyr^{104}-Gly^{105} bond in the sequence -Glu-Val-Tyr-Gly-Met- in the repeat 11 domain (Table 7) which is located in the central region of α-fodrin and involves the calmodulin binding site. Thus the cleavage site has Tyr at the P_1 position and Val at the P_2 position, and agrees well with the subsite sepcificity deduced with peptide substrates. Although the sequences, X-Tyr and X-Arg, where X is Leu, Val or Ile, occur many times within α-fodrin, they are absent from the repeat 11 domain. Moreover, the region encompassing the site of cleavage was predicted to be nonhelical according to the model of Chow and Fasman. Thus, the conformation of the cleavage site is thought to be an important determinant of the specificity as well as the specific residues around the cleavage site.

Myelin basic protein was shown to be susceptible to cleavage by calpain.[108] Tsubata and Takahashi[112] investigated the action of calpain II purified from bovine spinal cord on the myelin basic protein obtained from the same source, and found that the protein was cleaved mainly at the two peptide bonds Val^{93}-Thr^{94} in the sequence -Asn-Ile-Val-Thr-Pro- and Arg^{96}-Thr^{97} in the sequence -Thr-Pro-Arg-Thr-Pro- to similar extents and in a mutually exclusive way (Table 7). These results also agree partly, but not perfectly, with the subsite specificity deduced with synthetic peptides. In bovine myelin basic protein, however, the sequences X-Tyr and X-Arg, where X is Leu, Val, or Ile, are not present. Furthermore, minor cleavages occurred at the Ala^{16}-Ser^{17} bond in the sequence -Ala-Ser-Ala-Ser-Thr- and at the Gly^{68}-Ser^{69} bond in the sequence -His-Tyr-Gly-Ser-Leu-. These minor cleavage sites do not agree with the deduced major substrate specificity. Since four proline residues are clustered in the vicinity of the two major scissile bonds, this region should have a non-α-helical structure, and may extend out from the surface of the protein molecule so that it becomes the most sensitive site for calpain action.

In contrast with the above cases, calf thymus histones (H2A, H2B, and H3) appeared to be hydrolyzed by calpain II purified from rabbit muscle, rather randomly and incompletely.[121] There are ten X-Tyr and X-Arg sequences, where X is Leu, Val or Ile, in these histones, but the cleavage at the peptide bonds preceded by these sequences occurred in only three cases. On the other hand, additional cleavages took places at several peptides bonds apparently rather nonspecifically, as can be seen in Table 7. Thus, many peptide fragments were produced, but the yields of these peptides were generally low, ranging from 1 to 23%. However, among the eleven major cleavage sites, where cleavage occurred in a yield of over 10%, five sites have Leu or Val at the P_2 position and/or Arg or Tyr at the P_1 position (Table 7). Thus these cleavages are partially in accord with the deduced subsite specificity. The fragments released from histones appeared to undergo no further degradation, although many of the released fragments still contained the sites that would be susceptible to calpain if they were present in the intact histones. Further, the susceptible bonds appeared to be located in the vicinity of the boundaries between hydrophilic and hydrophobic clusters. These results suggest that the enzyme may recognize a certain conformation pertaining to the structural integrity of the substrate molecule.

More recently, the sites of cleavage of rat brain protein kinase C by limited proteolysis with rat kidney calpain have been reported.[90] All sites of cleavage are located in the variable region, which connects the regulatory domain (NH_2-terminal half) and the protein kinase domain (COOH-terminal half). In type I protein kinase C, both calpains I and II cleaved the second Ser-Ser bond in the sequence -Pro-Ser-Ser-Ser-Pro- and the Phe-Phe bond in the sequence -Arg-Cys-Phe-Phe-Gly-. In the type II enzyme, calpain I cleaved the Gly-Gln bond in the sequence -Lys-Ile-Gly-Gln-Gly- and the Lys-Thr bond in the sequence -Glu-Glu-Lys-Thr-Ala-, while calpain II cleaved only the former bond. In the type III enzyme, calpain I cleaved the Lys-Val bond in the sequence -Gly-Asn-Lys-Val-Ile- and the Arg-Lys bond in the sequence -Glu-Asp-Arg-Lys-Gln, whereas calpain II failed to cleave either of these bonds, but cleaved the Lys-Leu bond in the sequence -Lys-Ala-Lys-Leu-Gly-. Thus, the results show that calpains I and II have somewhat different specificity and that the cleavages agree only partially with the subsite specificity deduced with peptide substrates.

IV. CONCLUDING REMARKS

As described in the preceding sections, calpains are endopeptidases capable of cleaving various naturally occurring peptides and synthetic peptides with apparently rather distinct subsite specificity. On the other hand, they hydrolyze several native proteins by way of limited proteolysis. Although the exact cleavage sites have been elucidated with only a limited number of proteins, it seems that they do not necessarily agree with the subsite specificity deduced from studies using oligopeptides. Thus, conformational factors of proteins appear to influence considerably the action of calpains and hence the sites of cleavage in the limited proteolysis of native proteins. The deduced subsite specificity indicates clearly that the substrate specificity of calpains is unique and different from those of other cysteine proteinases, such as papain, cathepsin B which has peptidyldipeptidase activity,[128-132] and cathepsin L.[133,134] The subsite specificity, however, has been deduced from the results obtained with a limited kind and number of peptides. Therefore, further detailed and more systematic studies on both the subsite specificity and the conformational factors are necessary to fully understand the substrate specificity of calpain. For these studies, it will be necessary to use many more different kinds of peptides and proteins in both native and denatured forms. Furthermore, each calpain isozyme appears to have some preference for specific types of proteins and this specificity appears to be not always identical among the same calpain isozymes isolated from various sources. The structural basis for this difference should also be answered in future studies. These studies will help to fully define the native, physiological substrates of the calpains.

REFERENCES

1. **Murachi, T.,** Calpain and calpastatin, *Trends Biochem. Sci.,* 8, 167, 1983.
2. **Pontremoli, S. and Melloni, E.,** Extralysosomal protein degradation, *Annu. Rev. Biochem.,* 55, 455, 1986.
3. **Murachi, T., Hatanaka, M., and Hamakubo, T.,** Calpain and neuropeptide metabolism, in *Neuropeptides and Their Peptidases,* Turner, A. J., Ed., Ellis Horwood, Chichester, England, 1987, chap. 11.
4. **Guroff, G.,** A neutral, calcium-activated proteinase form the soluble fraction of rat brain, *J. Biol. Chem.,* 239, 149, 1964.
5. **Waxman, L.,** Characterization of two unique protease inhibitors and a calcium-activated protease from bovine cardiac muscle, in *Protein Turnover and Lysosome Function,* Segal, H. L. and Doyl, D. J., Eds., Academic Press, New York, 1978, 363.
6. **Ishiura, S., Sugita, H., Suzuki, K., and Imahori, K.,** Studies of a calcium-activated neutral protease from chicken skeletal muscle. II. Substrate specificity, *J. Biochem. (Tokyo),* 86, 579, 1979.

7. **Ishiura, S., Tsuji, S., Murofushi, H., and Suzuki, K.,** Purification of an endogenous 68,000-dalton inhibitor of Ca^{2+}-activated neutral protease from chicken skeletal muscle, *Biochim. Biophys. Acta,* 701, 216, 1982.
8. **Hiraro, T., Hara, K., and Takahashi, K.,** Degradation of neuropeptides by calcium-activated neutral protease, *J. Biochem. (Tokyo),* 94, 2071, 1983.
9. **Hirao, T. and Takahashi, K.,** Purification and characterization of a calcium-activated neutral protease from monkey brain and its action on neuropeptides, *J. Biochem. (Tokyo),* 96, 775, 1984.
10. **Sasaki, T., Kikuchi, T., Yumoto, N., Yoshimura, N., and Murachi, T.,** Comparative specificity and kinetic studies on porcine calpain I and calpain II with naturally occurring peptides and synthetic fluorogenic substrates, *J. Biol. Chem.,* 259, 12489, 1984.
11. **Hayashi, M., Inomata, M., Nakamura, M., Imahori, K., and Kawashima, S.,** Hydrolysis of protamine by calcium-activated neutral protease (CANP), *J. Biochem. (Tokyo),* 97, 1363, 1985.
12. **McDermott, J. R., Mantle, D., Biggins, J. A., Kidd, A. M., Davison, K., Lauffart, B., and Pennington, R. J. T.,** Specificity of neuropeptide degradation by two calcium-activated neutral proteases from human skeletal muscle, *Life Sci.,* 37, 725, 1985.
13. **Tsubata, T., Ishizaki, Y., and Takahashi, K.,** Degradation of neuropeptides by cytosolic and cytoskeleton-bound calcium-activated neutral proteases from bovine spinal cord, *Biomed. Res.,* 8, 281, 1987.
14. **Hatanaka, M., Sasaki, T., Kikuchi, T., and Murachi, T.,** Amidase-like activity of calpain I and calpain II on substance P and its related peptides, *Arch. Biochem. Biophys.,* 242, 557, 1985.
15. **Sasaki, T., Kikuchi, T., Fukui, I., and Murachi, T.,** Inactivation of calpain I and calpain II by specificity-oriented tripeptidyl chloromethyl ketones, *J. Biochem. (Tokyo),* 99, 173, 1986.
16. **Murachi, T., Tanaka, K., Hatanaka, M., and Murakami, T.,** Intracellular Ca^{2+}-dependent protease (calpain) and its high-molecular-weight endogenous inhibitor (calpastatin), *Adv. Enzyme Regul.,* 19, 407, 1981.
17. **Hara, K. and Takahashi, K.,** Inhibition of calcium-activated neutral protease of monkey cardiac muscle by epoxysuccininc acid derivatives, *Biomed. Res.,* 4, 121, 1983.
18. **Sugita, H., Ishiura, S., Suzuki, K., and Imahori, K.,** Inhibition of epoxide derivatives on chicken calcium-activated neutral protease (CANP) *in vitro* and *in vivo, J. Biochem. (Tokyo),* 87, 339, 1980.
19. **Dayton, W. R., Reville, W. J., Goll, D. E., and Stromer, M. H.,** A Ca^{2+}-activated protease possibly involved in myofibrillar protein turnover. Partial characterization of the purified enzyme, *Biochemistry,* 15, 2159, 1976.
20. **Reville, W. J., Goll, D. E., Stromer, M. H., Robson, R. M., and Dayton, W. R.,** A Ca^{2+}-activated protease possibly involved in myofibrillar protein turnover. Subcellular localization of the protease in porcine skeletal muscle, *J. Cell. Boil.,* 70, 1,1976.
21. **Busch, W. A., Stromer, M. H., Goll, D. E., and Suzuki, A.,** Ca^{2+}-specific removal of Z lines from rabbit skeletal muscle, *J. Cell. Biol.,* 52, 367, 1972.
22. **Reddy, M. K., Etlinger, J. D., Rabinowitz, M., Fischman, D. A., and Zak, R.,** Removal of Z-lines and α-actinin from isolated myofibrils by a calcium-activated neutral protease, *J. Biol. Chem.,* 250, 4278, 1975.
23. **Azanza, J.-L., Raymond, J., Robin, J.-M., Cottin, P., and Ducasting, A.,** Purification and some physico-chemical and enzymic properties of a calcium ion-activated neutral proteinase from rabbit skeletal muscle, *Biochem. J.,* 183, 339, 1979.
24. **Sugita, H., Ishiura, S., Suzuki, K., and Imahori, K.,** Ca-activated neutral protease and its inhibitors: *in vitro* effect on intact myofibrils, *Muscle and Nerve,* 3, 355, 1980.
25. **Ohtsuki, I., Shiraishi, F., Suenaga, N., Miyata, T., and Tanokura, M.,** A 26K fragment of troponin T from rabbit skeletal muscle, *J. Biochem. (Tokyo),* 95, 1337, 1984.
26. **Ishiura, S., Murofushi, H., Suzuki, K., and Imahori, K.,** Studies of a calcium-activated neutral protease from chicken skeletal muscle. I. Purification and characterization, *J. Biochem. (Tokyo),* 84, 225, 1978.
27. **Hara, K., Ichihara, Y., and Takahashi, K.,** Purification and characterization of a calcium-activated neutral protease from monkey cardiac muscle, *J. Biochem. (Tokyo),* 93, 1435, 1983.
28. **Toyo-oka, T. and Masaki, T.,** Calcium-activated neutral protease from bovine ventricular muscle: Isolation and some of its properties, *J. Mol. Cell. Cardiol.,* 11, 769, 1979.
29. **Davies, P. J. A., Wallach, D., Willingham, M. C., and Pastan, I.,** Filamin-actin interaction. Dissociation of binding from gelation by Ca^{2+}-activated proteolysis, *J. Biol. Chem.,* 253, 4036, 1978.
30. **Ishizaki, Y., Kurokawa, M., and Takahashi, K.,** A calcium-dependent protease associated with the neural cytoskeleton. Purification and partial characterization, *Eur. J. Biochem.,* 146, 331, 1985.
31. **Schlaepfer, W. W. and Micko, S.,** Chemical and structural changes of neurofilaments in transected rat sciatic nerve, *J. Cell. Biol.,* 78, 369, 1978.
32. **Kamakura, K., Ishiura, S., Sugita, H., and Toyokura, Y.,** Identification of Ca^{2+}-activated neutral protease in the peripheral nerve and its effects on neurofilament degeneration, *J. Neurochem.,* 40, 908, 1983.

33. **Schlaepfer, W. W., Lee, C., Lee, V. M.-Y., and Zimmerman, U.-J. P.**, An immunoblot study of neurofilament degradation *in situ* and during calcium-activated proteolysis, *J. Neurochem.*, 44, 502, 1985.
34. **Schlaepfer, W. W. and Micko, S.**, Calcium-dependent alterations of neurofilament proteins of rat peripheral nerve, *J. Neurochem.*, 32, 211, 1979.
35. **Schlaepfer, W. W. and Freeman, L. A.**, Calcium-dependent degradation of mammalian neurofilaments by soluble tissue factor(s) from rat spinal cord, *Neuroscience*, 5, 2305, 1980.
36. **Tashiro, T., and Ishizaki, Y.**, A calcium-dependent protease selectively degrading the 160,000 M_r component of neurofilaments is associated with the cytoskeletal preparation of the spinal cord and has an endogenous inhibitory factor, *FEBS Lett.*, 141, 41, 1982.
37. **Ishizaki, Y., Tashiro, T., and Kurokawa, M.**, A calcium-activated protease which preferentially degrades the 160-kDa component of the neurofilament triplet, *Eur. J. Biochem.*, 131, 41, 1983.
38. **Banik, N. L., McAlhaney, W. W., and Hogan, E. L.**, Is Ca^{++}-activated neutral proteinase a myelin enzyme?, *J. Neurochem.*, 41(Suppl.), 130A, 1983.
39. **Banik, N. L., McAlhaney, W. W., and Hogan, E. L.**, Calcium-stimulated proteolysis in myelin: Evidence for a Ca^{2+}-activated neutral proteinase associated with purified myelin of rat CNS, *J. Neurochem.*, 45, 581, 1985.
40. **Nixon, R. A., Brown, B. A., and Marotta, C. A.**, Posttranslational modification of a neurofilament protein during axoplasmic transport: Implications for regional specialization of CNS axons, *J. Cell. Biol.*, 94, 150, 1982.
41. **Nixon, R. A., Brown, B. A., and Marotta, C. A.**, Limited proteolytic modification of a neurofilament protein involves a proteinase activated by endogenous levels of calcium, *Brain Res.*, 275, 384, 1983.
42. **Kubota, S., Onaka, T., Murofushi, H., Ohsawa, N., and Takaku, F.**, Purification and characterization of high Ca^{2+}-requiring neutral proteases from porcine and bovine brains, *Biochemistry*, 25, 8396, 1986.
43. **Malik, M. N., Meyers, L. A., Igbal, K., Sheikh, A. M., Scotto, L., and Wisniewski, H. M.**, Calcium activated proteolysis of fibrous proteins in central nervous system, *Life Sci.*, 29, 795, 1981.
44. **Malik, M. N., Fenko, M. D., Igbal, K., and Wisniewski, H. M.**, Purification and characterization of two forms of Ca^{2+}-activated neutral protease from calf brain, *J. Biol. Chem.*, 258, 8955, 1983.
45. **Zimmerman, U.-J. P. and Schlaepfer, W. W.**, Characterization of a brain calcium-activated protease that degrades neurofilament proteins, *Biochemistry*, 21, 3977, 1982.
46. **Zimmerman, U.-J. P. and Schlaepfer, W. W.**, Multiple forms of Ca-activated protease from rat brain and muscle, *J. Biol. Chem.*, 259, 3210, 1984.
47. **DeArmond, S. J., Fajardo, M., Naughton, S. A., and Eng, L. F.**, Degradation of glial fibrillary acidic protein by a calcium dependent proteinase: an electroblot study, *Brain Res.*, 262, 275, 1983.
48. **Baudry, M., Bundman, M. C., Smith, E. K., and Lynch, G. S.**, Micromolar calcium stimulates proteolysis and glutamate binding in rat brain synaptic membranes, *Science*, 212, 937, 1981.
49. **Nelson, W. J. and Traub, P.**, Purification and further characterization of the Ca^{2+}-activated proteinase specific for the intermediate filament proteins vimentin and desmin, *J. Biol. Chem.*, 257, 5544, 1982.
50. **Nelson, W. J. and Traub, P.**, Proteolysis of vimentin and desmin by the Ca^{2+}-activated proteinase specific for these intermediate filament proteins, *Mol. Cell. Biol.*, 3, 1146, 1983.
51. **Ireland, M. and Maisel, H.**, Evidence for a calcium activated protease specific for lens intermediate filaments, *Current Eye Res.*, 3, 423, 1984.
52. **Yoshida, H., Murachi, T., and Tsukahara, I.**, Degradation of actin and vimentin by calpain II, a Ca^{2+}-dependent cysteine proteinase, in bovine lens, *FEBS Lett.*, 170, 259, 1984.
53. **Fischer, S., Vandekerckhove, J., Ampe, C., Traub, P., and Weber, K.**, Protein-chemical identification of the major cleavage sites of the Ca^{2+} proteinase on murine vimentin, the mesenchymal intermediate filament protein, *Biol. Chem. Hoppe-Seyler*, 367, 1147, 1986.
54. **Harris, A. S., Croall, D. E., and Morrow, J. S.**, The calmodulin-binding site in α-fodrin is near the calcium-dependent protease-I cleavage site, *J. Biol. Chem.*, 263, 15754, 1988.
55. **Siman, R., Baudry, M., and Lynch, G.**, Brain fodrin: substrate for calpain I, and endogenous calcium-activated protease, *Proc. Natl. Acad. Sci. U.S.A.*, 81, 3572, 1984.
56. **Sandoval, I. V. and Weber, K.**, Calcium-induced inactivation of microtubule formation in brain extracts. Presence of a calcium-dependent protease acting on polymerization-stimulating microtubule-associated proteins, *Eur. J. Biochem.*, 92, 463, 1978.
57. **Klein, I., Lehotay, D., and Gondek, M.**, Characterization of a calcium-activated protease that hydrolyzes a microtubule-associated protein, *Arch. Biochem. Biophys.*, 208, 520, 1981.
58. **Burgoyne, R. D. and Cumming, R.**, Evidence for the presence of high-M_r microtubule-associated proteins and their Ca^{2+}-dependent proteolysis in synaptosomal cytosol, *FEBS Lett.*, 146, 273, 1982.
59. **Yoshimura, N., Tsukahara, I., and Murachi, T.**, Calpain and calpastatin in porcine retina. Identification and action on microtubule-associated proteins, *Biochem. J.*, 223, 47, 1984.
60. **Croall, D. E., Morrow, J. S., and DeMartino, G. N.**, Limited proteolysis of the erythrocyte membrane skeleton by calcium-dependent proteinases, *Biochim. Biophys. Acta*, 882, 287, 1986.

61. **Pant, H. C., Virmani, M., and Gallant, P. E.**, Calcium-induced proteolysis of spectrin and band 3 protein in rat erythrocyte membranes, *Biochem. Biophys. Res. Commun.*, 117, 372, 1983.
62. **Murakami, T., Hatanaka, M., and Murachi, T.**, The cytosol of human erythrocytes contains a highly Ca^{2+}-sensitive thiol protease (calpain I) and its specific inhibitor protein (calpastatin), *J. Biochem. (Tokyo)*, 90, 1809, 1981.
63. **Hatanaka, M., Yoshimura, N., Murakami, T., Kannagi, R., and Murachi, T.**, Evidence for membrane-associated calpain I in human erythrocytes. Detection by an immunoelectrophoretic blotting method using monospecific antibody, *Biochemistry*, 23, 3272, 1984.
64. **Anderson, D. R., Davis, J. L., and Carraway, K. L.**, Calcium-promoted changes of the human erythrocyte membrane. Involvement of spectrin, transglutaminase, and a membrane-bound protease, *J. Biol. Chem.*, 252, 6617, 1977.
65. **Gache, Y., Landon, F., Touitou, H., and Olomucki, A.**, Susceptibility of platelet α-actinin to a Ca^{2+}-activated neutral protease, *Biochem. Biophys. Res. Commun.*, 124, 877, 1984.
66. **Truglia, J. A. and Stracher, A.**, Purification and characterization of a calcium-dependent sulfhydryl protease from human platelets, *Biochem. Biophys. Res. Commun.*, 100, 814, 1981.
67. **Fox, J. E. B., Goll, D. E., Reynolds, C. C., and Phillips, D. R.**, Identification of two proteins (actin-binding protein and P235) that are hydrolyzed by endogenous Ca^{2+}-dependent protease during platelet aggregation, *J. Biol. Chem.*, 260, 1060, 1984.
68. **Beckerle, M. C., Burridge, K., DeMartino, G. N., and Croall, D. E.**, Colocalization of calcium-dependent protease II and one of its substrates at sites of cell adhesion, *Cell*, 51, 569, 1987.
69. **Puca, G. A., Nola, E., Sica, V., and Bresciani, F.**, Estrogen-binding proteins of calf uterus. Interrelationship between various forms and identification of a receptor-transforming factor, *Biochemistry*, 11, 4157, 1972.
70. **Puca, G. A., Nola, E., Sica, V., and Bresciani, F.**, Estrogen binding proteins of calf uterus. Molecular and functional characterization of the receptor transforming factor: a Ca^{2+}-activated protease, *J. Biol. Chem.*, 252, 1358, 1977.
71. **Sherman, M. R., Atienza, S. B. P., Shansky, J. R., and Hoffman, L. M.**, Progesterone receptors of chick oviduct. Steroid-binding "SUBUNIT" formed with divalent cations, *J. Biol. Chem.*, 249, 5351, 1974.
72. **Vedeckis, W. V., Freeman, M. R., Schrader, W. T., and O'Malley, B. W.**, Progesterone-binding components of chick oviduct: partial purification and characterization of a calcium-activated protease which hydrolyzes the progesterone receptor, *Biochemistry*, 19, 335, 1980.
73. **Baudry, M. and Lynch, G.**, Regualtion of hippocampal glutamate receptors: evidence for the involvement of a calcium-activated protease, *Proc. Natl. Acad. Sci. U.S.A.*, 77, 2298, 1980.
74. **Lynch, C. J., Sobo, G. E., and Exton, J. H.**, An endogenous Ca^{2+}-sensitive proteinase converts the hepatic α_1-adrenergic receptor to guanine nucleotide-insensitive forms, *Biochim. Biophys. Acta*, 885, 110, 1986.
75. **Cassel, D. and Glaser, L.**, Proteolytic cleavage of epidermal growth factor receptor. A Ca^{2+}-dependent, sulfhydryl-sensitive proteolytic system in A431 cells, *J. Biol. Chem.*, 257, 9845, 1982.
76. **Gates, R. E. and King, L. E., Jr.**, Proteolysis of the epidermal growth factor receptor by endogenous calcium-activated neutral protease from rat liver, *Biochem. Biophys. Res. Commun.*, 113, 255, 1983.
77. **Yeaton, R. W., Lipari, M. T., and Fox, C. F.**, Calcium-mediated degradation of epidermal growth factor receptor in dislodged A431 cells and membrane preparations, *J. Biol. Chem.*, 258, 9254, 1983.
78. **Ek, B. and Heldin, C.-H.**, Specific cleavage of the fibroblast receptor for platelet-derived growth factor by an endogenous Ca^{2+}-dependent thiol protease, *Eur. J. Biochem.*, 155, 409, 1986.
79. **Yoshida, N., Weksler, B., and Nachman, R.**, Purification of human platelet calcium-activated protease. Effect on platelet and endothelial function, *J. Biol. Chem.*, 258, 7168, 1983.
80. **McGowan, E. B., Yeo, K.-T., and Detwiler, T. C.**, The action of calcium-dependent protease on platelet surface glycoproteins, *Arch. Biochem. Biophys.*, 227, 287, 1983.
81. **Meyer, W. L., Fischer, E. H., and Krebs, E. G.**, Activation of skeletal muscle phosphorylase b kinase by Ca^{2+}, *Biochemistry*, 3, 1033, 1964.
82. **Huston, R. B. and Krebs, E. G.**, Activation of skeletal muscle phosphorylase kinase by Ca^{2+}. II. Identification of the kinase activating factor as a proteolytic enzyme, *Biochemistry*, 7, 2116, 1968.
83. **Drummond, G. I. and Duncan, L.**, On the mechanism of activation of phosphorylase b kinase by calcium, *J. Biol. Chem.*, 243, 5532, 1968.
84. **Drummond, G. I. and Duncan, L.**, The action of calcium ion on cardiac phosphorylase b kinase, *J. Biol. Chem.*, 241, 3097, 1966.
85. **Corbin, J. D., Brostrom, C. O., King, C. A., and Krebs, E. G.**, Studies on the adenosine 3′,5′-monophosphate-dependent protein kinases of rabbit skeletal muscle, *J. Biol. Chem.*, 247, 7790, 1972.
86. **Beer, D. G., Butley, M. S., and Malkinson, A. M.**, Developmental changes in the endogenous Ca^{2+}-stimulated proteolysis of mouse lung cAMP-dependent protein kinases, *Arch. Biochem. Biophys.*, 228, 207, 1984.

87. **Inoue, M., Kishimoto, A., Takai, Y., and Nishizuka, Y.**, Studies on a cyclic nucleotide-independent protein kinase and its proenzyme in mammalian tissues. II. Proenzyme and its activation by calcium-dependent protease from rat brain, *J. Biol. Chem.*, 252, 7610, 1977.
88. **Kishimoto, A., Kajikawa, N., Shiota, M., and Nishizuka, Y.**, Proteolytic activation of calcium-activated, phospholipid-dependent protein kinase by calcium-dependent neutral protease, *J. Biol. Chem.*, 258, 1156, 1983.
89. **Takai, Y., Yamamoto, M., Inoue, M., Kishimoto, A., and Nishizuka, Y.**, A proenzyme of cyclic nucleotide-independent protein kinase and its activation by calcium-dependent neutral protease from rat liver, *Biochem. Biophys. Res. Commun.*, 77, 542, 1977.
90. **Kishimoto, A., Mikawa, K., Hashimoto, K., Yasuda, I., Tanaka, S., Tominaga, M., Kuroda, T., and Nishizuka, Y.**, Limited proteolysis of protein kinase C subspecies by calcium-dependent neutral protease (calpain), *J. Biol. Chem.*, 264, 4088, 1989.
91. **Melloni, E., Pontremoli, S., Michetti, M., Sacco, O., Sparatore, B., Salamino, F., and Horecker, B. L.**, Binding of protein kinase C to neutrophil membranes in the presence of Ca^{2+} and its activation by a Ca^{2+}-requiring proteinase, *Proc. Nat. Acad. Sci. U.S.A.*, 82, 6435, 1985.
92. **Melloni, E., Pontremoli, S., Michetti, M., Sacco, O., Sparatore, B., and Horecker, B. L.**, The involvement of calpain in the activation of protein kinase C in neutrophils stimulated by phorbol myristic acid, *J. Biol. Chem.*, 261, 4101, 1986.
93. **Tapley, P. M. and Murray, A. W.**, Platelet Ca^{2+}-activated, phospholipid-dependent protein kinase: evidence for proteolytic activation of the enzyme in cells treated with phospholipase C, *Biochem. Biophys. Res. Commun.*, 118, 835, 1984.
94. **Tapley, P. M. and Murray, A. W.**, Modulation of Ca^{2+}-activated phospholipid-dependent protein kinase in platelets treated with a tumor-promoting phorbol ester, *Biochem. Biophys. Res. Commun.*, 122, 158, 1984.
95. **Juskevich, J. C., Kuhn, D. M., and Lovenberg, W.**, Calcium enhanced inactivation of calmodulin dependent protein kinase from synaptosomes, *Biochem. Biophys. Res. Commun.*, 108, 24, 1982.
96. **Ito, M., Tanaka, T., Nunoki, K., Hidaka, H., and Suzuki, K.**, The Ca^{2+}-activated protease (calpain) modulates Ca^{2+}/calmodulin dependent activity of smooth muscle myosin light chain kinase, *Biochem. Biophys. Res. Commun.*, 145, 1321, 1987.
97. **Belocopitow, E., Appleman, M. M., and Torres, H. N.**, Factors affecting the activity of muscle glycogen synthetase. II. The regulation by Ca^{2+}, *J. Biol. Chem.*, 240, 3473, 1965.
98. **Hiraga, A. and Tsuiki, S.**, Activation of a D-form of rabbit muscle glycogen synthase by Ca^{2+}-activated protease, *FEBS Lett.*, 205, 1, 1986.
99. **Au, K. S.**, Activation of erythrocyte membrane Ca^{2+}-ATPase by calpain, *Biochim. Biophys. Acta*, 905, 273, 1987.
100. **Tallant, E. A., Brumley, L. M., and Wallace, R. W.**, Activation of a calmodulin-dependent phosphatase by a Ca^{2+}-dependent protease, *Biochemistry*, 27, 2205, 1988.
101. **Mellgren, R. L., Aylward, J. H., Killilea, S. D., and Lee, E. Y. C.**, The activation and dissociation of a native high molecular weight form of rabbit skeletal muscle phosphorylase phosphatase by endogenous Ca^{2+}-dependent proteases, *J. Biol. Chem.*, 254, 648, 1979.
102. **Low, M. G., Carroll, R. C., and Weglicki, W. B.**, Multiple forms of phosphoinositide-specific phospholipase C of different relative molecular masses in animal tissues. Evidence for modification of the platelet enzyme by Ca^{2+}-dependent proteinase, *Biochem. J.*, 221, 813, 1984.
103. **Hamon, H., Bourgoin, S., Artaud, F., and Hery, F.**, Rat brain stem tryptophan hydroxylase: mechanism of activation by calcium, *J. Neurochem.*, 28, 811, 1977.
104. **Togari, A., Ichikawa, S., and Nagatsu, T.**, Activation of tyrosine hydroxylase by Ca^{2+}-dependent neutral protease, calpain, *Biochem. Biophys. Res. Commun.*, 134, 749, 1986.
105. **Ando, Y., Imamura, S., Yamagata, Y., Kitahara, A., Saji, H., Murachi, T., and Kannagi, R.**, Platelet factor XIII is activated by calpain, *Biochem. Biophys. Res. Commun.*, 144, 484, 1987.
106. **Inomata, M., Imahori, K., and Kawashima, S.**, Autolytic activation of calcium-activated neutral protease, *Biochem. Biophys. Res. Commun.*, 138, 638, 1986.
107. **Crawford, C., Willis, A. C., and Gagnon, J.**, The effect of autolysis on the structure of chicken calpain II, *Biochem. J.*, 248, 579, 1987.
108. **Yanagisawa, K., Sato, S., Amaya, N., and Miyatake, T.**, Degradation of myelin basic protein by calcium-activated neutral protease (CANP) in human brain and inhibition by E-64 analogue, *Neurochem. Res.*, 8, 1285, 1983.
109. **Sato, S. and Miyatake, T.**, Degradation of myelin basic protein by calcium-activated neutral protease (CANP)-like enzyme in myelin and inhibition by E-64 analogue, *Biomed. Res.*, 3, 461, 1982.
110. **Sato, S., Quarles, R. H., and Brady, R. O.**, Suseptibility of the myelin-associated glycoprotein and basic protein to a neutral protease in highly purified myelin from human and rat brain, *J. Neurochem.*, 39, 97, 1982.

111. **Sato, S. and Miyatake, T.**, Degradation of myelin basic protein and myelin-associated glycoprotein by calcium-activated neutral protease like enzyme, *J. Neurochem.*, 41(Suppl.), 46A, 1983.
112. **Tsubata, T. and Takahashi, K.**, Limited proteolysis of bovine myelin basic protein by calcium-dependent proteinase from bovine spinal cord, *J. Biochem. (Tokyo)*, 105, 23, 1989.
113. **Yanagisawa, K., Sato, S., O'Shannessy, D. J., Quarles, R. H., Suzuki, K., and Miyatake, T.**, Myelin-associated calpain II, *J. Neurochem.*, 51, 803, 1988.
114. **Phillips, D. R. and Jakábová, M.**, Ca^{2+}-dependent protease in human platelets. Specific cleavage of platelet polypeptides in the presence of added Ca^{2+}, *J. Biol. Chem.*, 252, 5602, 1977.
115. **Kōsaki, G., Tsujinaka, T., Kambayashi, J., Morimoto, K., Yamamoto, K., Yamagami, K., Sobue, K., and Kakiuchi, S.**, Specific cleavage of calmodulin-binding proteins by low Ca^{2+}-requiring form of Ca^{2+}-activated neutral protease in human platelets, *Biochem. Int.*, 6, 767, 1983.
116. **Yoshida, H., Yumoto, N., Tsukahara, I., and Murachi, T.**, The degradation of α-crystallin at its carboxyl-terminal portion by calpain in bovine lens, *Invest. Ophthalmol. Vis. Sci.*, 27, 1269, 1986.
117. **Hutton, J. C., Davidson, H. W., and Peshavaria, M.**, Proteolytic processing of chromogranin A in purified insulin granules. Formation of a 20 kDa N-terminal fragment (betagranin) by the concerted action of a Ca^{2+}-dependent endopeptidase and carboxypeptidase H (EC 3.4.17.10), *Biochem. J.*, 244, 457, 1987.
118. **Kunicki, T., Montgomery, R. R., and Schullek, J.**, Cleavage of human von Willebrand factor by platelet calcium-activated protease, *Blood*, 65, 352, 1985.
119. **Lang, R. D. A., Wickenden, C., Wynne, J., and Lucy, J. A.**, Proteolysis of ankyrin and of band 3 protein in chemically induced cell fusion. Ca^{2+} is not mandatory for fusion, *Biochem. J.*, 218, 295, 1984.
120. **Yoshihara, Y., Ueda, H., Imajoh, S., Takagi, H., and Satoh, M.**, Calcium-activated neutral protease (CANP), a putative processing enzyme of the neuropeptide, kyotorphin, in the brain, *Biochem. Biophys. Res. Commun.*, 155, 546, 1988.
121. **Sakai, K., Akanuma, H., Imahori, K., and Kawashima, S.**, A unique specificity of a calcium activated neutral protease indicated in histone hydrolysis, *J. Biochem. (Tokyo)*, 101, 911, 1987.
122. **Melloni, E., Salamino, F., Sparatore, B., Michetti, M., and Pontremoli, S.**, Characterization of the single peptide generated from the amino-terminus end of α-and β-hemoglobin chains by the Ca^{2+}-dependent neutral proteinase, *Biochim. Biophys. Acta*, 788, 11, 1984.
123. **Nakamura, M., Inomata, M., Imajoh, S., Suzuki, K., and Kawashima, S.**, Fragmentation of an endogenous inhibitor upon complex formation with high- and low-Ca^{2+}-requiring forms of calcium-activated neutral proteases, *Biochemistry*, 28, 449, 1989.
124. **Pontremoli, S., Melloni, E., Salamino, F., Sparatore, B., Michetti, M., and Horecker, B. L.**, Cytosolic Ca^{2+}-dependent neutral proteinases from rabbit liver: activation of the proenzymes by Ca^{2+} and substrate, *Proc. Natl. Acad. Sci. U.S.A.*, 81, 53, 1984.
125. **Siezen, R. J., and Hoenders, H. J.**, The quaternary structure of bovine α-crystallin. Surface probing by limited proteolysis *in vitro*, *Eur. J. Biochem.*, 96, 431, 1979.
126. **Geisler, N., and Weber, K.**, The amino acid sequence of chicken muscle desmin provides a common structural model for intermediate filament proteins, *EMBO J.*, 1, 1649, 1982.
127. **Harris, A. S., and Morrow, J. S.**, Proteolytic processing of human brain alpha spectrin (fodrin): identification of a hypersensitive site, *J. Neurosci.*, 8, 2640, 1988.
128. **Aronson, N. N., Jr., and Barrett, A. J.**, The specificity of cathepsin B. Hydrolysis of glucagon at the *C*-terminus by a peptidyldipeptidase mechanism, *Biochem. J.*, 171, 759, 1978.
129. **Bond, J. S., and Barrett, A. J.**, Degradation of fructose-1,6-bisphosphate aldolase by cathepsin B. A further example of peptidyldipeptidase activity of this proteinase, *Biochem. J.*, 189, 17, 1980.
130. **Barrett, A. J., and Kirschke, H.**, Cathepsin B, cathepsin H, and cathepsin L, *Methods Enzymol.*, 80, 535, 1981.
131. **Hirao, T., Hara, K., and Takahashi, K.**, Purification and characterization of cathepsin B from monkey skeletal muscle, *J. Biochem. (Tokyo)*, 95, 871, 1984.
132. **Takahashi, T., Dehdarani, A. H., Yonezawa, S,, and Tang, J.**, Porcine spleen cathepsin B is an exopeptidase, *J. Biol. Chem.*, 261, 9375, 1986.
133. **Kärgel, H.-J., Dettmer, R., Etzold, G., Kirschke, H., Bohley, P., and Langner, J.**, Action of cathepsin L on the oxidized B-chain of bovine insulin, *FEBS Lett.*, 114, 257, 1980.
134. **Towatari, T. and Katsunuma, N.**, Selective cleavage of peptide bonds by cathepsins L and B from rat liver, *J. Biochem. (Tokyo)*, 93, 1119, 1983.

Chapter 5

PROTEIN AND PEPTIDE INHIBITORS OF CALPAINS

Catherine Crawford

TABLE OF CONTENTS

I. INTRODUCTION

Investigation of the inhibitors of an enzyme can provide a great deal of useful information. Identification and characterization of natural inhibitors may suggest how the enzyme is regulated *in vivo*. For proteinases, use of active-site directed synthetic inhibitors allows the class of the proteinase to be assigned. Experimentation with a range of peptide inhibitors gives an indication of the primary sequence specificity of the site cleaved by a proteinase. Also, use of specific inhibitors *in vivo* facilitates understanding of the physiological roles of enzymes. For calpains, both natural and synthetic inhibitors have been identified. In all cases inhibition is only observed in the presence of Ca^{2+}, the conditions under which enzyme activity is expressed; and generally little difference has been observed in the inhibition characteristics of calpains I and II. In this chapter, I first describe how the characteristics of a particular inhibitor can be studied, then discuss the known protein and peptide inhibitors of calpains, and finally describe how inhibitors have been used to investigate the physiological roles of calpain.

II. CHARACTERIZATION OF AN INHIBITOR

An inhibitor is a molecule which, when bound to an enzyme in the presence of substrate, decreases the turnover of the substrate. Once an inhibitor has been identified it can be characterized. The first question to consider is whether the inhibitor is reversible or irreversible. Following reversible inhibition, enzyme activity can be recovered by gel filtration, dilution or dialysis, provided sufficient time is allowed for dissociation. If enzyme activity is not recovered in these situations then inhibition is irreversible and the enzyme has been inactivated. It is not always easy to distinguish between reversible and irreversible inhibitors particularly in the case of tight binding inhibitors. The next question to address is how good is the inhibitor? This can be quantitated either by determining the equilibrium binding constant (K_i) for reversible inhibitors or the rate constant of inactivation for irreversible inhibitors. Measurement of IC_{50} values, that is, the concentration of inhibitor required to decrease the activity of the enzyme by 50%, is generally uninformative. IC_{50} values give no information about enzyme mechanism; equilibrium binding constants cannot be calculated from them; and, in the case of irreversible or tight binding inhibitors, the IC_{50} value approximates to 50% of the enzyme concentration used in the assay.

Procedures for quantitating inhibition have been excellently reviewed recently.[1] Briefly, for irreversible inhibitors of proteinases the reaction is usually assumed to proceed via a noncovalent enzyme-inhibitor complex according to Scheme 1:

$$\mathrm{E} + \mathrm{I} \overset{\mathrm{K_i}}{\rightleftharpoons} \mathrm{EI} \xrightarrow{\mathrm{k_2}} \mathrm{E_i}$$

Scheme 1

where K_i is the dissociation constant for the intermediate complex and k_2 is the first order rate constant for the formation of the inactive enzyme (E_i). Values of K_i, k_2, and for the overall second order rate constant for the inactivation can be found by the method of Kitz and Wilson[2] or more conveniently by the method of Tian and Tsou.[3] In both procedures it is important to insure that inhibitor depletion does not occur, and that the relative levels of the other components in the assay system are consistent with the assumptions made in the kinetic analyses. For reversible inhibitors, and when Michaelis-Menten kinetics are observed in the presence of inhibitor, the inhibition can be classified as competitive if it increases K_M, or as noncompetitive if it decreases k_{cat}. Scheme 2 shows a model for reversible inhibition.

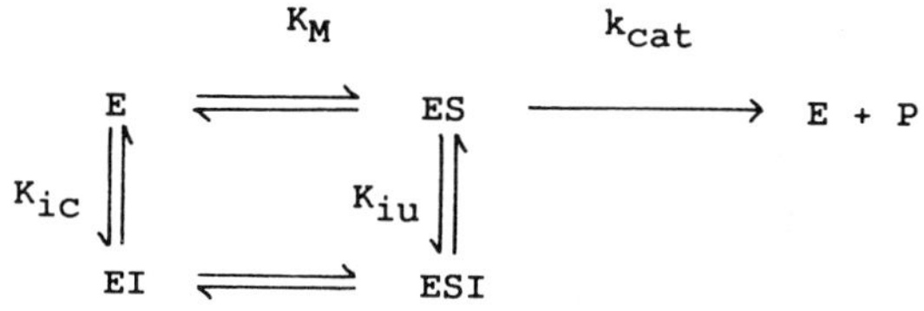

Scheme 2

Reversible inhibitors are frequently analyzed by observing the effects of increasing the inhibitor concentration on the velocity of inhibition at fixed values of enzyme and substrate concentration and analyzing the results using Dixon plots.[4] It is then possible to obtain values of K_{ic} and K_{iu} (the smaller K_i dominates), and to assign the type of inhibition. There are two situations frequently encountered in the analysis of reversible proteinase inhibitors where the above analysis is not valid and Michaelis-Menten kinetics are not observed in the presence of inhibitor. This occurs when an inhibitor is either tight-binding or slow-binding (or both). For tight-binding inhibitors, when significant inhibitor depletion usually occurs during assays, kinetic parameters can be determined using the method of Ackermann and Potter[5] or Henderson.[6] For slow-binding inhibitors, analysis is complicated by the fact that a significant time is taken to reach equilibrium between enzyme, substrate, and inhibitor. These problems have been discussed by Morrison[7] and Cha.[8]

The final question to consider when characterizing an inhibitor is what is its specificity? For proteinase inhibitors, it is pertinent to determine whether the inhibitor is effective against only one proteinase, against a class of proteinases, or against all proteinases.

The characteristics of all the calpain inhibitors described in this chapter are summarized in Table 1.

III. PROTEIN INHIBITORS OF CALPAIN

A. CALPASTATIN

Calpastatin is a specific inhibitor of calpains (see Reference 9 for a review). It occurs in various molecular forms; for example in the pig, a 107,000-M_r form is present in the heart muscle and a 68,000-M_r form is present in the erythrocytes. There have been considerable advances recently in our understanding of the structure of calpastatin. These advances are described in Chapter 3 of this volume, and the physiological aspects of calpain/calpastatin interactions are presented in Chapter 8. This section will therefore concentrate on the molecular aspects of calpain/calpastatin interactions.

Calpastatin inhibits both calpains I and II.[10,11] The inhibitor does not show any species or tissue specificity.[11-13] Calpastatin is only inhibitory against calpains; it does not inhibit other cysteine proteinases such as papain, cathepsin B, bromelain, or ficin, nor proteinases from other classes including trypsin, chymotrypsin, plasmin, thrombin, subtilisin, elastase, pepsin, cathepsin D, collagenase, or thermolysin.[11,14-18] Calpain is only active in the presence of Ca^{2+}. However, calpastatin does not inhibit calpain by functioning as a Ca^{2+} chelator since high concentrations of Ca^{2+} do not reverse the inhibition,[10,13] calpastatin does not alter the Ca^{2+} concentration required for half maximal enzyme activity,[19,20] and calpastatin does not bind Ca^{2+}.[21] Calpastatin acts as an inhibitor by complexing with calpain. Interaction is thought to occur through the large subunit of the enzyme.[10,21] Ca^{2+} has been shown to be required for this complex formation, as calpain only binds to calpastatin immobilized on Sepharose in the presence of Ca^{2+}[22,23] and calpain/calpastatin complexes are only observed on gel filtration in the presence of Ca^{2+}.[16,24] It is assumed that the role of Ca^{2+} in the complexes is to bind to calpain. This results in a conformational change in the enzyme allowing it to express activity and associate with calpastatin. The inhibitory complex formed between calpain and calpastatin is dissociated by addition of EDTA;[24,25] however, the reversibility of the enzyme inhibitor interaction in the presence of Ca^{2+} has not been formally

TABLE 1
Summary of the Properties of Calpain Inhibitors

Inhibitor	Specificity	Reversible or irreversible	K_i (M) or rate constant ($M^{-1}s^{-1}$)
Calpastatin	Calpains only	Reversible(?) Competitive(?)	3.0×10^{-9} M
α_2-macroglobulin	All proteinases	Irreversible	3.0×10^{4} $M^{-1}s^{-1}$
Kininogen	Cysteine proteinases	Reversible	1.0×10^{-9} M
Peptide aldehydes	Serine and cysteine proteinases	Reversible Competitive(?)	4.0×10^{-7} M for leupeptin
Epoxysuccinyl peptides	Cysteine proteinases	Irreversible	7.5×10^{3} $M^{-1}s^{-1}$ for E-64
Peptide chloromethanes	Serine and cysteine proteinases	Irreversible	$> 10^{5}$ $M^{-1}s^{-1}$ for Leu-Leu-Phe-CH_2Cl
Peptide diazomethanes	Cysteine proteinases	Irreversible	2.3×10^{5} $M^{-1}s^{-1}$ for Z-Leu-Leu-Tyr-CHN_2

Note: Ca^{2+} is required for all the inhibitors. The inhibitors react in a very similar way with calpains I and II.

demonstrated. It has been reported that calpastatin is digested by calpain[22,24,26] and that calpain autolyzes even in the presence of the inhibitor.[27] These observations are relevant when considering the outcome of calpain/calpastatin interactions.

It is clear from the recent sequence analysis of calpastatins that the inhibitors contain multiple inhibitory domains.[28-31] These domains (M_r 14,000) can act in isolation as inhibitory units. This has been shown by analysis of proteolytic fragments of calpastatin[32] and by analysis of domains obtained by expression of cDNA fragments of the inhibitor.[33-35] The domains each contain a central consensus sequence (e.g., GERDDTIPPEYR for pig domain 4) that has been postulated to be important for inhibition. Analysis of the domains by mutagenesis has shown that a region 52 amino acids long, including the central consensus sequence, is necessary for inhibition.[36] Whether this region acts as an inhibitor simply by forming a tight complex with calpain, or whether the inhibitor is cleaved by calpain during complex formation, remains to be elucidated.

Quantitation of the inhibition of calpain by calpastatin has proved difficult. The two parameters that could be determined are the equilibrium binding constant (K_i) and the number of molecules of calpain inhibited by each molecule of calpastatin. Initial attempts to determine the K_i used Dixon plots, which are inappropriate for tight-binding inhibitors, and also led to the misleading conclusion that inhibition was noncompetitive.[20,23] More recently a K_i of 3.0 nM for one inhibitory domain of pig calpastatin with calpain I has been determined, and inhibitor shown to be competitive.[36] Measuring correct K_i values for the whole calpastatin molecule will be complicated because the individual domains may have different K_i's and the domains may show cooperativity. Finding the stoichiometry of calpain calpastatin complexes has also been difficult. Values ranging up to ten molecules of calpain bound per molecule of calpastatin have been reported from analysis of complexes by titration or gel filtration.[10,23,24,37] The diversity of values obtained is partly due to the difficulty of determining accurate values for the concentration of active molecules in enzyme and inhibitor solutions. Calpastatin has a very low extinction coefficient[17,38] and runs with anomalous M_r on gel filtration and SDS/PAGE.[29] Now that the structure of calpastatins has been elucidated, it is clear that no more than three or four molecules of calpain can bind per inhibitor molecule, the number depending on the number of inhibitory domains in a particular calpastatin. It has been reported that the inhibitor will bind up to the maximum number of one calpain per inhibitory domain,[31,34] suggesting that no steric hindrance occurs between the domains.

B. ALPHA-2-MACROGLOBULIN

Alpha-2-macroglobulin (α_2-M) is a tetrameric plasma protein (M_r 725,000) that inhibits all classes of proteinase (see Reference 39 for a review). The molecule has the appearance of a cyrillic H in electron micrographs.[40] The upper and lower halves of the H are thought to act as independent inhibitory units. The reaction of α_2-M with proteinases involves several steps. First, the proteinase complexes with α_2-M and cleaves a peptide bond in the "bait-region" of the inhibitor. This causes a conformational change in α_2-M which entraps the proteinase within the arms of the H.[41,42] The conformational change also activates a thiol-ester bond that may react with proteinase lysine residues to form a covalent interaction. This covalent bond formation is not necessary for irreversible inhibition. The final consequence of the conformational change is the exposure of receptor recognition sites on α_2-M for hepatocytes and macrophages. One of the unique features of the reaction of α_2-M is that the proteinase retains activity against low M_r substrates and inhibitors but reaction with high M_r molecules is completely abolished.

Calpain is inhibited by α_2-M.[43-45] The reaction is irreversible with a rate constant of the order of $3 \times 10^4\ M^{-1}s^{-1}$. This is intermediate in the range of values that have been found for the reaction of proteinases with α_2-M. For example, the rate constant for bovine trypsin is $1.5 \times 10^7\ M^{-1}s^{-1}$ and that for thrombin is $5 \times 10^2\ M^{-1}s^{-1}$. The calpain complexed with α_2-M showed only limited reactivity with the protein substrate azocasein, but was fully reactive with the peptide substrate Suc-Leu-Tyr-aminomethylcoumarin or with the peptide inhibitor E-64. Calpain trapped by α_2-M was at least partially protected from loss of activity due to autolysis, consistent with the observation that some of the autolytic reactions are intermolecular.[46,47] α_2-M traps a heterodimeric form of calpain, M_r 98,000 to 110,000. α_2-M is capable of trapping other large proteinases such as plasmin (M_r 77,000) and kallikrein (M_r 88,000), but calpain is the largest proteinase so far found to be inhibited by α_2-M.

C. THE CYSTATIN SUPERFAMILY

The cystatin superfamily comprises proteins that are homologous with chicken cystatin. The majority of the proteins are cysteine proteinase inhibitors (see Reference 48 for a review). The superfamily is divided into three families. The members of family 1, which includes stefins and cystatins A and B, have M_r values of about 11,000 and no disulfide bonds. Family 2, which includes chicken cystatin and human cystatin C, have M_r about 13,000 and two disulfide bonds. The third family comprises the kininogens, which are now known to be the same proteins as α-cysteine proteinase inhibitors. Two forms of kininogen are found in the plasma of most species; high M_r kininogen (H-kininogen, M_r 110,000) and low M_r kininogen (L-kininogen, M_r 60,000). These molecules are products of the same gene. In rat additional kininogens have been identified, termed T-kininogens. These are expressed in response to acute inflammation. The kininogens are single-chain glycoproteins which, when the kinin peptides are released after cleavage by kallikreins, are composed of an N-terminal heavy chain disulfide bonded to the C-terminal light chain. The heavy chains of L-kininogen and H-kininogen are identical and are constructed from three cystatin segments, with a total of nine disulfide bonds.

Cystatins inhibit cysteine proteinases by forming tight-binding reversible complexes. The K_i values for the papain-like enzymes are typically of the order of 0.01 to 1.0 nM. The precise nature of the interaction between enzyme and inhibitor is uncertain. The cystatins have a conserved QVVAG sequence[49,50] and a conserved glycine residue near the N-terminus. It has been proposed that both of these sites are important in the interaction with cysteine proteinases.[51] Hopefully the molecular details of the interaction will soon be elucidated because human cystatin C is available in quantity from an *E. coli* expression system[52] and the crystal structure of chicken cystatin has been determined at 2.0Å resolution.[53]

Investigation of the reaction of calpain with members of the cystatin superfamily has revealed some surprising results. Cystatins from families 1 and 2 do not inhibit calpains (rat cystatin β,[54] chicken cystatin C,[55] human amniotic-fluid cystatin,[56] human cystatins A, B, and C,[45] puff adder cystatin[57]). Calpain is, however, inhibited by L- and H-kininogens,[43,44,58,59] but only the central cystatin domain of the kininogen heavy chain is inhibitory.[60] This domain has a K_i of 1.0 n*M* for calpain II. Under conditions where an excess of calpain is incubated with kininogens, both inhibition and kinin release from the kininogen have been observed.[61] In contrast with the results for L- and H-kininogens, T-kininogens are not calpain inhibitors.[42,62] It would be interesting to understand why, of all the cystatin domains, only the central domain of L- and H-kininogens is a calpain inhibitor.

D. REGULATION OF CALPAIN ACTIVITY *IN VIVO*

The naturally occurring protein inhibitors of calpain described above are presumed to be involved in the regulation of calpain activity *in vivo*. Calpain is activated by a rise in intracellular Ca^{2+} levels and this activity is enhanced by interaction with phospholipid. The proteinase could then be inactivated by decreased Ca^{2+} levels, by interaction with calpastatin, and by autolysis.[47] In addition, other regulators may have a role. A cytoplasmic calpain activator protein has been described,[63-65] and isovalerylcarnitine is also reported to be a specific activator of calpain.[66] Another important factor in regulation might be the redox state of the cytoplasm as calpain is inactivated by oxidation.[67,68] Furthermore, if calpain were exposed to the extracellular environment, perhaps following cell death, then inhibition by α_2-M and kininogens would become important.

IV. PEPTIDE INHIBITORS OF CALPAIN

A. PEPTIDE ALDEHYDES

Leupeptin and antipain are peptides that have aldehyde derivatives of arginine (argininal) at their C-terminals. Leupeptin is Ac-Leu-Leu-argininal and antipain is *N*-[(*S*-1-carboxy-2-phenylethyl]-carbamoyl-Arg-Val-argininal. They are reversible inhibitors of both serine and cysteine proteinases. The aldehyde group of the inhibitors reacts with the active-site thiol of cysteine proteinases to form a hemithio-acetal as shown in Scheme 3.[69]

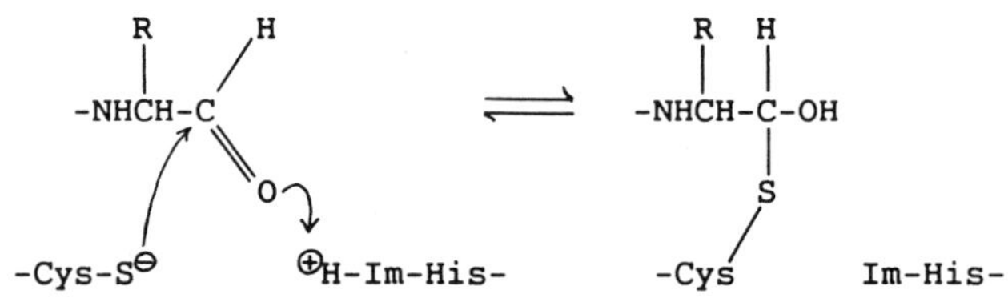

Scheme 3

The compounds are considered as transition state analogues and are tight-binding. Slow-tight-binding has been observed for the inhibition of cathepsin B by leupeptin.[70]

Leupeptin and antipain inhibit calpains.[71] Inhibition was competitive, with K_i values for leupeptin of 3.2×10^{-7} *M* for calpain I and 4.3×10^{-7} *M* for calpain II, and for antipain of 1.4×10^{-6} *M* for both calpains I and II.[72] Attempts have been made to find peptide aldehyde inhibitors with greater affinity and specificity for calpains. Strepin P-1 was isolated from *Streptomyces tanabeensis* and found to have the structure N-isoVal-Tyr-Val-argininal.[73] This compound inhibited calpain, papain, and also the serine proteinase trypsin. Strepin P-1 and antipain, with Val at P_2, inhibited papain more strongly than calpain, while leupeptin, with Leu at P_2, inhibited calpain more strongly than papain. This suggests calpain has a preference for Leu over Val at P_2. Additional peptide aldehydes have been isolated from another organism, *Staphylococcus tanabeensis,* called staccopin P1 (H-Val-Val-Val-Val-phenylalaninal) and staccopin P2 (H-Val-Val-Val-Val-tyrosinal).[74] These compounds inhibited calpains I and II and papain. They did not inhibit trypsin, but it would be surprising if they did not inhibit chymotrypsin-like serine proteinases. Two synthetic peptide aldehydes have also been tested; Ac-Leu-Leu-methioninal and Ac-Leu-Leu-norleucinal. They inhibited both papain and calpains. Modification of these synthetic peptide aldehydes by replacing the N-terminal Ac-Leu-with benzyloxycarbonyl or phenylbutyryl produced compounds that would penetrate the cell membrane.[75] Despite this attractive property, these compounds have limited usefulness for *in vivo* studies because they are not calpain specific. The investigations described above have revealed some interesting new peptide aldehyde inhibitors. Unfortunately, only IC_{50} values for the inhibition of various enzymes have been reported, so it is difficult to quantitate the effectiveness of these inhibitors.

B. EPOXYSUCCINYL PEPTIDES

Irreversible inhibition of cysteine proteinases by epoxysuccinyl peptides was initially discovered by Hanada et al.[76] The first to be isolated from *Aspergillus japonicans* was E-64, [*N*-(L-3-*trans*-carboxy-oxiran-2-carbonyl)-L-leucyl]-amido(4-guanido)butane, (abbreviated here as Eps-Leu-NH-$[CH_2]_4$-NH-CH-NH_2NH). Epoxysuccinyl peptides are presumed to inhibit cysteine proteinases by alkylating the active-site cysteine. The reaction shown in Scheme 4 has been proposed[77] where the leucine of the inhibitor binds in the S_2' pocket of the enzyme.

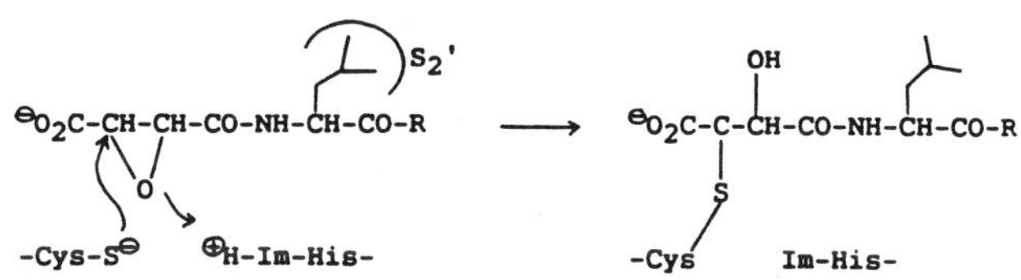

Scheme 4

Epoxysuccinyl peptides inhibit calpains.[78-80] Determination of the rate constants for inactivation for a series of epoxysuccinyl peptides with different peptide side chains showed that the fastest rates of inactivation were achieved with Ep-460 (Eps-Leu-NH$(CH_2)_4$-NH-Z, 23,300 $M^{-1}s^{-1}$), E-64 (Eps-Leu-NH$(CH_2)_4$-NH-CH-NH_2NH, 7,500 $M^{-1}s^{-1}$) and Ep-475 (Eps-Leu-NH-CH_2-$CH(CH_3)_2$, 7,400 $M^{-1}s^{-1}$).[46] It was found that the rate constants dropped significantly if the D rather than the L isomer of E-64 was used, if the epoxide was replaced by fumarate, a weaker alkylating agent, or if the free carboxyl of the epoxysuccinyl group was esterified. The rate constants observed for calpain are slow relative to the rate constants of inactivation of the lysosomal cysteine proteinases by epoxysuccinyl peptides.[81]

C. PEPTIDE CHLOROMETHANES

Peptide chloromethanes are irreversible inhibitors of cysteine and serine proteinases. For cysteine proteinases the active site cysteine is alkylated by one of the two routes shown in Scheme 5.

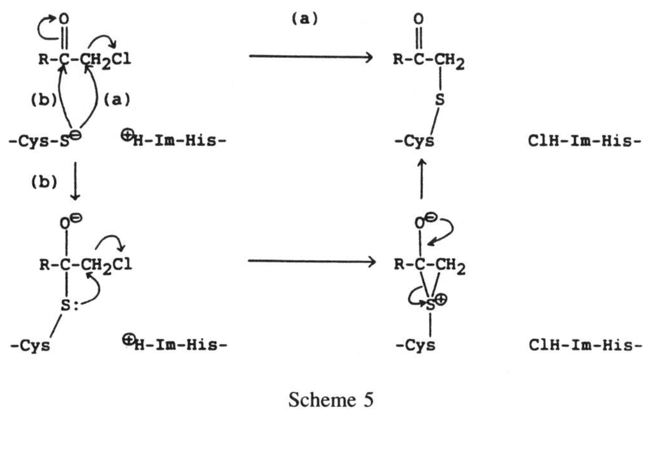

Scheme 5

Calpains I and II are inhibited by peptide chloromethanes. The fastest rate constants for inactivation are achieved with compounds that have peptide regions analogous to good calpain substrate, i.e., Phe, Tyr, Met or Lys in P_1 and Leu or Val in P_2 and P_3.[72] Thus Pro-Phe-Arg-CH_2Cl reacts slowly with calpain,[46] whereas Leu-Leu-Phe-CH_2Cl, Leu-Leu-Tyr-CH_2Cl and Leu-Leu-Lys-CH_2Cl are good inhibitors.[82] Second order rate constants for inactivation of greater than 100,000 $M^{-1}s^{-1}$ were found for Z-Leu-Leu-Phe-CH_2Cl and Leu-Leu-Phe-CH_2Cl.[83] Unfortunately these compounds are not specific for calpains; they react much faster with the lysosomal cysteine proteinases.

D. PEPTIDE DIAZOMETHANES

Peptide diazomethanes are irreversible inhibitors of cysteine proteinases. They do not inactivate serine proteinases in contrast to peptide chloromethanes.[84] This is an attractive advantage when considering the best inhibitors to use for investigating the *in vivo* function of proteinases. Recently, however, it has been found that Z-Phe-Arg-CHN_2 inhibits plasma kallikrein, but not trypsin.[85] Thus the specificity of peptide diazomethanes for cysteine proteinases may not be absolute. It has been possible, by varying the peptide structure of diazomethanes, to achieve selectivity between individual cysteine proteinases.[86] The probable mechanism for the reaction of cysteine proteinases with peptide diazomethanes is shown in Scheme 6.

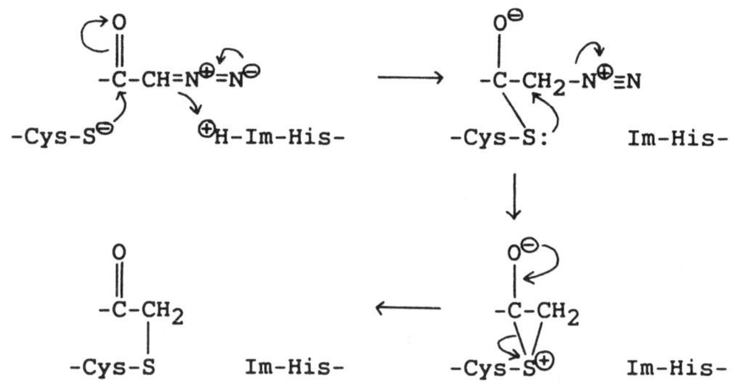

Scheme 6

Peptide diazomethanes inactivate calpains.[83] The fastest reacting inhibitor was Z-Leu-Leu-Tyr-CHN_2 with a rate constant of 2.3×10^5 $M^{-1}s^{-1}$. Investigation of a series of diazomethanes with different peptide structures showed that specificity for calpains is difficult to achieve. Using a range of inhibitors with the structure Z-Leu-X-CHN_2, the P_1 specificity of calpain was found to be: Leu<Trp<iodoTyr<Tyr<Met<homoPhe, in order of increasing rate constant of inactivation. However, these compounds were better inhibitors of the lysosomal enzymes cathepsins B and L. Increasing the length of the inhibitor, with Leu in P_2 since this is preferred by calpains, increased the rate of reaction and led to the very fast reacting compound Z-Leu-Leu-Tyr-CHN_2. Unfortunately, this compound was a better inhibitor of cathepsin L than of calpain. Examination of other compounds revealed that only Boc-Val-Lys(ε-Z)-Leu-Tyr-CHN_2 was a faster inactivator of calpain (rate constant 2.1×10^4 $M^{-1}s^{-1}$) than cathepsin L (rate constant 1.3×10^4 $M^{-1}s^{-1}$).

Recently a new class of proteinase inhibitors with specificity for cysteine proteinases have been described, peptidyl sulfonium salts.[87] Investigation of these compounds may prove useful in the search for specific calpain inhibitors.

E. NONPEPTIDE CYSTEINE MODIFYING REAGENTS

Calpains have a reactive cysteine in the active site that can interact with general thiol modifying reagents. As with reaction of the enzyme with any other inhibitor, the interaction will only occur in the presence of Ca^{2+}, since it is only under these conditions that the active site is accessible. Thus calpain is irreversibly inactivated by iodoacetic acid,[79] by mersalyl (*O*-[(3-hydroxymercuri-2-methoxypropyl)-carbamoly]-phenoxyacetic acid) and by *N*-ethylmaleimide.[88] Calpain is also inactivated by oxidation of the active site cysteine.[67,68]

V. USE OF INHIBITORS *IN VIVO*

The main reason for developing peptide inhibitors of calpain is to obtain reagents that could be used in the study of the role of calpains *in vivo*. For this purpose the most important characteristic of the inhibitor is that it should be specific for calpains. It is interesting to consider the difference in K_i values or second order rate constants required to achieve specificity. For a reversible inhibitor in the presence of two enzymes at equal concentrations, almost complete inhibition of one enzyme with only 1% inhibition of the other is achieved if there is a 10^4-fold difference in the K_i values for the two enzymes. For an irreversible inhibitor in the presence of two enzymes at equal concentrations, achieving only 1% inhibition of the first enzyme at the time the second enzyme is almost completely inhibited requires the second order rate constant for the inactivation of the second enzyme to be 500 times that of the first enzyme (assuming no depletion of the inhibitor). Obviously at longer times the first enzyme will show progressively greater inhibition, and this should be considered in experimental design. In addition to specificity, the other properties of an inhibitor to consider before it is used *in vivo* are whether it is more advantageous to use a reversible or an irreversible inhibitor; what is the K_i or rate constant of inhibition (in other words, how good is the inhibitor?); whether the inhibitor will penetrate the cell membrane; in which compartment(s) of the cell the inhibitor accumulates and at what concentration; whether the inhibitor is cytotoxic; and finally whether the inhibitor can be easily radiolabeled. Not all of these parameters are easy to determine.

Reference to Table 1 will show that the only inhibitor currently available that is specific for calpains is calpastatin. The major difficulty with using calpastatin *in vivo* is that it is a protein, rather than a peptide; thus there are problems of stability and cell penetration. Perhaps it will be possible to develop inhibitors which are easier to use, based on the minimum functional segment of one calpastatin domain. Another possibility for using calpastatin in *in vivo* studies is to administer the inhibitor by microinjection techniques.[89]

Peptide aldehydes have been used extensively in the study of the physiological roles of calpains. Leupeptin injected into dystrophic chickens reduced muscle wasting; it was suggested that calpain inhibition was the main reason for the improvement.[90] A role for calpain in the degradation of neruofilaments in synaptic termini was proposed, because leupeptin injected into the optic tectum of goldfish resulted in the accumulation of neruofilaments.[91] Calpains have been implicated in memory since leupeptin blocked both olfactory discrimination learning and spatial maze performance in rats.[92,93] Leupeptin has also been widely used to investigate the role of calpains in platelet activation. Calpain is thought to degrade cytoskeletal components during platelet activation allowing shape change,[88] fibrinogen receptor expression,[94] and aggregation[95] to occur. A role for calpain in platelet activation has also been suggested using Ac-Leu-Leu-norleucinal, Ac-Leu-Leu-methioninal and Leu-Leu-Phe-CH_2Cl.[96] The problem of using nonspecific inhibitors in this type of study was however amply demonstrated by Brass and Shatti.[97] These authors showed that the inhibition of thrombin-induced platelet activation by leupeptin in their experiments was due to leupeptin interacting with thrombin and not due to calpain inhibition.

Epoxysuccinyl peptides have also been used extensively to investigate the physiological roles of calpain. These inhibitors are probably more useful than the peptide aldehydes since they are irreversible and specific for cysteine proteinases. The major problem with their use is that they inhibit the lysosomal cysteine proteinases much faster than calpains. The membrane permeant epoxysuccinyl peptide E-64-d (ethyl(2*S*,3*S*)-3-[(*S*)-3-methyl-1-(3-methylbutylcarbamoly)-butylcarbamoyl]oxirane-2-carboxylate) was found to arrest A431 cells at mitotic metaphase and inhibition of calpain was thought to contribute to the effect.[98] The major area, however, where epoxysuccinyl peptides have been used, is in the investigation of the role of proteinase in muscle degradation.[99,100] It has been suggested that there might be a therapeutic use for calpain inhibitors in this area. Calpain is thought to degrade Z-lines, since the loss of Z-line in muscle slices incubated in the presence of Ca^{2+} and A23187 was inhibited by epoxysuccinyl peptides, but not by lysosomotrophic agents.[101,102] Injection of E-64 into dystrophic chickens reduced muscle wasting and the effect was attributed to calpain inhibition.[78] By contrast, the reduction in tissue damage achieved using epoxysuccinyl peptides in myocardial infarcts was attributed to inhibition of cathepsins B and L.[103]

The current lack of good specific low M_r inhibitors of calpains makes it difficult to design experiments using inhibitors to study the roles of calpains. With the presently available inhibitors it is really necessary to test for inhibition of all the types of proteinase that react with a particular compound, before the effects of an inhibitor can be attributed to inhibition of a particular proteinase. Experience with peptide diazomethanes has shown the cathepsin L is the major cysteine proteinase likely to cross-react with calpain inhibitors. Perhaps the answer to the specificity problem is to develop inhibitors based on calpastatin.

NOTE ADDED IN PROOF

Since writing this chapter a number of publications concerning calpain inhibitors have appeared. Further work with peptide aldehydes has identified 4-phenyl-butyryl-Leu-methioninal as a good inhibitor of calpains I and II;[104] however, the coupound is also a strong inhibitor of cathespin L. Benzyloxycarbonyl-Val-phenylalaninal has also been shown to be a calpain inhibitor.[105] Details of the interaction between E-64 and papain have been found from X-ray crystallographic analysis.[106,107] This work suggests E-64 binds in the S subsites of the enzyme rather than the S′ subsites, and that His-159 of papain plays a binding role in the inactivation process, rather than the catalytic role suggested in scheme 4. The calmodulin antagonists melittin, calmidazolium, and trifluoperazine have been found to inhibit calpain activity.[108,109] Work with calpastatin has provided further evidence for inhibitor fragmentation by calpain,[110] and considerable progress has been made towards defining the minimum calpastatin sequence (approximately 25 amino acids) necessary for inhibition.[111-113]

REFERENCES

1. **Knight, C. G.,** The characterization of enzyme inhibition, in *Proteinase Inhibitors,* Barrett, A. J. and Salvesen, G., Eds., Elsevier, Amsterdam, 1986, 23.
2. **Kitz, R. and Wilson, I. B.,** Esters of methanesulphonic acid as irreversible inhibitors of acetylcholinesterase, *J. Biol. Chem.,* 237, 3245, 1962.
3. **Tian, W.-X. and Tsou, C.-L.,** Determination of the rate constant of enzyme modification by measuring the substrate reaction in the presence of modifier, *Biochemistry,* 21, 1028, 1982.
4. **Dixon, M.,** The determination of enzyme inhibitor constants, *Biochem. J.,* 55, 170, 1953.
5. **Ackermann, W. W. and Potter, V. R.,** Enzyme inhibition in relation to chemotherapy, *Proc. Soc. Exp. Biol. Med.,* 72, 1, 1949.
6. **Henderson, P. J. F.,** A linear equation that describes the steady-state kinetics of enzymes and subcellular particles interacting with tightly bound inhibitors, *Biochem. J.,* 127, 321, 1972.
7. **Morrison, J. F.,** The slow-binding and slow, tight-binding inhibition of enzyme-catalyzed reactions, *Trends Biochem. Sci.,* 7, 102, 1982.
8. **Cha, S.,** Tight-binding inhibitors I. Kinetic behavior, *Biochem. Pharmacol.,* 24, 2177, 1975. See also correction *Biochem. Pharmacol.,* 25, 1561, 1976.

9. **Parkes, C.**, Calpastatins, in *Proteinase Inhibitors,* Barrett, A. J. and Salvesen, G., Eds., Elsevier, Amsterdam, 1986, 571.
10. **DeMartino, G. N. and Croall, D. E.**, Purification and characterization of a protein inhibitor of calcium-dependent proteases from rat liver, *Arch. Biochem. Biophys.*, 232, 713, 1984.
11. **Mellgren, R. L. and Carr, T. C.**, The protein inhibitor of calcium-dependent proteases. Purification from bovine heart and possible mechanism of regulation, *Arch Biochem. Biophys.*, 225, 779, 1983.
12. **Nishiura, I., Tanaka, K., and Murachi, T.**, A high molecular weight inhibitor of Ca^{2+}-dependent neutral protease in rat brain, *Experientia,* 35, 1006, 1979.
13. **Takano, E. and Murachi, T.**, Purification and some properties of human erythrocyte calpastatin, *J. Biochem.*, 92, 2021, 1982.
14. **Nishiura, I., Tanaka, K., Yamato, S., and Murachi, T.**, The occurrence of an inhibitor of Ca^{2+}-dependent neutral protease in rat liver, *J. Biochem.*, 84, 1657, 1978.
15. **Waxman, L.**, Characterisation of two unique protease inhibitors and a calcium activated protease from bovine cardiac muscle, in *Protein Turnover and Lysosomal Function,* Segal, H. L. and Doyle, D. J., Eds., Academic Press, New York, 1978, 363.
16. **Sakon, M., Tsujinaka, T., Kambayashi, J., and Kosaki, G.**, A specific endogenous inhibitor of two forms of Ca^{2+}-activated neutral proteases in platelets, *Experientia,* 38, 1099, 1982.
17. **Nakamura, M., Inomata, M., Hayashi, M., Imahori, K., and Kawashima, S.**, Purification and characterisation of an inhibitor of calcium-activated neutral protease from rabbit skeletal muscle: purification of 50,000 dalton inhibitor, *J. Biochem.*, 96, 1399, 1984.
18. **Lepley, R. A., Pampusch, M., and Dayton, W. R.**, Purification of a high molecular weight inhibitor of the calcium-activated proteinase, *Biochim. Biophys. Acta,* 828, 95, 1985.
19. **Ishiura, S., Tsuji, S., Murofushi, H., and Suzuki, K.**, Purification of an endogenous 68,000 dalton inhibitor of Ca^{2+}-activated neutral protease from chicken skeletal muscle, *Biochim. Biophys. Acta,* 701, 216, 1982.
20. **Shiba, E., Tsujinaka, T., Kambayashi, J., and Kosaki, G.**, Purification and characterisation of Ca^{2+}-activated neutral protease inhibitor from human platelets, *Thromb. Res.*, 32, 207, 1983.
21. **Cottin, P., Vidalenc, P. L., and Ducastaing, A.**, Ca^{2+}-dependent association between a Ca^{2+}-activated neutral proteinase (CaANP) and its specific inhibitor, *FEBS Lett.*, 136, 221, 1981.
22. **Shigeta, K., Yumoto, N., and Murachi, T.**, Fragmentation of a 70,000 dalton calpastatin molecule upon its complex formation with calpain, *Biochem. Int.*, 9, 327, 1984.
23. **Cottin, P., Vidalenc, P. L., Merdaci, N., and Ducastaing, A.**, Evidence for noncompetitive inhibiton between two calcium-dependent activated neutral proteinases and their specific inhibitor, *Biochim. Biophys.*, 743, 299, 1983.
24. **Murachi, T., Tanaka, K., Hatanaka, M., and Murakami, T.**, Intracellular Ca^{2+}-dependent protease (calpain) and its high molecular weight endogenous inhibitor (calpastatin), *Adv. Enzyme Regul.*, 19, 407, 1981.
25. **Imahjoh, S. and Suzuki, K.**, Reversible interaction between Ca^{2+}-activated neutral protease (CANP) and its endogenous inhibitor, *FEBS Lett.*, 187, 47, 1985.
26. **Mellgren, R. L., Mericle, M. L., and Lane, R. D.**, Proteolysis of the calcium-dependent protease inhibitor by myocardial calcium-dependent protease, *Arch. Biochem. Biophys.*, 246, 233, 1986.
27. **Mellgren, R. L., Repetti, A., Muck, T. C., and Easly, J.**, Rabbit skeletal muscle calcium-dependent protease requiring millimolar Ca^{2+}. Purification, subunit structure and Ca^{2+}-dependent autoproteolysis, *J. Biol. Chem.*, 257, 7203, 1982.
28. **Takano, E., Maki, M., Hatanaka, M., Mori, H, Zenita, K., Sakihama, T., Kannagi, R., Marti, T., Titani, K., and Murachi, T.**, Evidence for the repetitive domain structure of pig calpastatin as demonstrated by cloning of complementary DNA, *FEBS Lett.*, 208, 199, 1986.
29. **Takano, E., Maki, M., Mori, H., Hatanaka, M., Marti, T., Titani, K., Kannagi, R., Ooi, T., and Murachi, T.**, Pig heart calpastatin: identification of repetitive domain structures and anomalous behavior in polyacrylamide gel electrophoresis, *Biochemistry,* 27, 1964, 1988.
30. **Emori, Y., Kawasaki, H., Imajoh, S., Imahori, K., and Suzuki, K.**, Endogenous inhibitor for calcium-dependent cysteine protease contains four internal repeats that could be responsible for its multiple reactive sites, *Proc. Natl. Acad. Sci. U.S.A.*, 84, 3590, 1987.
31. **Imajoh, S., Kawasaki, H., Emori, Y., and Suzuki, K.**, Calcium-activated neutral protease inhibitor from rabbit erythrocytes lacks the N-terminal region of the liver inhibitor but retains three inhibitory units, *Biochem. Biophys. Res. Comm.*, 146, 630, 1987.
32. **DeMartino, G. N., Wachendorfer, R., McGuire, M. J., and Croall, D. E.**, Proteolysis of the protein inhibitor of calcium-dependent proteases produces lower molecular weight fragments that retain inhibitory activity, *Arch. Biochem. Biophys.*, 262, 189, 1988.
33. **Maki, M., Takano, E., Mori, H., Sato, A., Murachi, T., and Hatanaka, M.**, All four internally repetitive domains of pig calpastatin possess inhibitory activities against calpains I and II, *FEBS Lett.*, 223, 174, 1987.

34. **Imajoh, S., Kawasaki, H., Emori, Y., Ishiura, S., Minami, Y., Sugita, H., Imahori, K., and Suzuki, K.**, A fragment of an endogenous inhibitor produced in *Escherichia coli* for calcium-activated neutral protease (CANP) retains an inhibitory activity, *FEBS Lett.*, 215, 274, 1987.
35. **Emori, Y., Kawasaki, H., Imajoh, S., Minami, Y., and Suzuki, K.**, All four repeating domains of the endogenous inhibitor for calcium-dependent protease independently retain inhibitory activity, *J. Biol. Chem.*, 263, 2364, 1988.
36. **Maki, M., Takano, E., Osawa, T., Ooi, T., Murachi, T., and Hatanaka, M.**, Analysis of structure function relationship of pig calpastatin by expression of mutated cDNAs in *Escherichia coli, J. Biol. Chem.* 263, 10254, 1988.
37. **Otsuka, Y. and Goll, D. E.**, Purification of the Ca^{2+}-dependent proteinase inhibitor from bovine cardiac muscle and its interaction with the millimolar Ca^{2+}-dependent proteinase, *J. Biol. Chem.*, 262, 5839, 1987.
38. **Imajoh, S., Kawasaki, H., Kisaragi, M., Mukai, M., Sugita, H., and Suzuki, K.**, A 107-kDa inhibitor for calcium-activated neutral proteinase (CANP): purification from the human liver, *Biomed. Res.*, 5, 481, 1984.
39. **Travis, J. and Salvesen, G. S.**, Human plasma proteinase inhibitors, *Annu. Rev. Biochem.*, 52, 655, 1983.
40. **Schramm, H. J. and Schramm, W.**, Computer averaging of single molecules of α_2-macroglobulin and the α_2-macroglobulin/trypsin complex, *Hoppe-Seyler's Z. Physiol. Chem.*, 363, 803, 1982.
41. **Barrett, A. J. and Starkey, P. M.**, The interaction of α_2-M with other proteases, *Biochem. J.*, 133, 709, 1973.
42. **Feldman, S. R., Gonias, S. L., and Pizzo, S. V.**, Model of α_2-macroglobulin structure and function, *Proc. Natl. Acad. Sci. U.S.A.*, 82, 5700, 1985.
43. **Sasaki, M., Taniguchi, K., Suzuki, K., and Imahori, K.**, Human plasma α_1- and α_2-thiol proteinase inhibitors strongly inhibit Ca^{2+}-activated neutral protease form muscle, *Biochem. Biophys. Res. Comm.*, 110, 256, 1983.
44. **Schmaier, A. H., Bradford, H., Silver, L. D., Farber, A., Scott, C. F., Schutsky, D., and Colman, R. W.**, High molecular weight kininogen is an inhibitor of platelet calpain, *J. Clin. Invest.*, 77, 1565, 1986.
45. **Crawford, C.**, Inhibition of chicken calpain II by proteins of the cystatin superfamily and α_2-macroglobulin, *Biochem. J.*, 248, 589, 1987.
46. **Parkes, C., Kembhavi, A. A., and Barrett, A. J.**, Calpain inhibition by peptide epoxides, *Biochem. J.*, 230, 509, 1985.
47. **Crawford, C., Willis, A. C., and Gagnon, J.**, The effects of autolysis on the structure of chicken calpain II, *Biochem. J.*, 248, 579, 1987.
48. **Barrett, A. J., Rawlings, N. D., Davies, M. E., Machleidt, W., Salvesen, G., and Turk, V.**, Cysteine proteinase inhibitors of the cystatin superfamily, in *Proteinase Inhibitors,* Barrett, A. J. and Salvesen, G., Eds., Elsevier, Amsterdam, 1986, 515.
49. **Ohkubo, I., Kurachi, K., Takasawa, T., Shiokawa, H., and Sasaki, M.**, Isolation of a human cDNA for α_2-thiol proteinase inhibitor and its identity with low molecular weight kininogen, *Biochemistry,* 23, 5691, 1984.
50. **Teno, N., Tsuboi, S., Itoh, N., Okamoto, H, and Okada, Y.**, Significant effects of Z-Gln-Val-Val-OMe, common sequence of thiol proteinase inhibitors on thiol proteinases, *Biochem. Biophys. Res. Comm.*, 143, 749, 1987.
51. **Abrahamson, M., Ritonja, A., Brown, M. A., Grubb, A., Machleidt, W., and Barrett, A. J.**, Identification of the probable inhibitory reactive sites of the cysteine proteinase inhibitors human cystatin C and chicken cystatin, *J. Biol. Chem.*, 262, 9688, 1987.
52. **Abrahamson, M., Dalboge, H., Olafsson, I., Carlsen, S., and Grubb, A.**, Efficient production of native, biologically active human cystatin C by *Escherichia coli, FEBS Lett.*, 236, 14, 1988.
53. **Bode, W., Engh, R., Musil, D., Thiele, U., Huber, R., Karshikov, A., Brzin, J., Kos, J., and Turk, V.**, The 20 Å X-ray crystal structure of chicken egg white cystatin and its possible mode of interaction with cysteine proteinases, *EMBO J.*, 7, 2593, 1988.
54. **Kominami, E., Wakamatsu, N., and Katunuma, N.**, Purification and characterisation of thiol proteinase inhibitor from rat liver, *J. Biol. Chem.*, 257, 14648, 1982.
55. **Johnson, P., Parkes, C., and Barrett, A. J.**, The characterisation of calpains and calpain inhibitors from chicken gizzard smooth muscle, *Biochem. Soc. Trans.*, 12, 1106, 1984.
56. **Rohrlich, S. T., Levy, H., and Rifkin, D. B.**, Purification and characterisation of a low molecular mass cysteine proteinase inhibitor from human amniotic fluid, *Hoppe-Seyler's Z. Physiol. Chem.*, 366, 147, 1985.
57. **Evans, H. J. and Barrett, A. J.**, A cystatin-like cysteine proteinase inhibitor from venom of the African puff adder, *(Bitis arietans), Biochem. J.*, 246, 795, 1987.
58. **Higashiyama, S., Ohkubo, I., Ishiguro, H., Kunimatsu, M., Sawaki, K., and Sasaki, M.**, Human high molecular weight kininogen as a thiol proteinase inhibitor: presence of the entire inhibition capacity in the native form of heavy chain, *Biochemistry,* 25, 1669, 1986.

59. **Ishiguro, H., Higashiyama, S., Namikawa, C., Kunimatsu, M., Takano, E., Tanaka, K., Ohkubo, I., Murachi, T., and Sasaki, M.**, Interaction of human calpains I and II with high molecuar weight and low molecular weight kininogens and their heavy chains: mechanism of interaction and the role of divalent cations, *Biochemistry,* 26, 2863, 1987.
60. **Salvesen, G., Parkes, C., Abrahamson, M., Grubb, A., and Barrett, A. J.**, Human low-M_r kininogen contains three copies of a cystatin sequence that are divergent in structure and in inhibitory activity for cysteine proteinases, *Biochem. J.*, 234, 429, 1986.
61. **Higashiyama, S., Ishiguro, H., Ohkubo, I, Fujimoto, S., Matsuda, T., and Sasaki, M.**, Kinin release from kininogens by calpains, *Life Sci.*, 39, 1639, 1986.
62. **Moreau, T., Esnard, F., Gutman, N., Degand, P., and Gauthier, F.**, Cysteine proteinase-inhibiting function of T kininogens and of its proteolytic fragments, *Eur. J. Biochem.*, 173, 185, 1988.
63. **DeMartino, G. N. and Blumenthal, D. K.**, Identification and partial purification of a factor that stimulates calcium-dependent proteases, *Biochemistry,* 21, 4297, 1982.
64. **Takeyama, Y., Nakanishi, H., Uratsuji, Y., Kishimoto, A., and Nishizuka, Y.**, A calcium-protease activator associated with brain microsomal-insoluble elements, *FEBS Lett.*, 194, 110, 1986.
65. **Pontremoli, S., Melloni, E., Michetti, M., Salamino, F., Sparatore, B., and Horecker, B. L.**, An endogenous activator of the Ca^{2+}-dependent proteinase of human neutrophils that increases its affinity for Ca^{2+}, *Proc. Natl. Acad. Sci. U.S.A.*, 85, 1740, 1988.
66. **Pontremoli, S., Melloni, E., Michetti, M., Sparatore, B., Salamino, F., Siliprandi, N., and Horecker, B. L.**, Isovalerylcarnitine is a specific activator of calpain of human neutrophils, *Biochem. Biophys. Res. Comm.*, 148, 1189, 1987.
67. **Di Cola, D. and Sacchetta, P.**, Irreversible inactivation of calcium-dependent proteinases from rat liver by biological disulphides, *FEBS Lett.*, 210, 81, 1987.
68. **Morelli, A., Grasso, M., and de Flora, A.**, Oxidative inactivation of the calcium-stimulated neutral proteinase from human red blood cells by divicine and intracellular protection by reduced glutathione, *Arch. Biochem. Biophys.*, 251, 1, 1986.
69. **Lewis, C. A. and Wolfenden, R.**, Thiohemiacetal formation by inhibitory aldehydes at the active site of papain, *Biochemistry,* 16, 4890, 1977.
70. **Baici, A. and Gyger-Marazzi, M.**, The slow, tight-binding inhibition of cathepsin B by leupeptin. A hysteretic effect, *Eur. J. Biochem.*, 929, 33, 1982.
71. **Toyo-Oka, T., Shimizu, T., and Masaki, T.**, Inhibition of proteolytic activity of calcium activated neutral proteinase by leupeptin and antipain, *Biochem. Biophys. Res. Comm.*, 82, 484, 1978.
72. **Sasaki, T., Kikuchi, T., Yumoto, N., Yoshimura, N., and Murachi, T.**, Comparative specificity and kinetic studies on porcine calpain I and calpain II with naturally occurring peptides and synthetic fluorogenic substrates, *J. Biol. Chem.*, 259, 12489, 1984.
73. **Ogura, K., Maeda, M., Nagai, M., Tanaka, T., Nomoto, K., and Murachi, T.**, Purification and structure of a novel cysteine proteinase inhibitor, strepin P1, *Agric. Biol. Chem.*, 49, 799, 1985.
74. **Saito, M., Kawaguchi, N., Hashimoto, M., Kodama, T., Higuchi, N., Tanaka, T., Nomoto, K., and Murachi, T.**, Purification and structure of novel cysteine proteinase inhibitors, staccopins P1 and P2 from *Staphylococcus tanabeensis, Agric. Biol. Chem.*, 51, 861, 1987.
75. **Tsujinaka, T., Kajiwara, Y., Kambayashi, J., Sakon, M., Higuchi, N., Tanaka, T., and Mori, T.**, Synthesis of a new cell penetrating calpain inhibitor (calpeptin), *Biochem. Biophys. Res. Comm.*, 153, 1201, 1988.
76. **Hanada, K., Tamai, M., Yamagishi, M., Ohmura, S., Sawada, J., and Tanaka, I.**, Studies on thiol protease inhibitors. Part I. Isolation and characterisation of E-64, a new thiol protease inhibitor, *Agric. Biol. Chem.*, 42, 523, 1978.
77. **Rich, D. H.**, Inhibitors of cysteine proteinases, in *Proteinase Inhibitors,* Barrett, A. J. and Salvesen, G., Eds., Elsevier, Amsterdam, 1986, 153.
78. **Sugita, H., Ishiura, S., Suzuki, K., and Imahori, K.**, Inhibition of epoxide derivatives on chicken calcium-activated neutral protease (CANP) *in vitro* and *in vivo, J. Biochem.*, 87, 339, 1980.
79. **Suzuki, K.**, Reaction of calcium-activated neutral protease (CANP) with an epoxysuccinyl derivative (E64c) and iodoacetic acid, *J. Biochem.*, 93, 1305, 1983.
80. **Hara, K. and Takahashi, K.**, Inhibition of calcium-activated neutral protease of monkey cardiac muscle by epoxysuccinic acid derivatives, *Biomed. Res.*, 4, 121, 1983.
81. **Barrett, A. J., Kembhavi, A. A., Brown, M. A., Kirschke, H., Knight, C. G., Tamai, M., and Hanada, K.**, L-*trans*-Epoxysuccinyl-leucylamido-(4-guanidino)butane (E64) and its analogues as inhibitors of cysteine proteinases including cathepsins B, H and L, *Biochem. J.*, 201, 189, 1982.
82. **Sasaki, T., Kikuchi, T., Fukui, I., and Murachi, T.**, Inactivation of calpain I and calpain II by specificity-oriented tripeptidyl chloromethyl ketones, *J. Biochem.*, 99, 173, 1986.
83. **Crawford, C., Mason, R. W., Wikstrom, P., and Shaw, E.**, The design of peptidyldiazomethane inhibitors to distinguish between the cysteine proteinase calpain II, cathepsin L and cathepsin B, *Biochem. J.*, 253, 751, 1988.

84. **Leary, R., Larsen, D., Watanabe, H., and Shaw, E.,** Diazomethyl ketone substrate derivatives as active-site directed inhibitors of thiol proteinases; Papain, Biochemistry, 16, 5857, 1977.
85. **Zumbrunn, A., Stone, S., and Shaw, E.,** Synthesis and properties of Cbz-Phe-Arg-CHN_2 (benzyloxy-carbonylphenylalanyldiazomethane) as a proteinase inhibitor, *Biochem. J.*, 250, 621, 1988.
86. **Shaw, E.,** The selective inactivation of thiol proteases in vitro and in vivo, *J. Prot. Chem.*, 3, 109, 1984.
87. **Shaw, E.,** Peptidyl sulphonium salts. A new class of protease inhibitors, *J. Biol. Chem.*, 263, 2768, 1988.
88. **Fox, J. E. B., Reynolds, C. C., and Phillips, D. R.,** Calcium-dependent proteolysis occurs during platelet aggregation, *J. Biol. Chem.*, 258, 9973, 1983.
89. **Schollmeyer, J. E.,** Calpain II involvement in mitosis, *Science*, 240, 911, 1988.
90. **Stracher, A., McGowan, E. B., and Shafiq, S. A.,** Muscular dystrophy: inhibition of degeneration *in vivo* with protease inhibitors, *Science*, 200, 50, 1978.
91. **Roots, B. I.,** Neurofilament accumulation induced in synapses by leupeptin, *Science*, 221, 971, 1983.
92. **Staubli, U., Baudry, M., and Lynch, G.,** Leupeptin, a thiol proteinase inhibitor, causes a selective impairment of spatial maze performance in rats, *Behav. Neurol. Biol.*, 40, 58, 1984.
93. **Staubli, U., Baudry, M., and Lynch, G.,** Olfactory discrimination learning is blocked by leupeptin, a thiol protease inhibitor, *Brain Res.*, 337, 333, 1985.
94. **Baldassare, J. J., Bakshian, S., Knipp, M. A., and Fisher, G. J.,** Inhibition of fibrinogen receptor expression and serotonin release by leupeptin and antipain, *J. Biol. Chem.*, 260, 10531, 1985.
95. **Ruggiero, M. and Lapetina, E. G.,** Protease and cyclo-oxygenase inhibitors synergistically prevent activation of human platelets, *Proc. Natl. Acad. Sci. U.S.A.*, 83, 3456, 1986.
96. **Kajiwara, Y., Tsujinaka, T., Sakon, M., Kambayashi, J., Ohshiro, T., Murachi, T., and Mori, T.,** Elucidation of calpain-dependent phosphorylation off myosin light chain in human platelets, *Biochem. Int.*, 15, 935, 1987.
97. **Brass, L. F. and Shattil, S. J.,** Inhibition of thrombin-induced platelet activation by leupeptin, *J. Biol. Chem.*, 263, 5210, 1988.
98. **Shoji-Kasai, Y., Senshu, M., Iwashita, S., and Imahori, K.,** Thiol protease-specific inhibitor E-64 arrests human epidermoid carcinoma A431 cells at mitotic metaphase, *Proc. Natl. Acad. Sci. U.S.A.*, 85, 146, 1988.
99. **Sugita, H., Ishiura, S., and Nonaka, I.,** Ca-activated neutral protease (CANP) and its inhibitors in pathological states, in *Proteinase Inhibitors: Medical and Biological Aspects,* Katunuma, N. et al., Eds., Springer-Verlag, Berlin, 1983, 69.
100. **Hanada, K., Tamai, M., Adachi, T., Oguma, K., Kashiwagi, K., Ohmura, S., Kominami, E., Towatari, T., and Katunuma, N.,** Characterisation of the three new analogues of E-64 and their therapeutic application, in *Proteinase Inhibitors: Medical and Biological Aspects,* Katunuma, N. et al., Eds., Springer-Verlag, Berlin, 1983, 25.
101. **Ishiura, S., Nonaka, I., and Sugita, H.,** Suppression of calcium-induced removal of the Z-line by a thiol protease inhibitor, E-64-c, *J. Biochem.*, 90, 283, 1981.
102. **Zeman, R. J., Kameyama, T., Matsumoto, K., Bernstein, P., and Etlinger, J. D.,** Regulation of protein degradation in muscle by calcium, *J. Biol. Chem.*, 260, 13619, 1985.
103. **Tsuchida, K., Aihara, H, Isogai, K., Hanada, K., and Shibata, N.,** Degradation of myocardial structural proteins in myocardial infarcted dogs is reduced by Ep-459, a cysteine proteinase inhibitor, *Biol. Chem. Hoppe-Seyler,* 367, 39, 1986.
104. **Sasaki, T., Kishi, M., Saito, M., Tanaka, T., Higuchi, N., Kominami, E., Katunuma, N., and Murachi, T.,** Inhibitory effect of di- and tripeptidyl aldehydes on calpains and cathepsins, *J. Enz. Inhib.*, 3, 195, 1990.
105. **Mehdi, S., Angelastro, M. R., Wiseman, J. S., and Bey, P.,** Inhibition of the proteolysis of rat erythrocyte membrane proteins by a synthetic inhibitor of calpain, *Biochem. Biophys. Res. Comm.*, 157, 1117, 1988.
106. **Varughese, K. I., Ahmed, F. R., Carey, P. R., Hasnain, S., Huber, C. P., and Storer, A. C.,** Crystal structure of a papain-E-64 complex. *Biochemistry,* 28, 1330, 1989.
107. **Matsumoto, K., Yamamoto, D., Ohishi, H., Tomoo, K., Ishida, T., Inoue, M., Sadatome, T., Kitamura, K., and Mizuno, H.,** Mode of binding of E-64-c, a potent thiol protease inhibitor, to papain as determined by X-ray crystal analysis of the complex, *FEBS Lett.*, 245, 177, 1989.
108. **Zhang, H. and Johnson, P.,** Inhibition of calpains by calmidazolium and calpastatin, *J. Enz. Inhib.*, 2, 163, 1988.
109. **Brumley, L. M. and Wallace, R. W.,** Calmodulin and protein kinase C antagonists also inhibit the Ca^{2+}-dependent protein protease, calpain I, *Biochem. Biophys. Res. Comm.*, 159, 1297, 1989.
110. **Nakamura, M., Inomata, M., Imajoh, S., Suzuki, K., and Kawashima, S.,** Fragmentation of an endogenous inhibitor upon complex formation with high- and low-Ca^{2+}-requiring forms of calcium-activated neutral proteases, *Biochemistry,* 28, 449, 1989.
111. **Kawasaki, H., Emori, Y., Imajoh-Ohmi, S., Minami, Y., and Suzuki, K.,** Identification and characterisation of inhibitory sequences in four repeating domains of the endogenous inhibitor for calcium-dependent protease, *J. Biochem.*, 106, 274, 1989.

112. **Maki, M., Bagci, H., Hamaguchi, K., Ueda, M., Murachi, T., and Hatanaka, M.,** Inhibition of calpain by a synthetic oligopeptide corresponding to an exon of the human calpastatin gene, *J. Biol. Chem.*, 264, 18866, 1989.
113. **Uemori, T., Shimojo, T., Asada, K., Asano, T., Kimizuka, F., Kato, I., Maki, M., Hatanaka, M., Murachi, T., Hanzawa, H., and Arata, Y.,** Characterisation of a functional domain of human calpastatin, *Biochem. Biophys. Res. Comm.*, 166, 1485, 1990.

Chapter 6

SUBUNIT AUTOLYSIS AND REGULATION OF CALPAIN II ACTIVITY

David R. Hathaway and Phadungchom McClelland

TABLE OF CONTENTS

I. BACKGROUND

Dayton et al.[1] were first to isolate a highly purified preparation of calpain from skeletal muscle and to propose the basic structure of the enzyme, which is a heterodimer consisting of 80-kDa and 30-kDa subunits present in a 1:1 molar ratio. Although a number of investigators have reported monomeric species of calpain, most often an 80-kDa variety, there are a number of reasons to believe that these are actually partially autolyzed forms. In particular, sodium dodecylsulfate polyacrylamide gel systems that poorly discriminate lower molecular weight proteins and incomplete autolysis of the 30-kDa subunit resulting in several fragments with $M_r < 30{,}000$ are potential sources of error.

Suzuki et al.[2] were first to report that avian skeletal muscle calpain II underwent autolysis of its high molecular weight subunit in the presence of Ca^{2+} and that autolysis enhanced the Ca^{2+} sensitivity of the enzyme. Subsequently, we demonstrated that both subunits of avian calpain II were cleaved during autolysis and, in agreement with Suzuki et al.,[2] that the Ca^{2+}-sensitivity of the enzyme was increased by a factor of approximately 20.[3] Initially, there followed some confusion regarding the relationship between native and autolyzed calpain II and the presence of two forms of mammalian calpain, calpain I and calpain II (or, μCANP and mCANP). It is now clear that calpain I and calpain II as originally reported by Mellgren[4] are different enzymes with identical regulatory subunits[5,6] but distinct catalytic (i.e., 75 to 89 kDa) subunits.[7-10] Moreover, both forms of calpain are subject to modulation of activity by autolysis.[11]

II. EFFECTS OF AUTOLYSIS ON SUBUNIT STRUCTURE

In addition to significant differences in amino acid sequence of their catalytic subunits, calpains I and II appear to differ with regard to subunit autolysis. Interestingly, avian calpain II is a structural chimera of calpains I and II.[12] Moreover, it elutes from DEAE-cellulose at 0.3 *M* salts, similar to calpain II, has a sensitivity to Ca^{2+} that is intermediate between mammalian calpains I and II, and undergoes autolysis of its 80-kDa catalytic subunit, similar to calpain I.[3] As shown in Figure 1, autolysis of avian calpain II results in the conversion of the 80 kDa/30 kDa heterodimer to a 76 kDa/18 kDa species. In the case of both subunits, a blocked amino terminus is cleaved yielding fragments that are amenable to sequencing as summarized in Table 1.

Mammalian calpain II isolated from vascular smooth muscle possesses a smaller catalytic subunit ($M_r = 75{,}000$) and the 30-kDa subunit is cleaved by autolysis (Figure 2). As summarized in Figure 2, only the 18-kDa autolytic fragment of the regulatory subunit yields sequence while the 75-kDa subunit remains blocked. Moreover, there is no obvious reduction in mass of the 75-kDa subunit. These results contrast with avian calpain II and with calpain I,[13,14] where autolysis of the catalytic subunit does result in both a measurable reduction in size as well as unblocking of the amino terminus.

III. EFFECTS OF AUTOLYSIS ON ENZYME ACTIVITY

An important effect of autolysis on calpain is to increase the apparent sensitivity of the enzyme to Ca^{2+}. This effect is best illustrated in Figure 3, where purified avian calpain II proenzyme (80-kDa/30-kDa species) is compared to the autolyzed product (76-kDa/18-kDa form). Despite this difference in sensitivity to Ca^{2+}, we have determined that no such large change in the Ca^{2+}-binding properties of avian calpain II occurs following autolysis as shown in Figure 4.

In order to simplify the problem, we have investigated mammalian vascular calpain II since only the regulatory subunit is cleaved during autolysis.[15] As shown in Figure 5, there

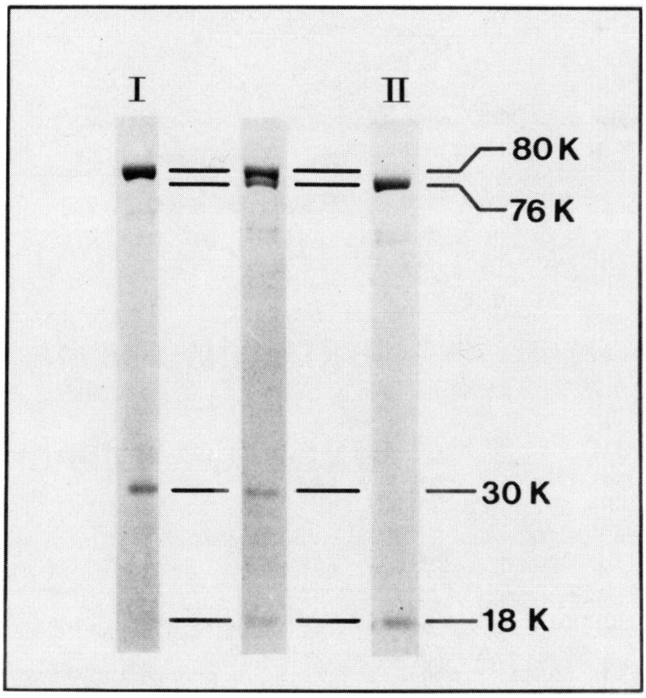

FIGURE 1. Sodium dodecylsulfate polyacrylamide gel electrophoresis (SDS-PAGE) of purified chicken gizzard calpain II and effects of autolysis on subunit structure. The enzyme was incubated with 500 μM Ca^{2+}. I, the unautolyzed enzyme; II, major autolytic subunits. The lane in the middle shows the transition between native and autolyzed forms. With longer incubations the 76-kDa subunit underwent additional autolysis. (From Hathaway, D., Werth, D., and Haeberle, J., *J. Biol. Chem.*, 257, 9072, 1982. With permission.)

TABLE 1
Partial Amino Terminal Sequences of the Native and Major Autolytic Cleavage Products of the Catalytic and Regulatory Subunits of Chicken Gizzard Calpain II

Subunit	Sequence[a]
Catalytic	
80 kDa	Blocked
76 kDa	AEGVGE
Regulatory	
30 kDa	Blocked
18 kDa	VDPNESEEVRQ

[a] Samples (100 pmol for each subunit) were subjected to SDS-PAGE, electroblotted onto polyvinylidene difluoride membranes, and sequenced by gas phase methodology.[27]

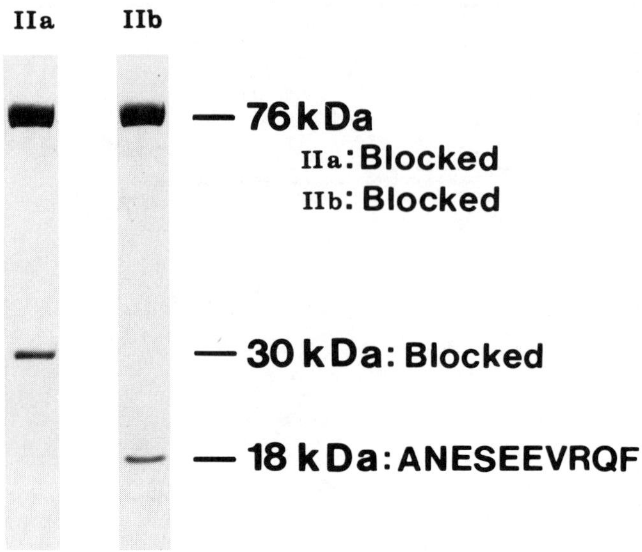

FIGURE 2. Autolysis of mammalian vascular calpain II. IIa, native enzyme: IIb, the autolyzed form resulting from incubation of the enzyme in 1.5 m*M* Ca^{2+} for 60 s. The subunits were electroblotted from gels onto polyvinylidene difluoride membranes and sequenced by gas phase methodology.

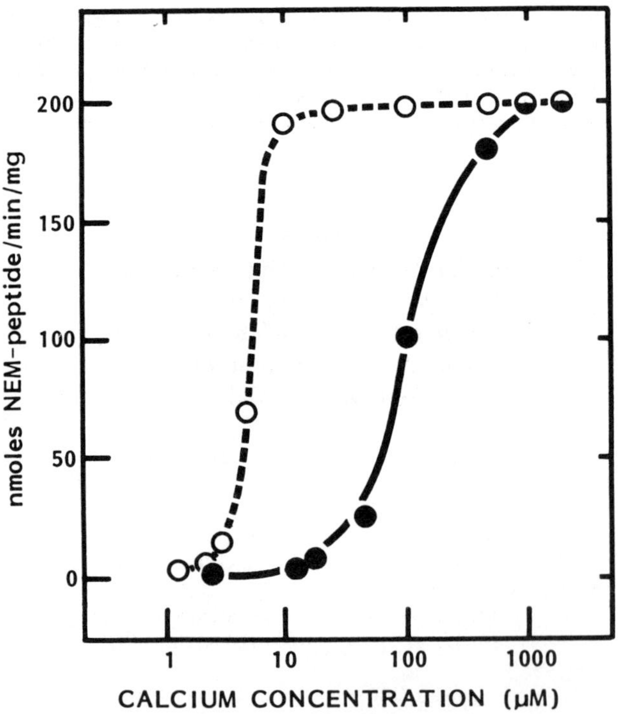

FIGURE 3. Ca^{2+} sensitivity of native and autolyzed forms of chicken gizzard calpain II; ● ● ●, the native form (I) of the enzyme as shown in Figure 1; ○ ○, the autolyzed form (II) as depicted in Figure 1. (From Hathaway, D., Werth, D., and Haeberle, J., *J. Biol. Chem.*, 257, 9072, 1982. With permission.)

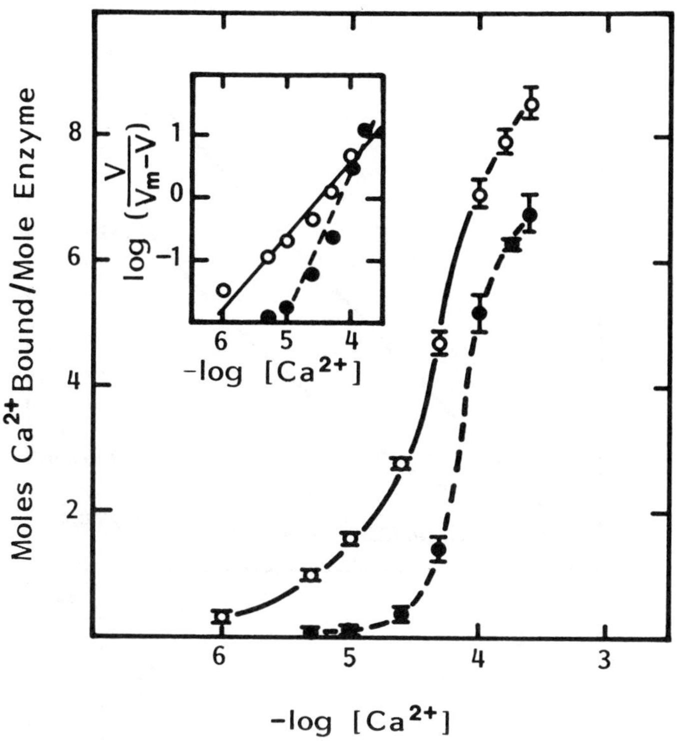

FIGURE 4. Ca^{2+}-binding to native and autolyzed forms of chicken gizzard calpain II. Binding studies were conducted at 4°C by equilibrium dialyses. All buffers contained 10 μM leupeptin to prevent intercurrent autolysis. The inset is a Hill plot of the binding data: ●-●, native calpain II; ○-○, autolyzed calpain II (see Figure 1). (From Coolican, S., Haiech, J., and Hathaway, D., *J. Biol. Chem.*, 261, 4170, 1986. With permission.)

is a fundamental difference in the kinetics of substrate degradation by native and autolyzed forms of calpain II with the native form exhibiting a lag phase inactivity giving rise to nonlinear kinetics of substrate degradation. When the time-dependent evolution of enzyme activity is compared to 30-kDa subunit autolysis (Figure 6) it is quite apparent that autolysis tends to precede activity and that substrate degradation achieves a linear rate when autolysis is essentially complete.

To link autolysis to the lag phase in substrate degradation, we applied a mathematical model for enzyme hysteresis (i.e., lag phase):

$$V_t = V_f + (V_i - V_f)e^{-kt}$$

where V_f = velocity as a function of time, V_i = initial velocity, V_f = final linear velocity, t = time and k = apparent rate constant of the transition from nonlinear to linear kinetics of substrate degradation.[16] This kind of analysis is useful because changes in k at varying Ca^{2+} concentrations should parallel changes in 30-kDa subunit autolysis if the latter is prerequisite for enzyme activity and, therefore, responsible for the lag phase in substrate degradation observed with native or unautolyzed calpain II. As shown in Figure 7, this proved to be the case. Thus, the high Ca^{2+} requirement of calpain II is actually due to a relatively low Ca^{2+} sensitivity of autolysis. Moreover, the latter can be regarded as a kind of "barrier" to enzyme activation and might well be the rate-limiting step in the proposed reaction scheme:

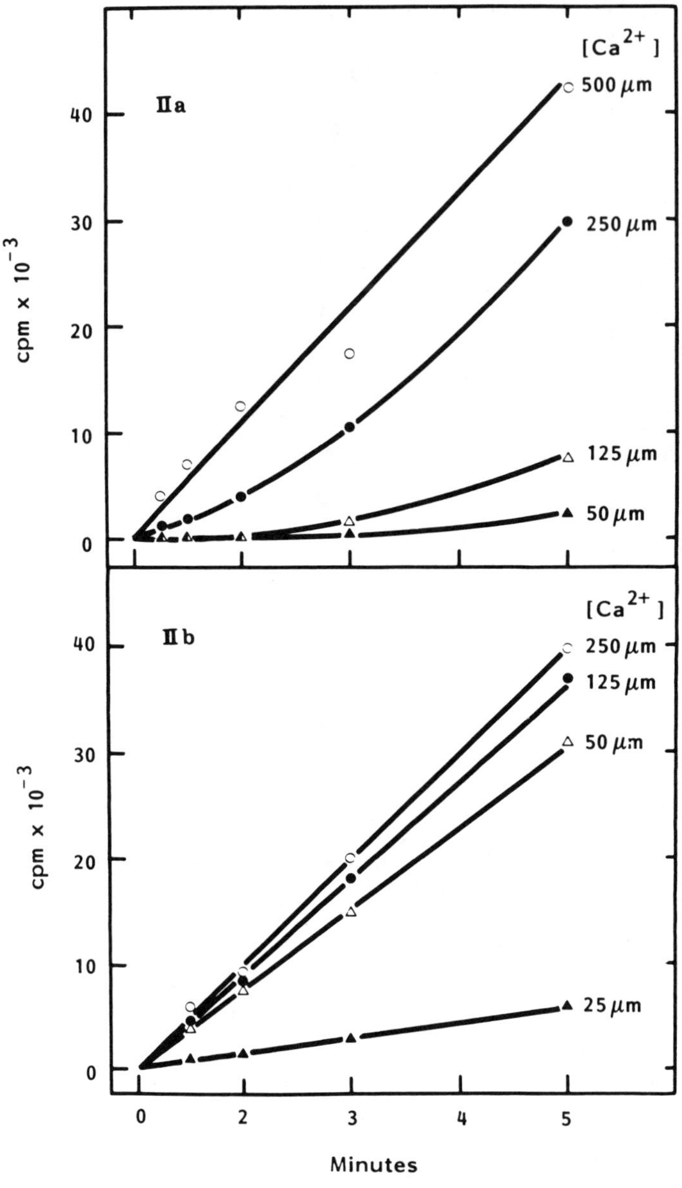

FIGURE 5. Time course of [^{3}H]-α-casein degradation by vascular calpain II: a, native enzyme; b, autolyzed enzyme (see Figure 2). The Ca^{2+} concentrations are given in the figure margins. (From Coolican, S., Haiech, J., and Hathaway, D., *J. Biol. Chem.*, 261, 4170, 1986. With permission.)

$$\text{CDP} \xrightarrow{Ca^{2+}} \text{CDP}' + \text{S} \xrightarrow{Ca^{2+}} \text{CDP}' + \text{P}$$

where CDP = native calpain, CDP′ = autolyzed calpain, S = protein substrate, and P = proteolytic products.

Since native calpain is a proenzyme, comparisons such as that depicted in Figure 3 have rather limited meaning. The Ca^{2+} sensitivity of proenzyme calpain II measured by substrate degradation (Figure 3) is more properly a measure of the Ca^{2+} sensitivity of autolysis. Moreover, autolysis and catalytic activity toward substrates subsequent to autolysis appear

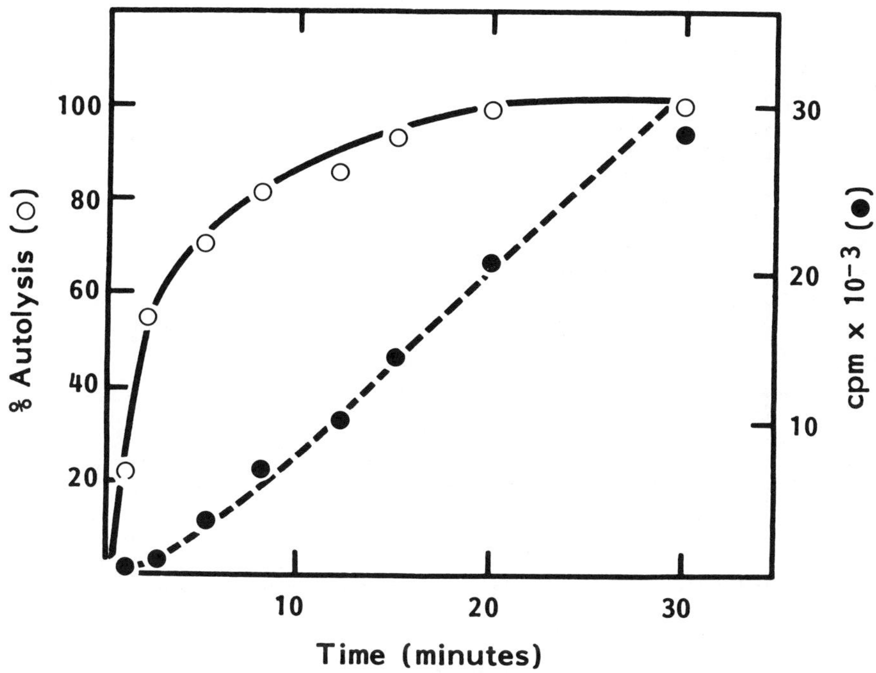

FIGURE 6. Time course of subunit autolysis and [^{3}H]-α-casein degradation by vascular calpain II: ●-●, casein degradation; ○-○, autolysis. Autolysis was measured as disappearance of the 30-kDa subunit from gel scans. (From Coolican, S., Haiech, J., and Hathaway, D., *J. Biol. Chem.*, 261, 4170, 1986. With permission.)

to have unique Ca^{2+} requirements suggesting that Ca^{2+} regulation of these two properties may involve distinct processes.

IV. PHOSPHOLIPIDS AND AUTOLYSIS

Once autolysis was recognized as a potentially important mechanism for regulation of calpain activity, a search for substances that might stimulate this process or modulate the Ca^{2+} sensitivity was undertaken. As summarized in Table 2, we found that phosphatidylinositol alone and in combination with diacylglycerol specifically stimulated autolysis but had no effect on the activity of the preautolyzed enzyme. The maximal stimulating effect of phosphatidylinositol was observed at a concentration of 10 μg/ml[17] and could be blocked entirely by the phenothiazine, trifluoperazine.[17]

Exploring the problem further, we found that phosphatidylinositol or phosphatidylinositol plus diacylglycerol could substantially increase the Ca^{2+} sensitivity of autolysis (Figure 8) by a factor of about 8 to 10. Although smaller effects were observed with other phospholipids, such as phosphatidylserine, in many cases thin layer chromatography of these lipids disclosed small amounts of phosphatidylinositol. Thus, we have concluded that this effect is specific for phosphatidylinositol.

Other investigators have confirmed these results[18] and have shown that calpain binds reversibly to membrane vesicles in the presence of Ca^{2+}.[19] Moreover, some studies have reported association of calpain with the plasmalemma in living cells[20] while others have shown a more diffuse distribution.[9] The latter results might be explained if translocation from cytoplasm to membrane was prerequisite for activation of calpain. If so, then factors initiating the translocation would have to be controlled or quite different results might be obtained. Schollmeyer microinjected calpain II into PtK_1 cells after growth-synchronizing

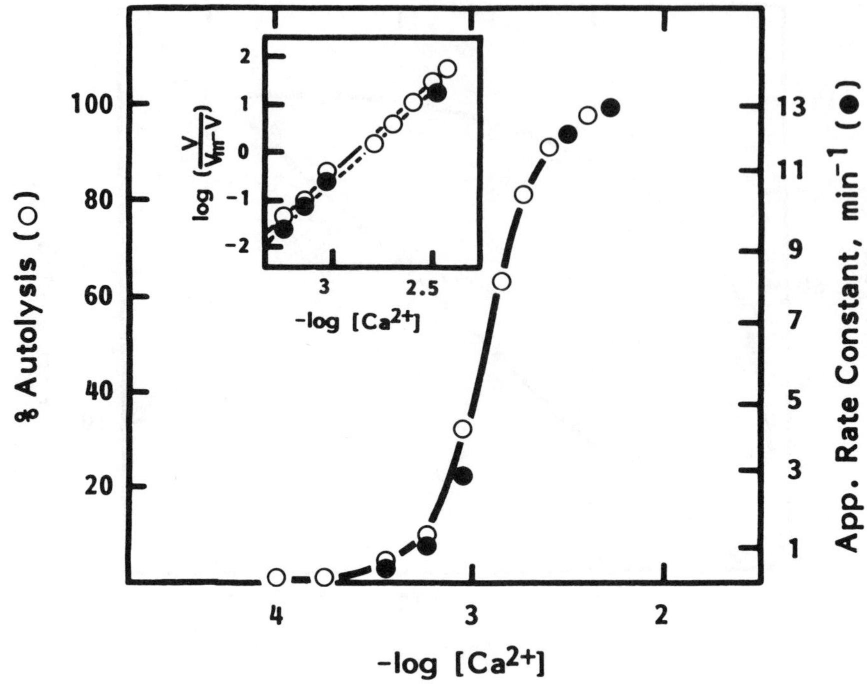

FIGURE 7. Comparison of the Ca^{2+}-dependence of autolysis and of the apparent rate constants of the transition of lag phase in substrate degradation by vascular calpain IIa (see Figure 2). The inset is a Hill plot of autolysis and rate constant data. The rate constants at varying concentrations of Ca^{2+} were determined as described by Coolican et al. (From Coolican, S., Haiech, J., and Hathaway, D., *J. Biol. Chem.*, 261, 4170, 1986. With permission.)

TABLE 2
Effects of Lipids and Phospholipid Derivatives on 30-kDa Subunit Autolysis of Vascular Calpain II[a]

Agent[b]	% stimulation[c]
Glycerol	0
Arachidonic acid	0
L-α-Phosphatidylinositol	500
L-α-Phosphatidylserine	12
L-α-Phosphatidylcholine	0
L-α-Phosphatidylethanolamine	0
Dioleoylglycerol	95
L-α-Phosphatidic acid	0
L-α-Phosphatidylinositol-4-monophosphate	0
L-α-Phosphatidylinositol-4,5-diphosphate	0
12-O-Tetradecanoylphorbol-13-acetate	0

[a] Autolysis was measured as disappearance of the 30-kDa regulatory subunit from scans of SDS-polyacrylamide gels.
[b] All agents were used at 50 μ*M*.
[c] The Ca^{2+} concentration was 500 μ*M*.

From Coolican, S. and Hathaway, D., *J. Biol. Chem.*, 48, 471, 1979. With permission.

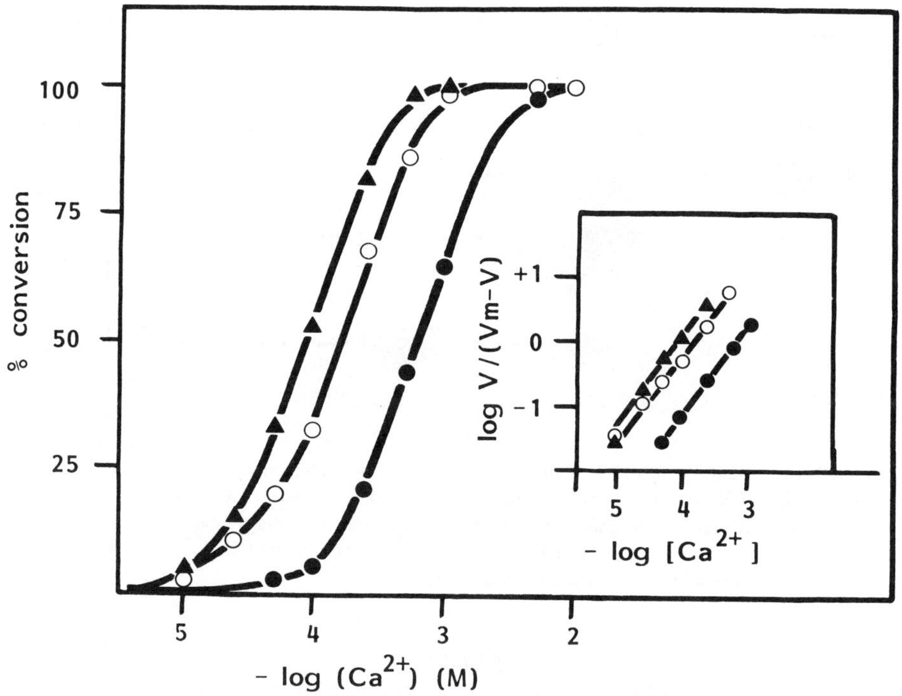

FIGURE 8. Effect of L-α-phosphatidylinositol and dioleoylglycerol on the Ca^{2+}-dependence of autolysis. Percent conversion refers to autolytic conversion of native vascular calpain II to the autolyzed forms as measured by disappearance of the 30-kDa subunit: ●, No lipid; ○, plus 50 μ*M* phosphatidylinositol; ▲, plus 50 μ*M* phosphatidylinositol and 50 μ*M* dioleoylglycerol. (From Coolican, S. and Hathaway, D., *J. Biol. Chem.*, 48, 471, 1979. With permission.)

these cells by thymidine block and noted plasmalemma localization of the rhodamine-labeled enzyme.[20] Since calpains may be activated during changes in cell growth or differentiation,[21] it is possible that the processes leading to translocation of calpain to the plasmalemma were activated in these experiments, whereas they were not in those studies indicating a more diffuse localization. These important issues will require more detailed investigation in the future in order to more thoroughly understand how calpain is activated in living cells.

It is noteworthy that other activators of calpain have been reported. DeMartino and Croall have isolated a small protein that increases the V_{max} of calpain[22] and Pontremoli et al.[23] have reported identification of a factor in neutrophils that activates calpain. The latter group has also shown that incubation of calpain I with hemoglobin increases its sensitivity to Ca^{2+}.[24] Finally, Zimmerman and Schlaepfer have described a marked enhancement of the Ca^{2+} sensitivity of both calpains I and II induced by certain carbohydrates in the presence of Ca^{2+} and leupeptin.[25] The relationship of all of these observations to the autolytic mechanism of calpain activation remains uncertain at this time.

V. THE AUTOLYTIC CLEAVAGE SITES IN THE REGULATORY SUBUNIT

Although there are a number of proteins which have been offered as likely endogenous substrates for calpain, the process of autolysis suggests that at least one substrate for calpain is its regulatory subunit. Thus, we have reasoned that precise determination of the autolytic cleavage sites might offer some insight into specific peptide sequences preferred by calpain. A similar but more limited approach has been reported by Crawford et al.[26] using avian

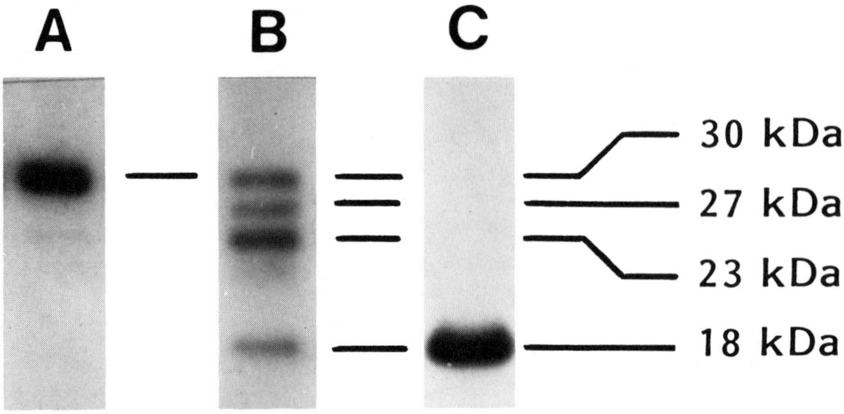

FIGURE 9. Stepwise autolysis of the 30-kDa subunit of vascular calpain II. Lane A, the native or unautolyzed 30-kDa regulatory subunit; lane B, partial autolysis of the regulatory subunit; lane C, the fully autolyzed and stable 18-kDa form of the regulatory subunit. Autolysis was conducted at 0.75 m*M* Ca^{2+}. The intermediates were electroblotted onto nitrocellulose and identified by autoradiography of radioimmunoblots.

TABLE 3
Autolytic Cleavage Sites in the 30-kDa Subunit of Vascular Calpain II

Subunit M_r	From cDNA[a] P_3-P_2-P_1		From Protein[b] P'_1-P'_2-P'_3	Peptide Bond
30 kDa		Blocked		
27 kDa	G-L-G	/	N-V-L	G/N
23 kDa	I-L-G	/	G-V-I	G/G
18 kDa	N-I-E	/	A-N-E	E/A

[a] Amino acids in positions 1—3 were inferred from the cDNA encoding the regulatory subunit.[28]

[b] The amino termini of the native 30-kDa subunit and its autolytic fragments were determined by gas phase sequencing of protein electroblotted onto polyvinylidene difluoride membranes.[27]

calpain II. As shown in Figure 9, autolysis of the regulatory subunit of vascular calpain II proceeds in stepwise fashion. Using newly developed technology for direct sequencing of proteins electroblotted onto polyvinylidene difluoride membranes,[27] we have determined the unique amino terminal sequences of the major autolytic fragments shown in Figure 9 (see Table 3). These studies confirm that autolysis proceeds in an amino to carboxyl direction since each intermediate does have a unique sequence. Although the entire amino acid sequence of the 30-kDa subunit of mammalian calpain has been deduced previously from cloning and sequencing of cDNA,[5,6] we have cloned and sequenced a cDNA from vascular muscle encoding this subunit as well.[28] This allowed us to predict the amino acid sequence upstream from the determined cleavage sites with certainty (Table 3). Overall, the sequence of autolytic cleavage sites does not indicate a high degree of primary sequence determinants. Nevertheless, there are some features that are noteworthy. First, a leucine (or isoleucine) residue in the upstream second position is a consistent feature of all cleavage sequences. This appears to be a common thread linking studies with synthetic inhibitors and substrates.[26,29,30] Indeed, the most potent synthetic inhibitor described to date is Z-leu-nleu-H, or calpeptin.[31] Moreover, with the exception of considerable variability in position 1, the remainder of residues upstream and downstream (i.e., P_3 to P'_3) are largely hydrophobic or neutral. Of course, the most obvious conclusion is that heretofore unidentified secondary or tertiary protein structural features are the prinicipal determinants of calpain specificity. It is possible that the information gleaned from sequencing of the autolytic sites may help in the design of more potent peptide inhibitors.

VI. CONCLUSIONS

At the present time, investigation of subunit autolysis of calpain has provided a framework for constructing models of enzyme regulation. However, there are several issues that remain incompletely answered. First, does autolysis account for calpain activation in living cells, and what are the proximate stimuli that activate this process? Second, are there alternative mechanisms that activate calpain without stimulating subunit autolysis? And third, what is the functional significance of catalytic subunit autolysis of calpain I? The answers to these questions will both require and yield a great deal of new information regarding the cellular functions of calpain.

REFERENCES

1. **Dayton, W. R., Goll, D. E., Zeece, M. C., Robson, R. M., and Reville, W. J.,** A Ca^{2+}-activated protease possibly inovlved in myofibrillar protein turnover. Purification from porcine muscle, *Biochemistry,* 15, 2150, 1976.
2. **Suzuki, K., Tsuji, S., Kubota, S., Kimura, Y., and Imahori, K.,** Limited autolysis of Ca^{2+}-activated neutral protease (CANP) changes its sensitivity to Ca^{2+} ions, *J. Biochem.,* 90, 275, 1981.
3. **Hathaway, D., Werth, D., and Haeberle, J.,** Limited autolysis reduces the calcium requirement of a smooth muscle Ca^{2+}-activated protease, *J. Biol. Chem.,* 257, 9072, 1982.
4. **Mellgren, R.,** Canine cardiac calcium-dependent proteases: resolution of two forms with different requirements for calcium, *FEBS Lett.,* 109, 129, 1980.
5. **Sakihama, T., Kakidani, H., Zenita, K., Yumoto, N., Kikuchi, T., Sasaki, T., Kannagi, R., Nakanishi, S., Ohmori, M., Takio, K., Titani, K., and Murachi, T.,** A putative Ca-binding protein: structure of the light subunit of porcine calpain elucidated by molecular cloning and protein sequence analysis, *Proc. Natl. Acad. Sci. U.S.A.,* 82, 6075, 1985.
6. **Emori, Y., Kawasaki, H., Imajoh, S., Kawashima, S., and Suzuki, K.,** Isolation and sequence analysis of cDNA clones for the small subunit of rabbit calcium-dependent protease, *J. Biol. Chem.,* 261, 9472, 1986.
7. **Emori, Y., Kawasaki, H., Sugihara, H., Imajoh, S., Kawashima, S., and Suzuki, K.,** Studies on a cyclic nucleotide-independent protein kinase and its proenzyme in mammalian tissues, *J. Biol. Chem.,* 261, 9465, 1986.
8. **Aoki, K., Imajoh, S., Ohno, S., Emori, Y., Koike, M., Kosaki, G., and Suzuki, K.,** Complete amino acid sequence of the large subunit of the low-$^{2+}$-requiring form of human Ca^{2+}-activated neutral protease (muCANP) deduced from its cDNA sequence, *FEBS Lett.,* 205, 313, 1986.
9. **Yoshimura, N., Hatanaka, M., Kitahara, A., Kawaguchi, N., and Murachi, T.,** Intracellular localization of two distinct Ca^{2+}-proteases (calpain I and calpain II) as demonstrated by using discriminative antibodies, *J. Biol. Chem.,* 259, 9847, 1984.
10. **Imajoh, S., Aoki, K., Ohno, S., Emori, Y., Kawasaki, H., Sugihara, H., and Suzuki, K.,** Molecular cloning of the cDNA for the large subunit of the high-calcium-requiring form of human calcium-activated protease, *Biochemistry,* 27, 8122, 1988.
11. **Suzuki, K.,** Calcium-activated neutral protease: domain structure and activity regulation, *Trends Biochem. Sci.,* 12, 103, 1987.
12. **Ohno, S., Emori, Y., Imajoh, S., Kawasaki, H., Kisaragi, M., and Suzuki, K.,** Evolutionary origin of a calcium-dependent protease by fusion of genes for a thiol protease and a calcium-binding protein, *Nature,* 312, 566, 1984.
13. **Dayton, W.,** Comparison of low- and high-calcium-requiring forms of the calcium-activated protease with their autocatalytic breakdown products, *Biochim. Biophys. Acta,* 709, 166, 1982.

14. **Imajoh, S., Kawasaki, H., and Suzuki, K.**, Limited autolysis of calcium-activated neutral protease (CANP): reduction of the Ca^{2+}-requirement is due to the NH_2-terminal processing of the large subunit, *J. Biochem.*, 100, 633, 1986.
15. **Coolican, S., Haiech, J., and Hathaway, D.**, The role of subunit autolysis in activation of smooth muscle Ca^{2+}-dependent proteases, *J. Biol. Chem.*, 261, 4170, 1986.
16. **Frieden, C.**, Slow transitions and hysteretic behavior in enzymes, *Annu. Rev. Biochem.*, 48, 471, 1979.
17. **Coolican, S. and Hathaway, D.**, Effect of L-alpha-phosphatidylinositol on a vascular smooth muscle Ca^{2+}-dependent protease, *J. Biol. Chem.*, 259, 11627, 1984.
18. **Imajoh, S., Kawasaki, H., and Suzuki, K.**, The amino-terminal hydrophobic region of the small subunit of calcium-activated neutral protease (CANP) is essential for its activation by phosphatidylinositol, *J. Biochem.*, 99, 1281, 1986.
19. **Pontremoli, S., Salamino, F., Sparatore, B., Michetti, M., Sacco, O., and Melloni, E.**, Following association to the membrane, human erytrocyte procalpain is converted and released as fully active calpain, *Biochim. Biophys. Acta*, 831, 335, 1985.
20. **Schollmeyer, J. E.**, Calpain II involvement in mitosis, *Science*, 240, 911, 1988.
21. **Schollmeyer, J. E.**, Possible role of calpain I and calpain II in differentiating muscle, *Exp. Cell Res.*, 163, 413, 1986.
22. **DeMartino, G. and Croall, D.**, Calcium-dependent proteases: a prevalent proteolytic system of uncertain function, *NIPS*, 2, 82, 1987.
23. **Pontremoli, S., Melloni, E., Michetti, M., Salamino, F., Sparatore, B. and Horecker, B. L.**, An endogenous activator of the Ca^{2+}-dependent proteinase of human neutrophils that increases its affinity for Ca^{2+}, *Proc. Nat. Acad. Sci. U.S.A.*, 85, 1740, 1988.
24. **Pontremoli, S. and Melloni, E.**, Extralysosomal protein degradation, *Annu. Rev. Biochem.*, 55, 455, 1986.
25. **Zimmerman, U. J. P. and Schlaepfer, W. W.**, Calcium-activated neutral proteases (calpains) are carbohydrate binding proteins, *J. Biol. Chem.*, 263, 11609, 1988.
26. **Crawford, C., Willis, A. C., and Gagnon, J.**, The effects of autolysis on the structure of chicken calpain II, *Biochem. J.*, 248, 579, 1987.
27. **Matsudaira, P.**, Sequence from picomole quantities of proteins electroblotted onto polyvinylidene difluoride membranes, *J. Biol. Chem.*, 262, 10035, 1987.
28. **McClelland, P., Lash, J. A., and Hathaway, D. R.**, Identification of autolytic cleavage sites in the regulatory subunit of calpain II: a comparison of partial amino terminal sequences to deduced sequences from complementary DNA, *J. Biol. Chem.*, 264, 17428, 1989.
29. **Sasaki, T., Kikuchi, T., Fukui, I., and Murachi, T.**, Comparative specificity and kinetic studies on porcine calpain I and calpain II with naturally occurring peptides and synthetic fluorogenic substrates, *J. Biol. Chem.*, 259, 12489, 1984.
30. **Sasaki, T., Kikuchi, T., Fukui, I., and Murachi, T.**, Inactivation of calpain I and calpain II by specificity-oriented tripeptidyl chloromethyl ketones, *J. Biochem.*, 99, 173, 1986.
31. **Tsujinaka, T., Kajiwara, Y., Kambayashi, J., Sakon, M., Higuchi, N., Tanaka, T., and Mori, T.**, Synthesis of a new cell penetrating calpain inhibitor (calpeptin), *Biochem. Biophys. Res. Commun.*, 153, 1201, 1988.

Chapter 7

REGULATION OF CALCIUM-DEPENDENT PROTEASE ACTIVITY *IN VITRO*: CLUES FOR DETERMINING THEIR PHYSIOLOGICAL FUNCTIONS

Dorothy E. Croall and George N. DeMartino

TABLE OF CONTENTS

I. INTRODUCTION

Biochemical purification and characterization of the mammalian calcium-dependent proteases, (CDPs) also known as calcium-activated neutral proteases, (CANPs) or calpains have revealed a variety of regulatory mechanisms that control proteolytic activity *in vitro*. These mechanisms include:

1. The absolute requirement for binding Ca^{2+};
2. Autoproteolytic processing of the protease subunits which activates the enzymes and results in an increased sensitivity to Ca^{2+};
3. Association of the proteases with membranes and/or phospholipid binding;
4. Inhibition of activation and/or subsequent proteolytic activity by a specific endogenous inhibitory protein (calpastatin); and
5. Stimulation of proteolytic activity by an endogenous stimulatory protein.

Each of the two distinct CANPs appears to be regulated similarly by these effectors. The Ca^{2+} concentration requirement for enzyme activation and subsequent proteolytic activity remains the sole major distinguishing catalytic feature between CANP-1 and CANP-2. It is not known how this complex proteolytic system is regulated in cells. Because of the multiple regulatory mechanisms it is unclear whether changes in intracellular Ca^{2+} per se result in CANP activation. However, the biochemical characterization of this proteolytic system suggests several approaches to address the difficult and important questions concerning the regulation and functions of these proteases *in vivo*.

Since other chapters in this book are devoted to specific regulatory mechanisms, e.g., the inhibitor protein and autoproteolysis, we will limit the focus of our discussion to the following issues. First, we will review the endogenous stimulatory protein, the least well-characterized component of the calpain system. Second, we will begin the transition from test tube to cells by asking if autoproteolytic modification of the CANP subunits occurs *in vivo* and therby provides an assay for enzyme activation. Last, we will discuss which signals may regulate activation of calcium-dependent proteases *in vivo*.

II. THE CALCIUM-DEPENDENT PROTEASE REGULATORY PROTEIN: A STIMULATORY FACTOR

Several groups have isolated a factor that stimulates the activity of the calcium-dependent proteases. The factor, presumed to be a protein, was first described by DeMartino and Blumenthal[1] in experiments designed to test the possible effect of calmodulin on the proteases. This work was conducted before it was clear that the proteases were themselves calcium-binding proteins and before their primary structures revealed intrinsic calmodulin-like domains. Although calmodulin had no effect on protease activity, some calmodulin preparations contained a contaminant that greatly increased protease activity. This factor has been isolated from bovine brain and partially purified. Many of its physical properties are similar to those of calmodulin, and this probably explains why calmodulin prepared by conventional protein purification techniques contained this protein. For example, the stimulatory factor has a native molecular weight of approximately 20,000, based on gel filtration chromatography, and is relatively heat stable. The factor increases the proteolytic activity of each calcium-dependent protease (i.e., type 1 and type 2) 5- to 25-fold, but has no effect on any other protease that has been tested (trypsin, chymotrypsin, papain and macropain). The stimulator has been identified in several species and tissues[1-3] and stimulates calcium-dependent proteases from all sources tested.

The mechanism of action of this stimulator is unclear. Kinetic data indicate that it does not influence the K_m of the proteases for substrates but does increase the V_{max} of proteolysis.

The stimulator has no effect on the calcium sensitivities of either type 1 or type 2 protease. This feature is distinct from the effect of calmodulin on its various target enzymes. Recent work indicates that the stimulator associates with the proteases in a calcium-dependent fashion,[68] similar to the interaction between the proteases and their inhibitor.[4] There is no evidence that the stimulatory protein is a calcium-binding protein, but it only interacts with the proteases after they have bound calcium.

The work of DeMartino and Blumenthal[1] used soluble tissue extracts as source material and made no attempt to examine other subcellular fractions for stimulatory activity. Recently, Takeyama et al.[2] have described the stimulatory factor in a Triton X-100 insoluble fraction from bovine brain and platelets. With the exception of its localization, most other physical and functional properties of the factor were very similar to the soluble form. This work indicated that the factor was much more abundant in the insoluble fraction than in the cytosolic fraction. Further support for a cytoskeletal location has been provided by Pontremoli et al.[3] in neutrophils. Some of the properties of the factor reported by this group, however, are at variance with other reports. Most notably, they indicated that the factor can increase the calcium sensitivity of the neutrophil calcium-dependent protease by 100-fold. Unfortunately, this protease appears unusual in comparison with typical mammalian CANPs because it is described as a monomer of 85,000 Da and has a calcium concentration requirement intermediate to those typical for the type 1 and type 2 proteases.[5,6]

The possible localization of the stimulatory factor in a putative cytoskeletal fraction (defined by Triton X-100 insolubility) has important functional implications in view of the possible association of both the proteases and their inhibitor with membrane and/or cytoskeletal fractions.[7-9] Because many cytoskeletal proteins are particularly susceptible to calcium-dependent proteolysis,[10-14] Takeyama et al. have speculated that the stimulatory factor may provide a mechanism to localize the protease to the cytoskeleton.[2] How the interaction between the stimulatory protein and proteases is regulated may contribute to our understanding of the apparent susceptibility of cytoskeletal proteins to these proteases. In addition to the complete purification and characterization of this stimulatory factor, it will also be important to determine its possible role in the interaction of the proteases with the inhibitory protein.

III. AUTOPROTEOLYTIC MODIFICATION OF CANP SUBUNITS: AN ASSAY FOR ENZYME ACTIVATION *IN VIVO*?

A. AUTOPROTEOLYSIS *IN VITRO*

We first suspected that the 80/30 heterodimeric forms of CANP were inactive proenzymes from studies comparing CANP-2, purified by conventional chromatography (reactive red agarose in the presence of Ca^{2+} chelators), with enzyme purified by Ca^{2+}-dependent substrate affinity chromatography.[15] Exposure of each CANP to conditions suitable for demonstrating proteolytic activity always resulted in partial or extensive proteolysis of the "30-kDa" subunit (Figure 1). This observation reconciled many reports of apparent monomeric forms of CANP where the small subunit (i.e., 17 kDa after modification) was present at insufficient concentration to be detected or electrophoresed at the ion front. Several laboratories have described autoproteolysis of each purified enzyme in the presence of calcium and in the absence of substrates[16-22] (see Chapter 6). Importantly for the type 2 enzyme, autoproteolysis is accompanied by a significant decrease in its Ca^{2+} concentration requirement.[16-18] This also occurs for type 1 enzyme.[72] In some species (e.g., chicken) and frequently for the type 1 enzymes (e.g., bovine, human), proteolytic modification of each subunit is easily detected by SDS-PAGE (84 kDa → 78 kDa; 26 kDa → 17 kDa).[23-26] A change in the type 2 80-kDa subunit is not usually observed by SDS-PAGE, but proteolysis of the smaller subunit occurs rapidly and is readily detected (Figure 1).[16-20]

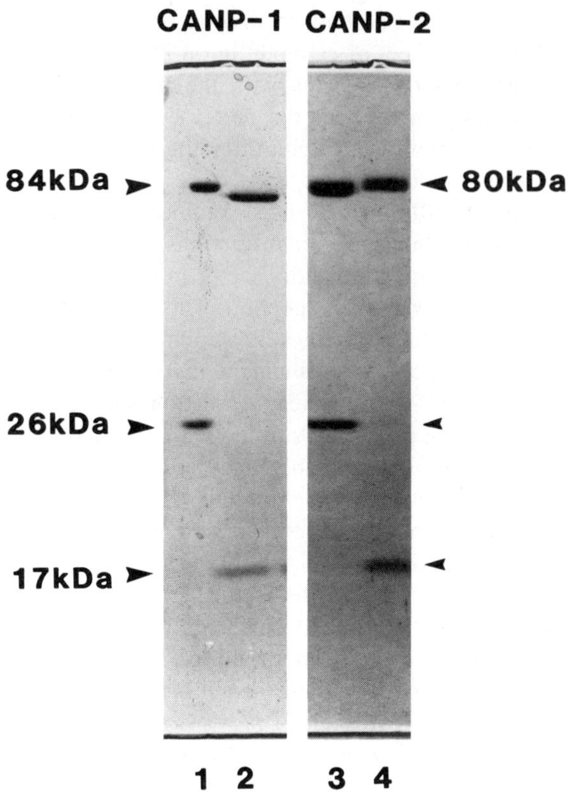

FIGURE 1. SDS-PAGE analysis of native and proteolytically modified calcium-dependent proteases *in vitro*. SDS-PAGE of native and autoproteolyzed calcium-dependent proteases 1 and 2 (CANP-1 and CANP-2) stained with coomassie blue. CANP-1 (lane 1) and CANP-2 (lane 3) were purified from bovine heart and electrophoresed on a 12% acrylamide gel using standard procedures. Each native protease was incubated under conditions known to allow autoproteolytic modification of each enzyme; 6 m*M* CaCl, 1 m*M* EDTA, 1 m*M* EGTA for 0.5 min at 4°C for CANP-1 (0.9 mg/ml); 9 m*M* CaCl, 1 m*M* EGTA for 1 min at 25°C for CANP-2 (0.16 mg/ml). Samples were then heated in SDS-PAGE sample buffer and electrophoresed (modified CANP-1, lane 2; modified CANP-2, lane 4). Each lane was loaded with 6 μg protein.

We and others have attempted to correlate specific proteolytic cleavages of the protease subunits with functional characteristics of the enzymes. What cleavage results in activation of the enzyme? Which forms of the enzyme demonstrate a reduced Ca^{2+} ion requirement? Using CANP-2 (from bovine heart) we examined autoproteolytic modification of the 26-kDa subunit and looked for changes in Ca^{2+} ion concentration requirement using ^{14}C casein hydrolysis as the assay for enzyme activity. By varying the enzyme and/or calcium concentrations and by preincubating the enzyme at lowered temperatures, proteolysis of the 26-kDa subunit is sufficiently slowed to identify numerous intermediate cleavages in the processing of 26 kDa to 17 kDa. A series of experiments revealed a correlation between the appearance of the 17-kDa fragment of CANP-26 kDa and the increase in enzyme sensitivity to Ca^{2+}.[16] By kinetic analysis of casein hydrolysis, it appears that the initial cleavages of the 26-kDa subunit are required for enzyme activation.[16,17] Because there is no change in the catalytic subunit of CANP-2 detectable by SDS-PAGE, we attributed the changes in catalytic properties to cleavages of the regulatory small subunit. Although it was clear at that time that the 80-kDa subunit was the catalytic subunit, alterations in its conformation due to Ca^{2+}-binding and its interaction with the modified regulatory subunit could be expected to alter its catalytic properties. Recently Suzuki et al. have documented proteolytic modification of the catalytic subunit of CANP-2 during autoproteolysis.[18] Eighteen residues of the amino terminus are removed with no resultant alteration of electrophoretic mobility of this subunit. They have also provided evidence that this small modification of the catalytic subunit is responsible for the decrease in the Ca^{2+} concentration requirement of the enzyme.[18] Regardless of which cleavage is (or if both cleavages are) required for enzyme activity and the reduction in Ca^{2+} required, if the modified protease subunits are present, the enzymes are (or have been) active.

Although no detailed studies of CANP-1 autoproteolysis are published, like CANP-2, its small subunit is rapidly proteolyzed through several intermediates to a 17-kDa form.[19,24] The NH_2-terminus of the catalytic subunit is also modified.[22-23,25-27,69] The proteolyzed enzyme (78 kDa/17 kDa) displays increased sensitivity to Ca^{2+} and extreme care is required to protect the enzyme from self-destruction. To date we have been unable to precisely determine if proteolysis of 84 kDa to 78 kDa precedes, accompanies, or follows proteolysis of 26 kDa to 17 kDa. As is true for the type 2 enzyme, the purification procedures used by us and others do not isolate any measurable, proteolytically modified enzyme from various sources. It is not clear if this indicates (1) a low, or nonexistent, steady state level of activated proteases; (2) lack of stability of the activated proteases during purification; or (3) loss of the modified enzymes during purification. To identify cellular conditions which result in enzyme activation we can utilize the fact that the CANP subunits are modified when (and after) the proteases become activated. Our initial studies examined proteolytic processing of the regulatory subunit. Although such experiments will not allow us to distinguish between the activation of CANP-1 and CANP-2, this modification is easily detected by SDS-PAGE in combination with immunoblotting and will indicate if either or both enzymes are activated. We can also exploit some of the biological diversity of the relative composition of this proteolytic system in various cells. By selecting cells or cell lines that have relatively low levels of the inhibitory protein, the probable role of the inhibitory protein in preventing and/or limiting enzyme activation can be minimized. We can also select cells such as erythrocytes[4,28] or platelets[10,25,28] that have only, or predominantly CANP-1, or such cells as vascular smooth muscle cells[17] and BHK21/C13[70] that have predominantly CANP-2.

At the present time it seems that direct examination of the protease subunits is the simplest approach to studying activation of this proteolytic system *in vivo*. However, there are several limitations to this approach. For example, it may be difficult (or impossible) to detect the small subunit of the proteases in whole cell extracts from sources where protease concentration is low. Partial purification of the enzymes could increase the probability of detection but, if necessary, it will then also be essential to assess the recovery of both the proenzyme and the modified, activated forms if results are to be meaningful. One possible way to control for partial loss of proteins during fractionation is to compare the relative amounts of 26 kDa to the 80 kDa catalytic subunits. When the enzymes are activated, the 26 kDa will decrease relative to the 78 kDa or 80 kDa catalytic subunits and thus, these may serve as an internal control to correct for the recovery of enzyme. After identifying the activated enzymes within cells the more complex issues such as (1) its possible membrane association and (2) the roles of the inhibitory protein and stimulatory protein in modulating enzyme activity can be examined. Additionally, identification of cellular conditions which activate the CANPs will also define conditions appropriate for continuing the search for bona fide physiological substrates.

B. RAT ERYTHROCYTES AS A MODEL CELL

Rat erythrocytes were chosen as a model system in which to examine intracellular proteolytic modification of CANP-1.[24] Erythrocytes are the only cell type known to contain one CANP (type 1 μCANP/CANP-1/CDP-1) but not type 2 (mCANP/CANP-2/CDP-2). Thus measuring the loss of 26 kDa and/or the appearance of the 17 kDa fragment will reflect proteolysis of the CANP-1. Erythrocytes from rats are reported to have a relative excess of protease with respect to its endogenous inhibitor.[28-30] This is in contrast to human erythrocytes where the inhibitory protein is present in excess of the protease.[28-31] The use of cells from rats should minimize the role of the inhibitory protein in altering protease activation and/or its subsequent activity. One disadvantage of using these cells is that there is no apparent change in the M_r of the catalytic subunit of the type 1 enzyme after autoproteolysis, a change that is detected for the human and bovine enzymes.[23-25]

1. Proteolytic Processing of Rat Erythrocyte CANP-1 in a Crude Soluble Fraction

Two preliminary experiments were performed to confirm the excess of CANP-1 relative to its endogenous inhibitor and to document autoproteolysis of rat CANP-1 *in vitro* in a crude soluble fraction.[24] In order to obtain a sample suitable for immunoblot analysis of the CANP-1 from erythrocytes it was necessary to separate hemoglobin from the protease. The protease and the endogenous inhibitor, but not hemoglobin, bind to DEAE cellulose at neutral pH. Step elution at 0.5 *M* NaCl recovers CANP-1 in both its proenzyme or activated form.[24] Assay of this fraction for Ca^{2+}-dependent caseinolytic activity (under appropriate assay conditions) confirmed the relative excess of CANP-1 to the inhibitory protein (data not shown). Immunoblot analysis of the CANP regulatory subunit in this fraction before and after the addition of Ca^{2+} demonstrated Ca^{2+}-dependent proteolytic precessing of the rat CANP-1 similar to that observed for the purified bovine CANP-2.[24] It is likely that even in the crude extract this processing is autocatalytic although this has not yet been proven.

2. Erythrocytes Treated with Ionomycin and Calcium

Incubation of erythrocytes with Ca^{2+} ionophores, such as ionomycin or A23187, in the presence of Ca^{2+} results in proteolysis of several prominent membrane skeletal proteins including ankyrin, spectrin, Band 3, and Band 4.1.[24,29-31] Several results suggest that this proteolysis is a result of CANP-1 activity. For example, proteolysis is inhibited by leupeptin and *N*-ethylmaleimide (known inhibitors of CANP) and cleavage patterns generated *in vivo* are similar to those produced by the action of purified CANP-1 on red blood cell ghost proteins *in vitro*.[24,30,31] The differences between the relative ratios of CANP-1 and its inhibitor protein in rat cells relative to human cells correlate with the extent of proteolysis observed in ionophore treated cells.[28,30,32] If CANP-1 is responsible for the observed proteolysis, the proteolytically modified CANP regulatory subunit should be detected in treated cells if such a modification occurs *in vivo*.

To directly measure the relative abundance of the proenzyme form of the CANP-regulatory subunit and its proteolytically modified form, we probed Western blots of red blood cell lysates with polyclonal antisera raised against CANP-2.[24] The use of antisera raised against CANP-2 minimizes the possibility of detecting fragments of the CANP-1 catalytic subunit which would complicate interpretation of these results. Probing erythrocyte lysates with antibody specific for the CANP-1 84 kDa confirmed that the quantified immunoblot-positive band is not a fragment of this protein and therefore must be the 17-kDa portion of the common 26-kDa regulatory subunit. If calcium was the signal for enzyme activation *in vivo* we expected that (1) increasing ionomycin concentration at a fixed extracellular Ca^{2+} concentration; or (2) increasing extracellular Ca^{2+} at a fixed ionophore concentration; and (3) increasing the time of incubation with ionomycin and Ca^{2+} would each result in an increase in the proteolytically modified subunit and corresponding loss of the 26 kDa. In each case the extent of conversion of the regulatory subunit was proportional to the increased alterations in Ca^{2+} homeostasis.[24] Appropriate controls demonstrated that preadsorption of antisera against purified CANP-2 eliminated 85 to 95% of the immunoreactive signal, and using antisera affinity purified against SDS-treated nitrocellulose-immobilized antigen gave results similar to that described for the whole serum.

3. A Comparison between CANP Activation in Rat Erythrocytes and Human Erythrocytes

Early studies by Lorand[32] and the Kosowers[29,30] described differences between the extent of proteolysis that occurs in human and rat erythrocytes in response to ionophores and Ca^{2+}. Proteolysis of the rat erythroid membrane skeleton proteins proceeds more quickly and to a greater extent than that observed in human erythrocytes treated with similar concentrations of ionophore and Ca^{2+}. Although there are undoubtedly other important species differences between these two cells, it is now clear that the relative abundance of CANP-1 to its inhibitory protein may contribute significantly to the difference in proteolysis. A more detailed study of proteolytic modification of CANP-1 within human red blood cells may provide important information with regard to the role of the inhibitory protein in either preventing CANP-1 activation and/or preventing activity of the activated protease. Contrasting the results from human and rat erythrocytes may also be significant in determining if the endogenous inhibitory protein is proteolyzed *in vivo,* as has been suggested by several investigators.[33,34]

C. PLATELET AGGREGATION: ACTIVATION OF CANP-1 AND CLEAVAGE OF TALIN (P235), FILAMIN (ACTIN-BINDING PROTEIN), AND FODRIN (SPECTRIN)

To date the best evidence for a specific role for calcium-dependent proteolysis catalyzed *in vivo* by either calpain is from studies of platelet aggregation (see Chapter 14).[10,11,25,26] Fox and Phillips described the appearance of two proteins in aggregated platelets and identified them as proteolytic fragments of filamin and talin. Considerable circumstantial evidence, including sensitivity to inhibitors, cleavage of these proteins by purified CANP-1 *in vitro* and the abundance of CANP-1 in platelet extracts, implicated the type-1 calcium-dependent protease in proteolysis of these proteins. Human platelets, like rat erythrocytes, have an excess of CANP-1 relative to the endogenous inhibitor protein[10] (Croall unpublished observation).[71] Platelets also contain CANP-2 but its quantity is perhaps only 5 to 10% that of CANP-1.[10,28] Recently two groups have provided direct evidence for proteolytic modification (implying activation) of the catalytic subunit of CANP-1 during platelet aggregation.[25,26] The catalytic subunit of human CANP-1 undergoes autocatalytic amino terminal processing upon exposure to Ca^{2+} which results in a detectable shift in M_r on SDS-PAGE (84 kDa $\rightarrow$ 78 kDa). Monoclonal antibodies raised against CDP-1 (by Elce[25]) were shown to be specific for its catalytic subunit. This antibody was used by Samis et al.[25] and by Okita et al.[26] to document proteolysis of the CDP-1 84 kDa and the appearance of the activated 76-78 kDa form during platelet aggregation. When aggregation was blocked, activation of CDP-1 was not observed,[25] in agreement with the lack of proteolysis under these conditions as described by Fox et al.[10,11] We have also recently documented the loss of the CANP regulatory subunit in platelets treated with ionomycin and calcium.[35] A separate or additional role for CANP-2 during platelet activation or aggregation cannot yet be excluded.

There are several consequences of the proteolysis of filamin including the generation of procoagulant activity,[36] the loss of GpIb-IX from the membrane skeleton fraction, and a decrease in the ability of GpIb-IX to bind to Von Willebrand factor.[37] Proteolysis of talin precedes its association with GpIIb-IIIa and the fractionation of these proteins with the Triton X-100 insoluble, actin-based membrane skeleton.[38] Presumably proteolysis of fodrin (spectrin)[11] has related consequences in altering the submembrane cytoskeleton and the transmembrane proteins associated with it.

D. OTHER CELL-TYPES TO EXAMINE FOR PROTEOLYTIC MODIFICATION OF THE SUBUNITS OF CALCIUM-DEPENDENT PROTEASES 1 AND 2

Clearly there are numerous cell types with well-described, Ca^{2+}-regulated processes which may be amenable to investigating the activation of calpains. Two additional models are suggested below (one for type 1 calpain and one for type 2). Each of these cell types has been studied previously with regard to the calcium-dependent proteases and Ca^{2+}-regulated signaling pathways.

1. Neutrophils

Neutrophils treated with the chemotactic peptide f-met-leu-phe (f-MLF) become activated through a Ca^{2+}-dependent signaling pathway.[39] From studies examining both f-MLF- and phorbol ester-treated neutrophils, Pontremoli et al. have proposed that a calcium-dependent protease (probably type 1) cleaves protein kinase C to its unregulated form, protein kinase M.[6,40-42] Additionally, phosphorylation of myosin light chains by protein kinase M is reported to increase degradation of the light chains, a process that may also be catalyzed by the endogenous calpain.[42] Thus with several known stimuli (f-MLF, phorbol esters, or ionomycin) and putative substrate proteins identified, this cell model should be appropriate for further study of calpain activation. Currently, however, there is some controversy as to the relative abundance and form of CANP-1 and 2 within these cells.[40-44] Pontremoli et al. have described the enzyme as the 85-kDa monomer with a Ca^{2+} requirement in between the typical type 1 and 2 enzymes.[40-42] Other groups however have described both heterodimeric proteases in this cell type.[43,44] The type 1 enzyme may be more abundant because approximately equal units of activity are recovered[43] and the type 1 enzyme has a lower specific activity against casein. It is not certain from previously published work if the inhibitory protein (calpastatin) is present in excess of the proteases in these cells, but this can readily be determined.

2. Smooth Muscle: Tissue and Cells

In smooth muscle, e.g., bovine aorta or trachealis, CANP-2 is abundant in comparison with the type 1 enzyme.[45,46] Calcium-dependent proteolytic activity is measurable in crude extracts of these tissues; thus CANP-2 is apparently in excess of its endogenous inhibitory protein. The regulation of Ca^{2+} homeostasis and its role in smooth muscle contraction have been studied extensively.[47-48] Thus we suggest that this model may be appropriate for determining if CDP-2 is activated in smooth muscle cells in response to any of the agonists known to alter intracellular Ca^{2+}. There are a number of potential physiological substrates to examine in this tissue, including myosin light chain kinase[49] and other calmodulin binding proteins,[50] myosin light chains, protein kinase C, as well as talin and filamin.

E. PROPOSED PHYSIOLOGICAL ROLES OF CANPs-1 AND 2

It is tempting to speculate that, as it does in platelets, CANP in other cells may also cleave proteins that link transmembrane proteins to the underlying membrane skeleton and thereby alter the cell surface. Cleavage of ankyrin by CANP-1 alters its ability to bind Band 3, the anion transporter of RBCs.[51] Ankyrin apparently interacts with Na^+,K^+-ATPase in other cell types.[52] How the proteolytically modified ankyrin interacts with Na^+,K^+-ATPase is not yet known. Talin, the protein which may link integrin type receptors with the actin cytoskeleton, is colocalized with CANP-2 in adhesion plaques in several cell types.[53] Cleavage of talin could modify its interaction with the fibronectin receptor *in vivo.* Conditions that activate CANP-2 and/or result in increased talin cleavage *in vivo,* however, have not yet been identified. A proteolytic fragment of talin, characteristic of CANP cleavage *in vitro,* is frequently observed in immunoprecipitates of talin from some cells.[54]

Several possible roles for CANP cleavage of fodrin/spectrin have been proposed, including membrane fusion,[29,30] stimulated exocytosis,[55,56] and alteration of neuronal cell surfaces in the generation of long-term potentiation.[52] Interestingly and perhaps importantly, the binding of Ca^{2+}-calmodulin to fodrin increases the susceptibility of fodrin to cleavage.[58,59] Furthermore, proteolysis of the α fodrin alters its ability to cross-link actin filaments.[59]

Nelson and Veshnock have recently described a model in which the role of CANP in the proteolysis of fodrin may be tested.[60,61] They have compared the degradation rates of newly synthesized ∝ fodrin in monolayers and cell contact-naive MDCK (Madin Darby canine kidney) epithelial cells. When the cells establish polarity the ∝ fodrin becomes insoluble in Triton X-100 and its half-life is increased four to sixfold ($>$ 70 h vs. 12 to 16 h previously.) By manipulating extracellular Ca^{2+} levels in the cell culture media, cell-cell contacts can be rapidly established or destroyed and the $t_{1/2}$ of newly synthesized fodrin is determined by the state of cell-cell contact during its synthesis.[60,61] As the protein linkages between transmembrane proteins and the submembrane cortical cytoskeleton become better defined, our ability to determine possible roles for calcium-dependent proteolysis in regulating these interactions will be greatly enhanced.

F. DOES Ca^{2+} REGULATE CANP ACTIVATION *IN VIVO*?

It is not clear from studies of platelet aggregation what the signals are for CANP-1 activation. Numerous experiments have documented a rapid increase in intracellular free Ca^{2+} in platelets treated with a variety of agonists.[62-67] The extent (maximum Ca^{2+} concentration reached) and duration of these changes is dependent upon the specific agonist used. The peak of $[Ca^{2+}]_i$ is reached prior to aggregation yet proteolysis of filamin or talin does not occur until much later.[10] In fact, proteolysis is not usually observed under conditions which prevent aggregation (e.g., absence of extracellular Ca^{2+} of lack of stirring.)[10,25] Interestingly, dibucaine, a local anesthetic that blocks aggregation of platelets, allows calcium-dependent proteolysis of talin and filamin to proceed in the absence of aggregation.[65] These observations suggest to us that either (1) some prior modification (e.g., phosphorylation) of talin or filamin must occur to make them susceptible to proteolysis *in vivo,* or (2) the activation of calpain -1 is stimulated by a Ca^{2+} signal that is different from the initial rise in $[Ca^{2+}]_i$, or (3) other factors regulate enzyme activation.

One possible explanation is that CANP-1 activation is linked to Ca^{2+} influx pathways. The initial spike of $[Ca^{2+}]_i$ is thought to result from the release of intracellular Ca^{2+} stores (from within the dense tubular system).[62-64] In platelets, and other cells as well, the sustained increase of Ca^{2+} results from increased Ca^{2+} influx from extracellular sources.[66] This suggestion is compatible with the knowledge that erythrocytes are the only cell to have just one CANP, i.e., CANP-1. More importantly, this hypothesis should be testable using the immunoblot analysis we have described to examine CANP activation. An obvious corollary to this hypothesis is that CANP-2 activation may be linked to the stimulated release of Ca^{2+} from intracellular stores. This too remains to be tested.

IV. SUMMARY AND CONCLUSIONS

From studies *in vitro* it has been established that the mammalian calcium-dependent proteases are synthesized and maintained intracellularly as inactive, heterodimeric proenzymes. In addition to binding Ca^{2+}, autoproteolytic processing of each subunit appears to be required for subsequent enzyme activity against, as yet unidentified, physiological substrates. To examine the regulation of enzyme activation *in vivo,* we have measured the disappearance of the proenzyme form of the regulatory subunit of these enzymes and the appearance of its proteolytically modified form. In rat erythrocytes, cells with exclusively type-1 CANP and low levels of the endogenous inhibitory protein (calpastatin), we have demonstrated a correlation between the appearance of proteolytically modified CANP-1 and conditions that increase intracellular Ca^{2+}. In human platelets, cells that also have predominantly CANP-1 and little inhibitory protein, Elce and colleagues[25] have documented the appearance of the proteolytically modified catalytic subunit of CANP-1 during platelet aggregation. These two model cells have thus provided the first direct evidence that this protease is proteolytically modified, and thereby activated, *in vivo*. Based on the currently available knowledge, we have proposed that activation of each calcium-dependent protease is regulated by specific Ca^{2+}-signals such as influx (e.g., via voltage regulated Ca^{2+}-channels) or the stimulated release of intracellular stores (calciosomes? [67]). Identification of the cellular conditions or signals that stimulate activation of either or both proteases will significantly enhance our ability to define physiological targets and functions of this proteolytic system.

REFERENCES

1. **DeMartino, G. N. and Blumenthal, D. K.,** Identification and partial purification of a factor that stimulates calcium-dependent proteases, *Biochemistry,* 21, 4297, 1982.
2. **Takeyama, Y., Nakanishi, H., Uratsuji, Y., Kishimoto, A., and Nishizuka, Y.,** A calcium-protease activator associated with brain microsomal-insoluble elements, *FEBS Lett.,* 194, 110, 1986.
3. **Pontremoli, S., Melloni, E., Michetti, M., Salamino, F., Sparatore, B., and Horecker, B. L.,** An endogenous activator of the Ca^{2+}-dependent proteinase of human neutrophils that increases its affinity for Ca^{2+}, *Proc. Natl. Acad. Sci. U.S.A.,* 85, 1988.
4. **DeMartino, G. N. and Croall, D. E.,** Purification and characterization of a protein inhibitor of calcium-dependent proteases from rat liver, *Arch. Biochem. Biophys.,* 232, 713, 1984.
5. **Pontremoli, S. and Melloni, E.,** Extralysosomal protein degradation, *Annu. Rev. Biochem.,* 55, 455, 1986.
6. **Pontremoli, S. Melloni, E., Damiani, G., Salamino, F., Sparatore, B., Michetti, M. and Horecker, B. L.,** Effects of a monoclonal anti-calpain antibody on responses of stimulated human neutrophils, *J. Biol. Chem.,* 263, 1015, 1988.
7. **Ishizaki, Y., Kurokawa, M., and Takahashi, K.,** A calcium-dependent protease associated with the neural cytoskeleton, *Eur. J. Biochem.,* 146, 331, 985.
8. **Mellgren, R. L.,** Calcium-dependent proteases: an enzyme system active at cellular membranes, *FASEB J.,* 1, 110, 1987.
9. **Mellgren, R. L., Lane, R. D., and Kakar, S. S.,** Isolated bovine myocardial sarcolemma and sarcoplasmic reticulum vesicles contain tightly bound calcium-dependent protease inhibitor, *Biochem. Biophys. Res. Commun.,* 142, 1025, 1987.
10. **Fox, J. E. B., Goll, D. E., Reynolds, C. C., and Phillips, D. R.,** Identification of two proteins (actin-binding protein and P235) that are hydrolyzed by endogenous Ca^{2+}-dependent protease during platelet aggregation, *J. Biol. Chem.,* 260, 1060, 1985.
11. **Fox, J. E. B., Reynolds, C. C., Morrow, J. S., and Phillips, D. R.,** Spectrin is associated with membrane-bound actin filaments in platelets and is hydrolyzed by the Ca^{2+}-dependent protease during platelet activation, *Blood,* 69, 537, 1987.
12. **Nelson, W. J. and Traub, P.,** Intermediate filament proteins and the calcium-activated proteinase specific for vimentin and desmin in the cells from fish to man: an example of evolutionary conservation, *J. Cell Sci.,* 57, 25, 1982.
13. **Zimmerman, U-J. P. and Schlaepfer, W.,** Characteristics of a brain calcium-activated protease that degrades neurofilament proteins, *Biochemistry,* 21, 3977, 1982.
14. **Klein, I., Lehotay, D., and Gondek, M.,** Characterization of a calcium-activated protease that hydrolyzes a microtubule-associated protein, *Arch. Biochem. Biophys.,* 2, 520, 1981.
15. **DeMartino, G. N. and Croall, D. E.,** Purification and characterization of a calcium-dependent protease from rat liver, *Biochemistry,* 22, 6287, 1983.
16. **DeMartino, G. N., Huff, C. A., and Croall, D. E.,** Autoproteolysis of the small subunit of calcium-dependent protease II activates and regulates protease activity, *J. Biol. Chem.,* 261, 12047, 1986.
17. **Coolican, S. A., Haiech, J., and Hathaway, D. R.,** The role of subunit autolysis in activation of smooth muscle Ca^{2+}-dependent proteases, *J. Biol. Chem.,* 261, 4170, 1986.
18. **Imajoh, S., Kawasaki, H., and Suzuki, K.,** Limited autolysis of calcium-activated neutral protease (CANP): reduction of the calcium requirement is due to the NH_2 terminal processing of the large subunit, *J. Biochem.,* 100, 633, 1986.
19. **Inomata, M., Hayashi, M., Nakamura, M., Imahori, K., and Kawashima, S.,** Hydrolytic and autolytic behavior of two forms of calcium-activated neutral protease (CANP), *J. Biochem.,* 98, 407, 1985.

20. **Mellgren, R. L., Repetti, A., Muck, T. C., and Easly, J.,** Rabbit skeletal muscle calcium-dependent protease requiring millimolar Ca^{2+} purification subunit structure and Ca^{2+}-dependent autoproteolysis, *J. Biol. Chem.*, 257, 7203, 1982.
21. **Suzuki, K., Tsuji, S., Ishiura, S., Kimura, Y., Kubota, S., and Imahori, K.,** Autolysis of calcium-activated neutral protease of chicken skeletal muscle, *J. Biochem.*, 90, 1787, 1981.
22. **Suzuki, K., Imajoh, S., Emori, Y., Kawasaki, H., Minami, Y., and Ohno, S.,** Calcium-activated neutral protease and its endogenous inhibitor, *FEBS Lett.*, 20, 271, 1987.
23. **Croall, D. E. and DeMartino, G. N.,** Comparison of two calcium-dependent proteinases from bovine heart, *Biochim. Biophys. Acta,* 788, 348, 1984.
24. **Croall, D. E.,** Proteolytic modification of calcium-dependent protease-1 in erythrocytes treated with ionomycin and calcium, *Biochemistry,* 28, 6882, 1989.
25. **Samis, J. A., Zboril, G., and Elce, J. S.,** Calpain I remains intact and intracellular during platelet activation, *Biochem. J.*, 246, 481, 1987.
26. **Okita, J. R., Wencel-Drake, J. D., and Kunicki, T. J.,** Redistribution of calpain I during thrombin-induced platelet activation, *J. Cell. Biol.*, 105, 290a, 1987.
27. **Suzuki, K., Imajoh, S., Emori, Y., Kawasaki, H., Minami, Y., and Ohno, S.,** Calcium-activated neutral protease and its endogenous inhibitor, *FEBS Lett.*, 220, 271, 1987.
28. **Murachi, T.,** Calpains and calpastatin, *TIBS,* 8, 167, 1983.
29. **Kosower, N. S., Glaser, T., and Kosower, E. M.,** Membrane mobility agent promoted fusion of erythrocytes: fusibility is correlated with attack of calcium-activated cytoplasmic proteases on membrane proteins, *Proc. Natl. Acad. Sci., U.S.A.*, 80, 7542, 1983.
30. **Glaser, T. and Kosower, N. S.,** Fusion of rat erythrocytes by membrane-mobility agent A_2C depends on membrane proteolysis by a cytoplasmic calpain, *Eur. J. Biochem.*, 159, 387, 1986.
31. **Croall, D. E., Morrow, J. S., and DeMartino, G. N.,** Limited proteolysis of the erythrocyte membrane skeleton by calcium-dependent proteinases, *Biochim. Biophys. Acta.*, 882, 287, 1986.
32. **Lorand, L., Bjerrum, D. J., Hawkins, M., Lowe-Krentz, L., and Siefring, G. E., Jr.,** Degradation of transmembrane proteins in Ca^{2+} enriched human erythrocytes, *J. Biol. Chem.*, 258, 5300, 1983.
33. **Mellgren, R. L., Mericle, M. T., and Lane, R. D.,** Proteolysis of the calcium-dependent protease inhibitor by myocardial calcium-dependent protease, *Arch. Biochem. Biophys.*, 246, 233, 1986.
34. **Nakamura, M., Inomata, M., Imajoh, S., Suzuki, K., and Kawashima, S.,** Fragmentation of endogenous inhibitor upon complex formation with high and low Ca^{2+} requiring forms of calcium-activated neutral proteases, *Biochemistry,* 28, 449, 1989.
35. **Croall, D. E.,** Regulation of calcium-dependent proteases: an assay for enzyme activation, *J. Cell. Biochem.*, 12b, 290, 1988.
36. **Verhallen, P. F. J., Bevers, E. M., Comfurius, P., and Zwaal, F. A.,** Correlation between calpain-mediated cytoskeletal degradation and expression of platelet procoagulant activity. A role for the platelet membrane-skeleton in the regulation of membrane lipid asymmetry?, *Biochim. Biophys. Acta,* 903, 206, 1987.
37. **Fox, J. E. B. and Boyles, J. K.,** The membrane skeleton: a distinct structure that regulates the function of cells, *Biol. Essays,* 8, 14, 1988.
38. **Isenberg, W., Fox, J. E. B., and Phillips, D. R.,** Platelet aggregation induces a redistribution of talin and the incorporation of talin and Gp IIb-IIIa into the actin cytoskeleton, *J. Cell. Biol.*, 107, 256a, 1988.
39. **Omann, G. M., Allen, R. A., Bokoch, G. M., Painter, R. G., Traynor, A. E., and Sklar, L. A.,** Signal transduction and cytoskeletal activation in the neutrophil, *Phys. Rev.*, 67, 285, 1987.
40. **Melloni, E., Pontremoli, S., Michetti, M., Sacco, O., Sparatore, B., Salamino, F., and Horecker, B. L.,** Binding of protein kinase C to neutrophil membranes in the presence of Ca^{2+} and its activation by Ca^{2+} by $Ca_2{}^{+}$-requiring proteinase, *Proc. Natl. Acad. Sci. U.S.A.*, 82, 6435, 1985.
41. **Melloni, E., Pontremoli, S., Michetti, M., Sacco, O., Sparatore, B., and Horecker, B. L.,** The involvement of calpain in the activation of protein kinase C in neutrophils stimulated by phorbol myristic acid, *J. Biol. Chem.*, 261, 4101, 1986.
42. **Pontremoli, S., Melloni, E., Michetti, M., Sparatore, B., Salamino, F., Sacco, O., and Horecker, B. L.,** Phosphorylation by protein kinase C of a 20 kDa cytoskeletal polypeptide enhances its susceptibility to digestion by calpain, *Proc. Natl. Acad. Sci. U.S.A.*, 84, 398, 1987.
43. **Fukui, I., Toyohara, H., Ito, K., Hamakubo, T., and Murachi, T.,** Molecular and catalytic characterization of intact heterodimeric and derived monomeric calpains isolated under different conditions from pig polymorphonuclear leukocytes, *Biochemistry,* 27, 3260, 1988.
44. **Legendre, J. L. and Jones, H. L.,** Purification and characterization of calpain from human polymorphonuclear leukocytes, *Inflammation,* 12, 51, 1988.
45. **Hathaway, D. R., Werth, D. K., and Haeberle, J. R.,** Limited autolysis reduces the Ca^{2+}-requirement of a smooth muscle Ca^{2+}-activated protease, *J. Biol. Chem.*, 257, 9072, 1982.
46. **Haeberle, J. R., Coolican, S. A., Evan, A., and Hathaway, D. R.,** The effects of a calcium-dependent protease on the ultrastructure and contractile mechanics of skinned uterine smooth muscle, *J. Muscle Res. Cell. Motil.*, 6, 347, 1985.

47. **Rasmussen, H., Takuwa, Y., and Park S.,** Protein kinase Ca^{2+} in the regulation of smooth muscle contraction, *FASEB J.*, 1, 177, 1987.
48. **Kamm, K. E. and Stull, J. T.,** The function of myosin and myosin light chain kinase phosphorylation in smooth muscle, *Annu. Rev. Pharmacol. Toxicol.*, 25, 593, 1985.
49. **Ito, M., Tanaka, T., Nunoki, K., Hidaka, H., and Suzuki, K.,** The Ca^{2+}-activated protease (calpain) modulates Ca^{2+}/calmodulin dependent activity of smooth muscle myosin light chain kinase, *Biochem. Biophys. Res. Commun.*, 145, 1321, 1987.
50. **Wallace, R. W., Tallant, E. A., and McManus, M. C.,** Human platelet calmodulin-binding proteins: identification and Ca^{2+}-dependent proteolysis upon platelet activation, *Biochemistry,* 26, 2766, 1987.
51. **Hall, T. G. and Bennett, V.,** Regulatory domains of erythrocyte ankyrin. *J. Biol. Chem.*, 262, 10549, 1987.
52. **Nelson, W. J. and Veshnock, P. J.,** Ankyrin binding to ($Na^{+}K^{+}$)ATPase and implications for the organization of membrane domains in polarized cells, *Nature,* 328, 533, 1987.
53. **Beckerle, M. C., Burridge, K., DeMartino, G. N., and Croall, D. E.,** Colocalization of calcium-dependent protease II and one of its substrates at sites of cell adhesion, *Cell,* 51, 569, 1987.
54. **O'Halloran, T. and Burridge, K.,** Purification of a 190 kDa protein from smooth muscle: relationship to talin, *Biochim. Biophys. Acta,* 869, 337, 1986.
55. **Perrin, D., Langley, O. K., and Aunis, D.,** Anti-$\propto$ fodrin inhibits secretion from permeabilized chromaffin cells, *Nature,* 326, 448, 1987.
56. **Burgoyne, R. D. and Cheek, T. R.,** Role of fodrin in secretion, *Nature,* 326, 448, 1987.
57. **Lynch, G. and Baudry, M.,** Brain spectrin, calpain and long-term changes in synaptic efficacy, *Brain Res. Bull.*, 18, 809, 1987.
58. **Harris, A. S., Croall, D. E., and Morrow, J. S.,** The calmodulin-binding site in $\propto$ fodrin is near the calcium-dependent protease-I cleavage site, *J. Biol. Chem.*, 263, 15754, 1988.
59. **Harris, A. S., Croall, D. E., and Morrow, J. S.,** Calmodulin regulates fodrin susceptibility to proteolysis by calcium-dependent protease-1, *J. Biol. Chem.*, 264, 17401, 1989.
60. **Nelson, W. J. and Veshnock, P. J.,** Dynamics of membrane skeleton (fodrin) organization during development of polarity in Madin-Darby canine kidney epithelial cells, *J. Cell. Biol.*, 103, 1751, 1986.
61. **Nelson, W. J. and Veshnock, P. J.,** Modulation of fodrin (membrane skeleton) stability by cell:cell contact in Madin-Darby canine kidney epithelial cells, *J. Cell. Biol.*, 104, 1527, 1987.
62. **Pollock W. K., Rink, T. J., and Irvine, R. F.,** Liberation of [^{3}H]arachidonic acid and changes in cytosolic free calcium in fura 2 loaded human platelets stimulated by ionomycin and collagen, *Biochem. J.*, 235, 869, 1986.
63. **Brass, L. F. and Joseph, S. K.,** A role for inositol triphosphate in intracellular Ca^{2+}-mobilization and granule secretion in platelets, *J. Biol. Chem.*, 260, 15172, 1985.
64. **Rink, T. J., Smith, S. W., and Tsien, R. Y.,** Cytoplasmic free Ca^{2+} in human platelets: Ca^{2+} thresholds and Ca^{2+} independent activation for shape change and secretion, *FEBS Lett.*, 148, 21, 1982.
65. **Fox, J. E. B.,** Platelet contractile proteins, in *Biochemistry of Platelets,* D. R. Phillips and M. A. Shuman, Eds., Academic Press, Orlando, FL, 1986, 116.
66. **Altin, J. G. and Bygrave, P. L.,** Second messengers and the regulations of Ca^{2+} fluxes by Ca^{2+} mobilizing agonists in rat liver, *Biol. Rev.*, 63, 551, 1988.
67. **Volpe, P., Krause, K-H., Hashimoto, S., Zorzato, F., Pozzan, T., Meldolesi, J., and Lew, D. P.,** "Calciosome" a cytoplasmic organelle: the inositol 1,4,5-triphosphate-sensitive Ca^{2+} store of nonmuscle cells? *Proc. Natl. Acad. Sci. U.S.A.*, 85, 1091, 1988.
68. **DeMartino, G. N. and Croall, D. E.,** unpublished observation.
69. **Croall, D. E.,** unpublished observation.
70. **Croall, D. E.,** unpublished observation.
71. **Croall, D. E.,** unpublished observation.
72. Unpublished observation.

Chapter 8

THE REGULATION OF CALPAINS BY INTERACTION WITH CALPASTATINS

Ronald L. Mellgren and Richard D. Lane

TABLE OF CONTENTS

I. INTRODUCTION

The calpains are proteolytic enzymes which coexist within animal cells with an infinite variety of potential substrates. The unregulated activity of the calpains would almost certainly result in serious damage to cells, if not cell death. It should not be surprising, then, that the calpains are subject to extremely tight regulation. The key regulatory elements defined thus far are a stringent requirement for high levels of free calcium ion and the presence of the highly specific and potent inhibitor proteins, the calpastatins. Indeed, calpain activity appears to be under such severe constraints that it is difficult to imagine how these proteases might be activated under any normal intracellular condition. In many cases, there is sufficient calpastatin to fully inhibit calpain activity in tissue extracts.[1] Nevertheless, the interactions of the calpains, calpastatin, and calcium ion in the intracellular environment are not yet fully appreciated, and it seems likely that the coordinated interplay of these three components is necessary to ensure a spatially and temporally controlled action of the calpains.

II. INTERACTION OF PURIFIED CALPAINS AND CALPASTATIN

As indicated in Chapter 3, calpastatins contain multiple copies of an inhibitory domain. Thus, the large forms of calpastatin (apparent M_r = 110 to 130 kDa on SDS-PAGE) contain four domains, each of which is capable of inhibiting calpains I or II.[2,3] The affinity of the calpastatin sites is sufficiently high to allow titration of calpain active sites by addition of stoichiometric amounts of calpastatin inhibitor domains.[4] Inhibition appears to be competitive, since low-affinity analogs of inhibitor domain 3 of porcine calpastatin, produced by site-directed mutagenesis, gave normal competitive inhibition characteristics on kinetic analysis.[5]

Calpastatins bind to calpains only in the presence of calcium ion. Recently, it has been shown that the calcium requirement for calpain II binding to calpastatin is close to the calcium requirement for its activation. On the other hand, calpain I activity is expressed at slightly lower concentrations of calcium ion than those required for binding to calpastatin (Chapter 1). Therefore, the latter calpain isozyme might be active in some cases in the presence of calpastatin.

Based on observations in a partially purified rat liver system, it was suggested early on that calpains might fragment calpastatin within the tertiary Ca^{2+}:calpain:calpastatin complex.[6] Later experiments with purified porcine calpains I and II and calpastatin confirmed cleavage of calpastatin by either of the calpains.[7] Calpastatins are excellent substrates for many proteolytic enzymes. One important consideration in the processing of calpastatin by calpains is the large amount of calpain required to initiate cleavage.[7] Substantial proteolysis of bovine myocardial calpastatin by calpain II,[8] or rabbit skeletal muscle calpastatin by calpain I or II[9] required at least a 1:1 mol ratio of calpain to calpastatin, i.e., occupation of at least one inhibitory domain. The rate of proteolysis was markedly increased by adding more calpain (Figure 1). In contrast, proteolysis of bovine myocardial calpastatin by trypsin was accomplished at much lower concentrations of the latter, and displayed a more usual dependence on trypsin concentration (Figure 1).

Another unusual feature of calpain-mediated calpastatin proteolysis is the relatively slow rate of cleavage. Even at a calpain to calpastatin mol ratio of 2:1, the half time for proteolysis of the large form of bovine myocardial calpastatin in the presence of 6 mM Ca^{2+} at 25°C was 30 min. In contrast, the same concentration of calpain II could completely degrade a 5-mg/ml solution of casein within 2 min.[8] These results suggest a highly unusual property of calpastatin: in a sense it can be regarded as a calpain substrate having extremely low K_m and V_{max} values.

Despite the evidence that calpastatins can be proteolyzed by calpains it is important to note that calpain cleavage, in contrast to degradation catalyzed by other proteases, results

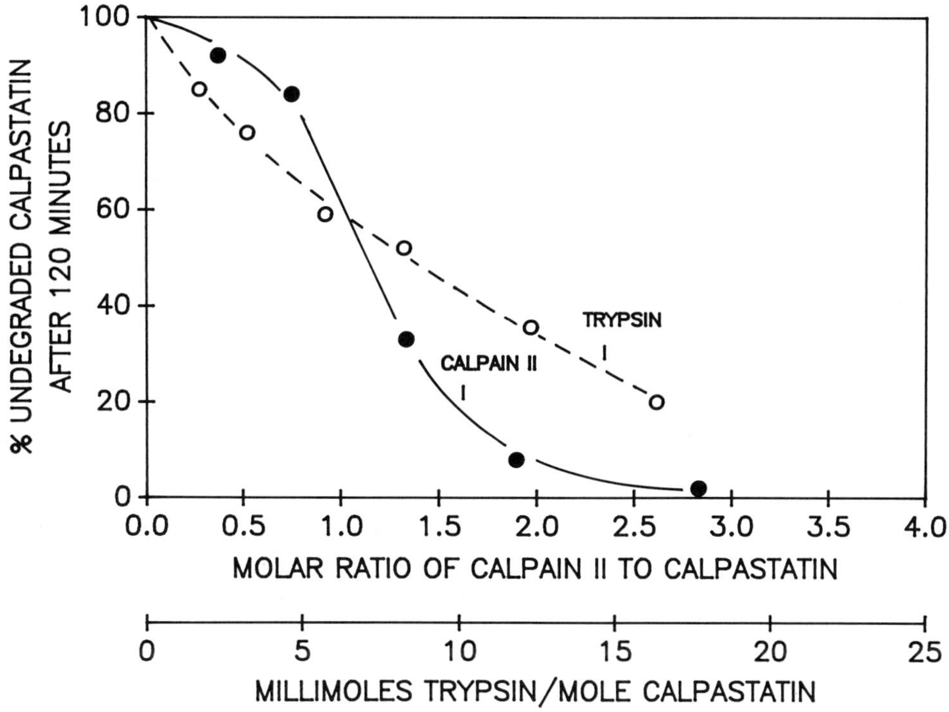

FIGURE 1. Proteolysis of bovine myocardial large calpastatin by calpain II or trypsin. Calpastatin was incubated for 2 h at 25°C with various amounts of calpain or trypsin and then subjected to SDS-PAGE, coomassie blue staining and densitometry to determine the amount of large (apparent M_r = 130,000 on SDS-PAGE) calpastatin remaining. (From data presented in Reference 8, but corrected for a calculated M_r of 81,000 for large calpastatin.[29])

in little or no loss of calpastatin inhibitory function. Presumably, the inhibitory domain is protected from calpain cleavage either by being intrinsically resistant to calpain hydrolysis, or by interacting strongly with a site on calpain which is adjacent to, but not part of, the catalytic domain. In the authors' laboratory, incubation of calpastatin with a tenfold molar excess of calpain II for 1 h, followed by heat-treatment to inactivate calpain, resulted in the subsequent recovery of about 50% of the initial inhibitory activity. This high ratio of calpain to calpastatin would not be likely to occur in most cells.

The features of calpain-mediated proteolysis of calpastatin outlined above suggest a highly unusual interaction between the protease and its specific inhibitor protein. The potential physiologic significance of this arrangement will be discussed later.

III. INTERACTION OF CALPAINS AND CALPASTATIN WITHIN CELLS

A full appreciation of the regulation of calpains by calpastatins requires an understanding of how these proteins interact within the intracellular environment. It is now fully established that calpastatin[10-12] and both forms of calpain are intracellular proteins.[11-16] The next question might well be: where within the cell are these different components localized? This question has been much more difficult to answer, and there is still considerable controversy regarding the localization of calpains and calpastatins in various cells.

Neither calpains nor calpastatins appear to be segregated inside intracellular organelles, and they are at least partially soluble within the cytosol. Some immunohistochemical studies have indicated that nearly all of the calpain immunoreactivity in bovine skeletal muscle,[13] porcine kidney,[17] rat liver,[18] and rat submandibular gland[18] appeared as diffuse staining in the cytoplasm. In other studies, however, there is evidence for localization of some calpain at the plasmalemma,[19-21] fibroblast substrate attachment plaques in various cell lines,[22] muscle cell I-band and Z-disk,[15,19,23] and cytoskeleton of nerve cells.[24] Although it is clearly necessary to know the intracellular localization of calpastatin as well as the calpains in order to predict the likely sites of calpain action, few studies have been conducted on localization of calpastatins. The authors' laboratory has investigated calpastatin localization in bovine[10] and canine[24,25] cardiac muscle, and found evidence for the association of calpastatin with membranes and intracellular organelles.

IV. EVIDENCE FOR THE ASSOCIATION OF CALPASTATINS WITH MYOCARDIAL MEMBRANES

Calpastatin activity is readily extractable from various tissues. As a result, it is tempting to assume that this protein is freely soluble and diffusible within the intracellular environment. However, the ease of extractability of proteins need not indicate their true disposition within the cell. For example, the cytoskeletal protein vinculin plays a key role in association of actin filaments with plasmalemma.[26] While vinculin has been demonstrated to bind to acidic phospholipids *in vitro,* this association is weak at physiologic ionic strength.[27] Spectrin is another cytoskeletal protein which can be removed from erythrocyte plasmalemma at low ionic strength.[28] Moreover, we have previously observed that a significant fraction of bovine or canine myocardial calpastatin is associated with sedimentable material in tissue homogenates.[29] In the case of rapidly prepared homogenates of freshly collected canine ventricle tissue, there appeared to be a larger fraction of sedimentable calpastatin relative to calpain[30] (Figure 2).

A. CALPASTATIN IN SARCOLEMMA AND SARCOPLASMIC RETICULUM.

Early immunofluorescence studies with bovine heart samples indicated that some of the calpastatin in myocardial myocytes was associated with the myofibril I-band and was near the sarcolemma.[10] Because of the limitations of the immunofluorescence technique, it was not possible to determine whether the calpastatin was actually associated with the sarcolemma. Further studies utilizing isolated bovine myocardial sarcolemma preparations demonstrated the association of calpastatin immunostaining material with the sarcolemma marker, oaubain-sensitive Na,K-ATPase[31,32] (Figure 3). Neither calpain activity nor calpain immunoreactivity was detectable in these preparations. Calpastatin was also associated with bovine myocardial sarcoplasmic reticulum (SR), although half of the endogenous calpastatin in this case was soluble calpastatin entrapped within the SR vesicles. In contrast, when sarcolemma vesicles were ruptured by repeated cycles of freezing and thawing, all of the endogenous calpastatin remained membrane associated and sedimentable.

The sarcolemma-associated calpastatin was capable of inhibiting the caseinolytic activity of exogenously added calpains I or II.[32] Freeze-thaw treatment of the vesicles resulted in a fivefold increase in inhibitory activity. Since the vesicles were predominantly right-side-out, the above results indicated that the calpastatin was associated with the inner face of the sarcolemma, i.e., in proper position to inhibit the intracellular calpains. In addition to decreasing calpain activity on the soluble substrate, casein, the sarcolemma-associated calpastatin could inhibit proteolysis of sarcolemma-associated protein kinase C.

B. MECHANISM OF CALPASTATIN BINDING TO MEMBRANES.

All of the SR-associated calpastatin, and about 50% of the sarcolemma-associated calpastatin could be removed by freeze-thaw treatment coupled with high salt washes. This indicated that ionic interactions were important in the association of calpastatin with mem-

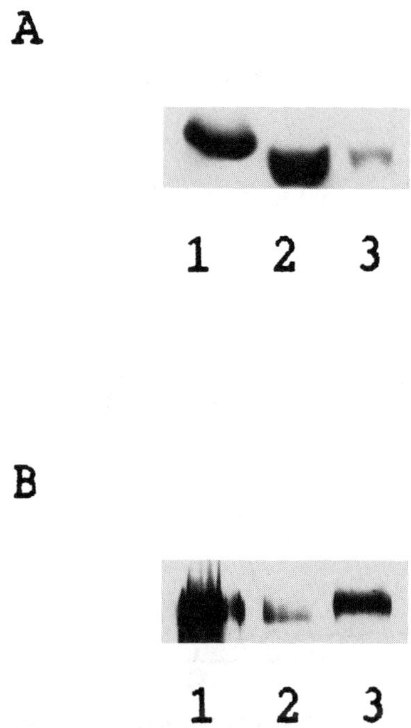

FIGURE 2. Differential distribution of calpains and calpastatin in canine myocardium. Samples of heart were collected within a few seconds post-mortem, and homogenized in a buffer of ionic strength 0.13. Following centrifugation at 10,000 × g, samples of supernatant were subjected to Western blot analysis for calpain small subunit (panel A, lane 2) or calpastatin (panel B, lane 2). The pellet was extracted with buffer at ionic strength = 0.53, and samples were immunostained for calpain (panel A, lane 3) or calpastatin (panel B, lane 3). Standard purified antigens were loaded in lane 1 of panels A and B. (From Mellgren, R. L., Lane, R. D., and Mericle, M. T., *Biochim. Biophys. Acta,* 999, 71, 1989. With permission.)

branes. Since calpastatins contain relatively small amounts of hydrophobic residues and are rich in charged residues, it was logical to postulate that the interaction with membranes might be through ionic bonding. Consequently, the binding of purified calpastatins to the polar phospholipids was investigated.[30,33] It was discovered that purified bovine myocardial calpastatin would bind to either of the two acidic phospholipids, phosphatidylserine (PS), or phosphatidylinositol (PI). In contrast, there was no detectable binding to the neutral phospholipids, phosphatidylethanolamine and phosphatidylcholine[30] (Figure 4). Canine myocardial and rabbit liver calpastatins would also bind to the acidic phospholipids. As predicted from the studies with biologic membranes, binding of myocardial calpastatin to the acidic phospholipids was highly dependent on ionic strength.[33] Treatment of sarcolemma with phospholipase C greatly diminished its capacity for binding myocardial calpastatin. Thus, it appears that binding of calpastatins to biologic membranes is at least partly dependent on association with acidic phospholipids.

Human and canine erythrocyte ghost membranes (plasmalemma) did not contain detectable calpastatin activity or immunoreactivity.[33] This was initially puzzling since, as discussed above, both bovine and canine sarcolemma appeared to have associated calpastatin.

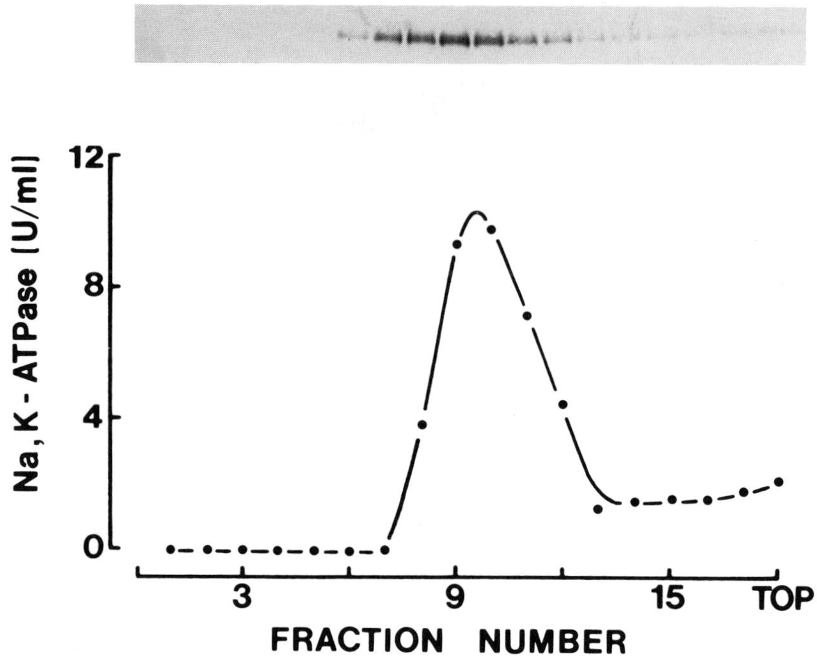

FIGURE 3. Co-migration of calpastatin and ouabain-sensitive Na,K-ATPase in isolated bovine myocardial sarcolemma. Isopycnic sucrose density gradient ultracentrifugation was performed on a sample of sarcolemma. The gradient was fractionated and assayed for ATPase, or subjected to Western blot analysis for calpastatin. (From Mellgren, R. L., Lane, R. D., and Kakar, S. S., *Biochim. Biophys. Acta,* 999, 71, 1989. With permission.)

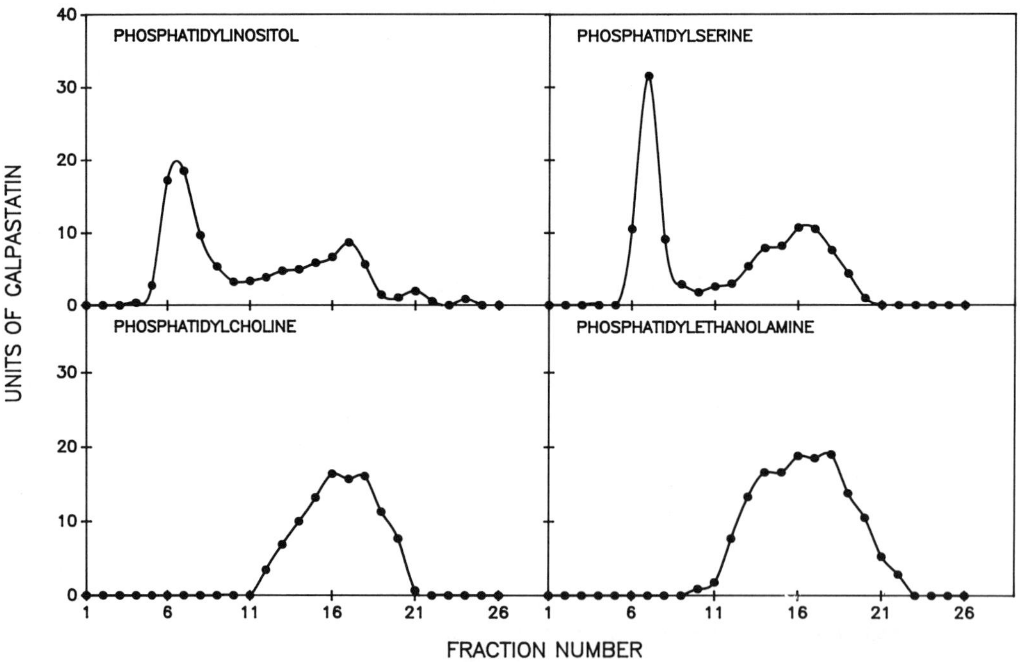

FIGURE 4. Binding of purified bovine myocardial calpastatin to phospholipids. Samples of calpastatin were incubated with a twofold weight excess of phospholipid, and chromatographed at room temperature on a column of Sepharose CL-2B. The column void volume and the position of migration of phospholipid-bound calpastatin were at tube 7. The peak position of calpastatin applied in the absence of phospholipids was at tube 16. (From Mellgren, R. L., Lane, R. D., and Mericle, M. T., *Biochim. Biophys. Acta,* 999, 71, 1989. With permission.)

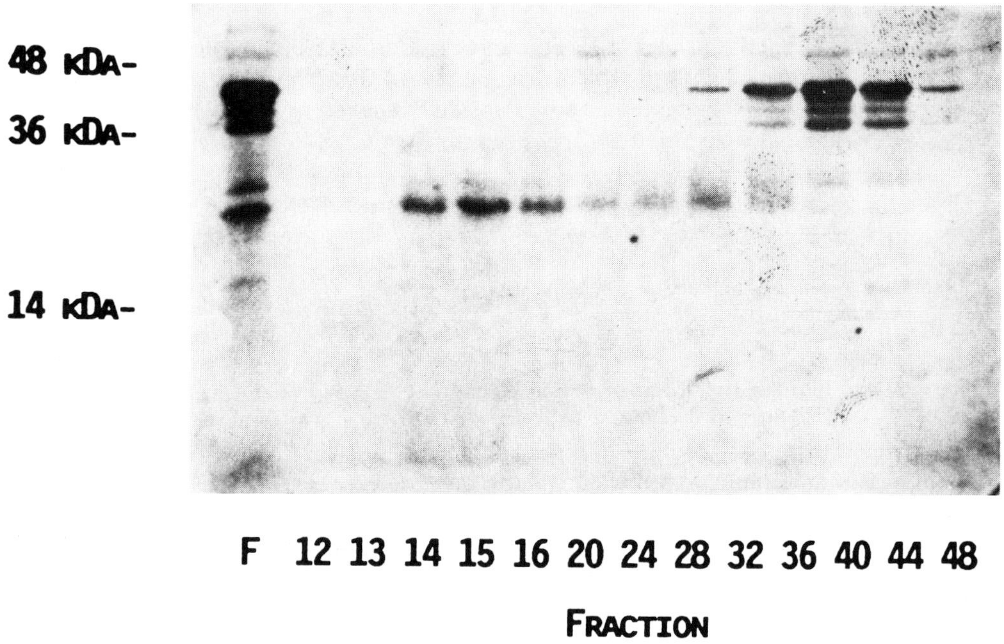

FIGURE 5. Isolation of a phospholipid-binding fragment from bovine myocardial calpastatin. A sample of calpastatin was digested with CNBr, and then mixed with phosphatidylinositol (PI), and subjected to gel filtration chromatography on Sepharose CL-2B as described in Figure 4. Fractions were then electrophoresed in a SDS-PAGE slab gel and stained with coomassie blue. The column void volume, containing calpastatin fragments bound to PI, was at fraction 15. (From Mellgren, R. L., Lane, R. D., and Mericle, M. T., *Biochim. Biophys. Acta,* 999, 71, 1989. With permission.)

It did not seem likely that the plasmalemma of the erythrocyte would differ greatly in phospholipid composition relative to heart sarcolemma. The other possiblity was that the calpastatin from erythrocytes did not bind well to membranes. Erythrocyte calpastatin is smaller than the calpastatin purified from most other cells,[34] and it could be missing a phospholipid binding region that is present in the larger forms. Indeed, calpastatin purified from human erythrocytes did not bind to PI and PS vesicles under the same experimental conditions used to demonstrate binding of the large calpastatin forms to these phospholipids.[33]

Rabbit erythrocyte calpastatin differs from rabbit liver calpastatin by lacking an amino-terminal sequence present in the latter.[35] It therefore seemed likely that the large calpastatin forms contained a specific region at or near the amino-terminal end which bound acidic phospholipids. To investigate this possibility, bovine myocardial calpastatin was fragmented with CNBr, and the mixture of fragments was subjected to Sepharose CL-2B chromatography in the presence of PI vesicles[30] (Figure 5). An approximately 20-kDa fragment was specifically bound to PI. It was identified as the amino-terminal fragment by partial sequence analysis and comparison with the predicted amino acid sequences of pig and rabbit calpastatin as determined from the sequences of the cDNAs (Table I).

V. IMMUNOELECTRON MICROSCOPIC STUDIES ON CALPAIN AND CALPASTATIN LOCALIZATION IN CANINE MYOCARDIUM

Using antimouse IgG-coated gold particles and specific antibodies against canine calpain large and small subunits and calpastatin, it was possible to investigate the localization of

TABLE 1
Amino Terminal Sequences of Rabbit and Pig Calpastatins Predicted by cDNA Sequences, Compared with the Partial Sequences of Cow Calpastatin Phospholipid-Binding CNBr (CB) Fragment, and a Peptide Prepared by Subdigesting the CB Fragment with *Staphylococcus aureus* Strain V8 Proteinase (V8).

	1 15		54 66
Rabbit	MNPAEAKAVPISKEM		GKPKEHTKPKSTH
	1 15		55 67
Pig	MNPTETKAIPVSKOL		VKPKEHPEPKSLP
Cow CB	GPTEAKAIPGSKOL	Cow V8	VKPKEHSEPKXLP

Note: **Bold** Residues: conserved in rabbit, pig and cow.
Underlined residues: conserved in pig and cow.

the proteinases and inhibitor in fresh canine heart.[39] Control experiments were performed by eliminating the first antibody incubation, or substituting nonimmune mouse IgG for the monoclonal antibodies. Fixation protocols were chosen which allowed ultrastructural preservation while maintaining antigenicity. For calpain immunolabeling, it was possible to fix the tissue samples with 1% paraformaldehyde and 4% glutaraldehyde. This produced excellent tissue preservation and allowed good specific labeling of calpain (Figure 6, panel A). For calpastatin immunolabeling, it was necessary to use a lighter fixative (0.5% paraformaldehyde) to allow preservation of calpastatin immunogenicity. Because of this, the tissue morphology was less well preserved (Figure 7, panel A). Nevertheless, it was possible to recognize binding of calpastatin to specific intracellular compartments. With either labeling protocol, there was little nonspecific labeling (Figures 6 and 7, panel B). Neither calpain nor calpastatin immunoreactivity was associated specifically with any one particular subcellular fraction. It was therefore necessary to perform morphometric analysis to establish the relative distribution of the two antigens in the different subcellular compartments (Figure 8). The results indicated that both calpains and calpastatin are located at many different sites in the cell, and that, in general, the patterns of distribution are much the same. There was surprisingly little calpain or calpastatin detected in the cytosolic compartment in these studies. However, it is impossible to rule out the potential loss of soluble (cytosolic) antigen during washing of the samples on the electron microscope sample grids. The important point from these studies was that calpastatin could be detected on the various organelles by the immunogold procedure. The binding of calpastatin to many of these sites, especially the SL, SR t-tubules, and intercalated disk, could be due in part to association with phospholipids.

VI. A MODEL FOR THE INTERACTION OF CALPAINS AND CALPASTATIN AT CELLULAR MEMBRANES

Since the discovery of the calpains, one of the major problems in determining how these enzymes function within the cell has been the extremely high concentrations of calcium required for their activation. This is especially troublesome when considering the activation of calpain II, a process which requires m*M* calcium concentrations in the test tube. A possible solution to this dilemma is that calpain II is exported out of the cell and functions as an extracellular protease. However, the calpains do not appear to be stored in secretory vesicles. Moreover, the carbohydrate groups characteristic of secreted proteins have not been detected in isolated calpains. It is conceivable that calpains are released from cells after cell death and serve some as yet undefined autacoid function. In this respect, it is of interest that some of the effects of calpains on platelet function may be brought about by their action on proteins located on the outside of the platelet plasmalemma (Chapter 15). However, there is little

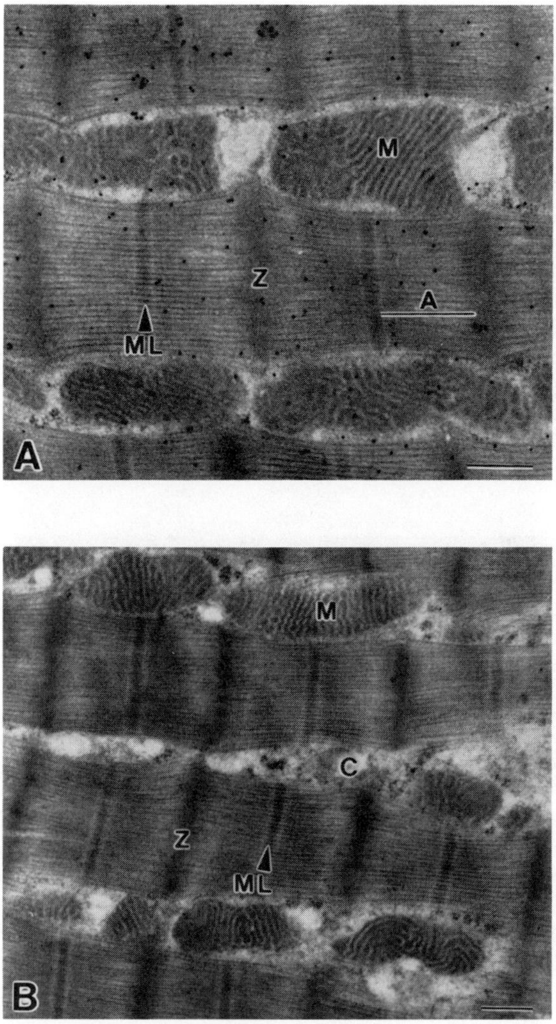

FIGURE 6. Immunolocalization of calpains in canine myocardium. Samples of canine myocardial papillary muscle were immunolabeled with antimouse IgG-coated gold particles, and a mixture of monoclonal antibodies for calpain large and small subunits (panel A). Panel B shows control immunolabeling (no added monoclonal antibodies). M = mitochondria, Z = muscle Z-line, ML = M-line, A = A-band.

evidence to suggest an extracellular function for calpains in other systems. If the calpains function inside cells, there must be some mechanism for greatly decreasing their calcium requirements. To date, the best evidence for such a mechanism involves binding of calpains to intracellular membranes, followed by an autoactivation process which results in a decreased calcium requirement.[34,36,37] The view that interaction of calpains with membranes is important in calpain function is strengthened by the observation that some of the best substrates for calpains are membrane-associated proteins.[38]

Another problem which arises in considering the activation of calpains in cells is the presence of endogenous calpastatin. Clearly, there must be a physical or functional separation of calpain and calpastatin activity in order for calpain to have significant activity in most cells. In this chapter, we have presented evidence for the association of calpastatin with

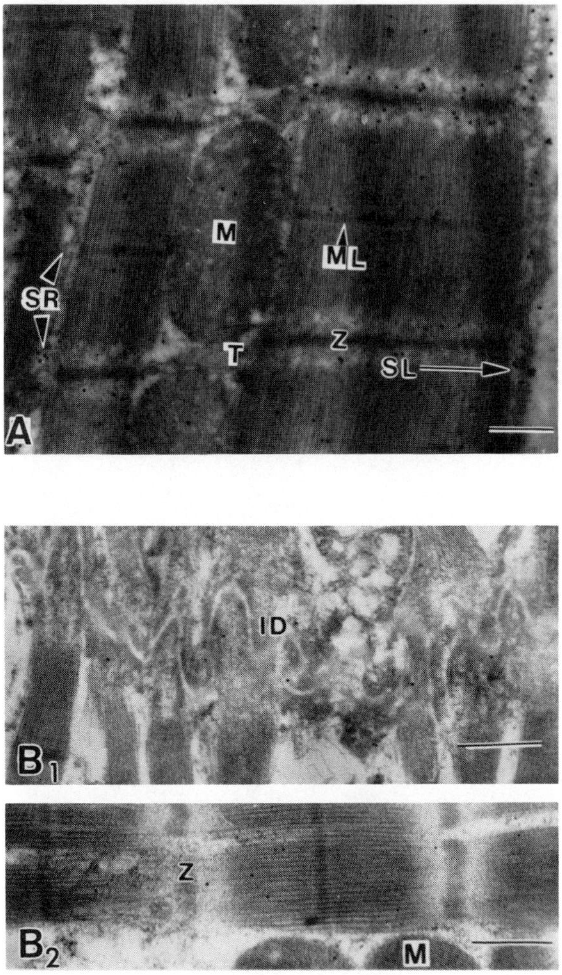

FIGURE 7. Immunolocalization of calpastatin in canine myocardium. Samples of papillary muscle were immunolabeled as in Figure 6, except that a mixture of monoclonal antibodies for canine calpastatin was employed as first antibody.

membranes. This would seem to place yet one more constraint on intracellular activity of the calpains. Specifically, if calpain I or II within a cell should associate with membranes prior to autoactivation, our studies suggest that there is membrane-associated calpastatin which is capable of inhibiting either isozyme. There is at least one scenario which would allow calpain to function at membranes in the presence of localized calpastatin: the possibility that calpastatin is less likely than calpains to diffuse to localized sites of high calcium concentrations near cell membranes. This would produce an imbalance of calpain/calpastatin and allow localized proteolysis. Some of our observations are consistent with this possibility. For instance, as noted above (Figure 2), in freshly prepared samples from canine heart, much of the calpastatin is associated with particulate material, while calpain immunostaining is predominantly extractable. It is also noteworthy that we have not detected calpain immunostaining in Western blots of SL or SR preparations, while calpastatin is readily detectable. On the other hand, the electron microscopic immunolocalization studies seem to indicate that much of the calpastatin and calpain are associated with various intracellular organelles. The possibility that much of the cytosolic calpain and calpastatin immunoreac-

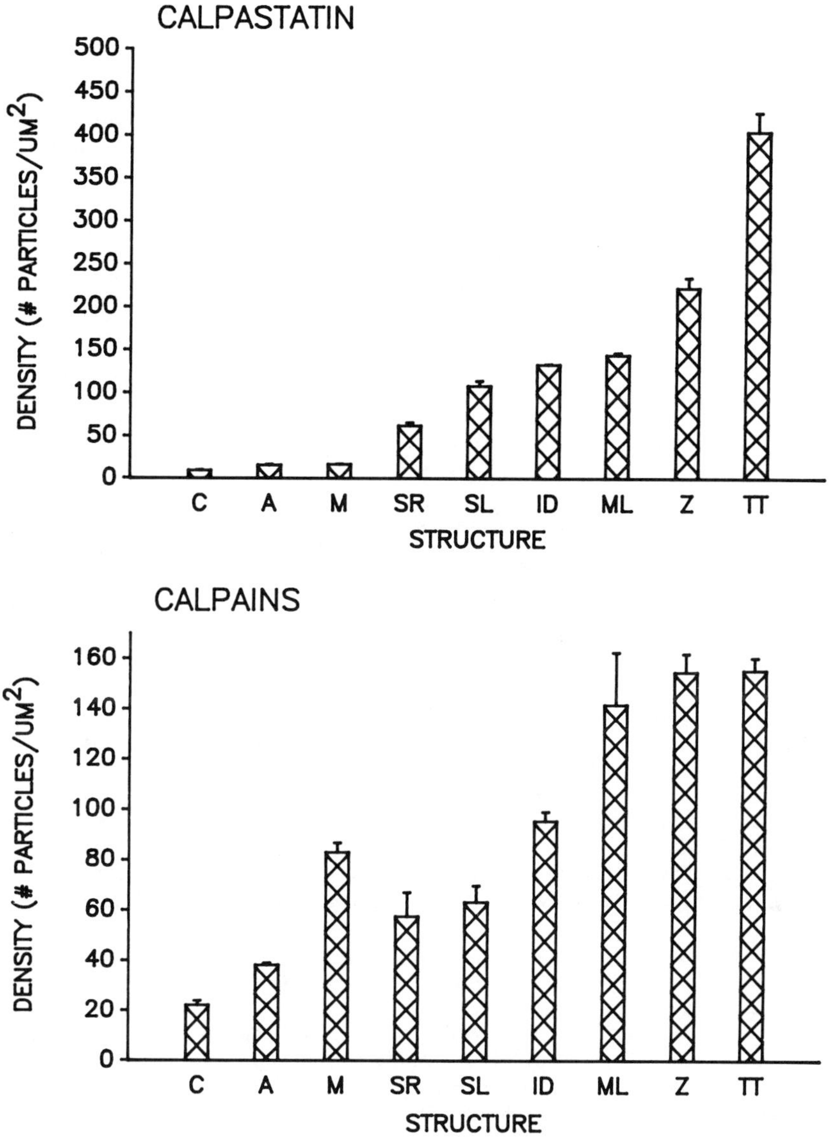

FIGURE 8. Density of labeling of canine myocardium for calpains and calpastatin in various compartments: C = cytosol, A = A-band, M = mitochondria, SR = sarcoplasmic reticulum, SL = sarcolemma, ID = intercalated disk, ML = M-line, Z = Z-disk, TT = transverse tubules.

tivity could be underestimated by this procedure has been discussed above. Therefore, the immunoelectron microscopy can only provide evidence for the presence of organelle-associated calpain and calpastatin, not for the absence of either protein in the cytosol, and the question of the relative distribution of the proteases and inhibitor between organelles and cytosol is not properly addressed by the electron microscopy studies.

Another potential factor in the regulation of calpain activity at membranes is the ability of calpains to proteolyze calpastatin (Figure 1). A single hydrolysis near the amino terminal of calpastatin could effectively remove it from membranes, and it is conceivable that localized calpain/calpastatin balance could be increased in this manner. The presence of multiple calpastatin forms, presumably proteolytic fragments, in fresh tissue samples is well documented,[8,39] and could be the result of calpain action on the inhibitor.[8]

In summary, the studies presented above suggest a mechanism for temporally and spatially controlled proteolysis at intracellular membranes in the presence of calcium. While the hypothetical interactions of calpains and calpastatin at biologic membranes are far from established at this time, our studies indicate that this is a potentially important mechanism for regulating the interaction between these proteases and their specific inhibitor.

ACKNOWLEDGMENTS

This work was supported in part by USPHS grant 36573 and was performed during the tenure of an American Heart Association Established Investigatorship to RM.

REFERENCES

1. **Murachi, T.,** Calcium-dependent proteinases and specific inhibitors: calpain and calpastatin, *Biochem. Soc. Symp.*, 49, 149, 1984.
2. **Emori, Y., Kawasaki, H., Imajoh, S., Imahori, K., and Suzuki, K.,** Endogenous inhibitor for calcium-dependent cysteine protease contains four internal repeats that could be responsible for its multiple reactive sites, *Proc. Natl. Acad. Sci. U.S.A.*, 84, 3590, 1987.
3. **Takano, E., Park, Y.-H., Kitahara, A., Yamagata, Y., Kannagi, R., and Murachi, T.,** Distribution of calpains and calpastatin in human blood cells, *Biochem. Int.*, 16, 391, 1988.
4. **Takahashi-Nakamura, M., Tsuji, S., Suzuki, K., and Imahori, K.,** Purification and characterization of an inhibitor of calcium-activated neutral protease from rabbit skeletal muscle, *J. Biochem.*, 90, 1583, 1981.
5. **Maki, M., Takano, E., Osawa, T., Ooi, T., Murachi, T., and Hatanaka, M.,** Analysis of structure-function relationships of pig calpastatin by expression of mutated cDNAs in *Escherichia coli, J. Biol. Chem.*, 263, 10254, 1988.
6. **Murachi, T., Tanaka, K., Hatanaka, M., and Murakami, T.,** Intracellular Ca^{2+}-dependent protease (calpain) and its high-molecular-weight endogenous inhibitor (calpastatin), *Adv. Enzyme Regul.*, 19, 407, 1981.
7. **Goll, D. E., Shannon, J. D., Edmunds, T., Sathe, S. K., Kleese, W. C., and Nagainis, P. A.,** Properties and regulation of the Ca^{2+}-dependent proteinase, in *Calcium-Binding Proteins 1983,* de Bernard, B., et al., Eds., Elsevier, Amsterdam, 1983, 19.
8. **Mellgren, R. L., Mericle, M. T., and Lane, R. D.,** Proteolysis of the calcium-dependent protease inhibitor by myocardial calcium-dependent protease, *Arch. Biochem. Biophys.*, 246, 233, 1986.
9. **Nakamura, M., Inomata, M., Imajoh, S., Suzuki, K., and Kawashima, S.,** Fragmentation of an endogenous inhibitor upon complex formation with high- and low-Ca^{2+}-requiring forms of calcium-activated neutral proteases, *Biochemistry,* 28, 449, 1989.
10. **Lane, R. D., Mellgren, R. L., and Mericle, M. T.,** Subcellular localization of bovine heart calcium-dependent protease inhibitor, *J. Mol. Cell Cardiol.*, 17, 863, 1985.
11. **Kleese, W. C., Goll, D. E., Edmunds, T., and Shannon, J. D.,** Immunofluorescent localization of the Ca^{2+}-dependent proteinase and its inhibitor in tissues of *Crotalus atrox, J. Exp. Zool.*, 241, 277, 1987.
12. **Kitahara, A., Takano, E., Ohtsuki, H., Kirihata, Y., Yamagata, Y., Kannagi, R., and Murachi, T.,** Reversed distribution of calpains and calpastatin in human pituitary gland and selective localization of calpastatin in adrenocorticotropin-producing cells as demonstrated by immunohistochemistry, *J. Clin. Endocrinol. Metab.*, 63, 343, 1986.
13. **Goll, D. E., Edmunds, T., Kleese, W. C., Sathe, S. K., and Shannon, J. D.,** Some properties of the Ca^{2+}-dependent proteinase, *Prog. Clin. Biol. Res.*, 180, 151, 1985.
14. **Murachi, T., and Yoshimura, N.,** Intracellular localization of low and high calcium-requiring forms of calpain, *Prog. Clin. Biol. Res.*, 180, 165, 1985.
15. **Ishiura, S., Sugita, H., Nonaka, I., and Imahori, K.,** Calcium-activated neutral protease: its localization in the myofibril, especially at the Z-band, *J. Biochem.*, 87, 343, 1980.
16. **Kamakura, K., Ishiura, S., Fujita, T., Nonaka, I., and Sugita, H.,** Localization of calcium-activated neutral protease (CANP) in the peripheral nerve, *Muscle & Nerve,* 8, 380, 1985.
17. **Yoshimura, N., Hatanaka, M., Kitahara, A., Kawaguchi, N., and Murachi, T.,** Intracellular localization of two distinct Ca^{2+}-proteases (calpain I and calpain II) as demonstrated by using discriminative antibodies, *J. Biol. Chem.*, 259, 9847, 1984.

18. **Yoshimura, N., Ohtsuki, H., Hamakubo, T., Kitahara, A., Kannagi, R., and Murachi, T.,** Distribution of calpain in various organs of the rat — an immunohistochemical study, *Biomed. Res.*, 5, 419, 1984.
19. **Dayton, W. R. and Schollmeyer, J. E.,** Immunocytochemical localization of a calcium-activated protease in skeletal muscle cells, *Exp. Cell Res.*, 136, 423, 1981.
20. **Hatanaka, M., Yoshimura, N., Murakami, T., Kannagi, R., and Murachi, T.,** Evidence for membrane-associated calpain I in human erythrocytes. Detection by an immunoelectrophoretic blotting method using monospecific antibody, *Biochemistry*, 23, 3272, 1984.
21. **Schollmeyer, J. E.,** Role of Ca^{2+} and Ca^{2+}-activated protease in myoblast fusion, *Exp. Cell Res.*, 162, 411, 1986.
22. **Beckerle, M. C., Burridge, K., DeMartino, G. N., and Croall, D. E.,** Colocalization of calcium-dependent protease II and one of its substrates at sites of cell adhesion, *Cell*, 51, 569, 1987.
23. **Yoshimura, N., Murachi, T., Heath, R., Kay, J., Jasani, B., and Newman, G. R.,** Immunogold electron-microscopic localisation of calpain I in skeletal muscle of rats, *Cell Tissue Res.*, 244, 265, 1986.
24. **Mellgren, R. L., Renno, W. M., and Lane, R. D.,** The nonlysosomal, calcium-dependent proteolytic system of mammalian cells, in *Current Trends in the Study of Intracellular Protein Degradation*, Knecht, E. and Grisolia, S., Eds., Springer, International, Vizcaya, Spain, 1989, 140.
25. **Renno, W. M., Mellgren, R. L., and Lane, R. D.,** unpublished data, 1989.
26. **Fuchtbauer, A., and Jockusch, B. M.,** Vinculin as a functional component of focal contacts in tissue culture cells, *J. Submicrosc. Cytol.*, 16, 109, 1984.
27. **Niggli, V., Dimitrov, D. P., Brunner, J., and Burger, M. M.,** Interaction of the cytoskeletal component vinculin with bilayer structures analyzed with a photoactivatable phospholipid, *J. Biol. Chem.*, 261, 6912, 1986.
28. **Tyler, J. M., Hargreaves, W. R., and Branton, D.,** Purification of two spectrin-binding proteins: biochemical and electron microscopic evidence for site-specific reassociation between spectrin and bands 2.1 and 4.1, *Proc. Natl. Acad. Sci. U.S.A.*, 76, 5192, 1979.
29. **Mellgren, R. L., Lane, R. D., Hegazy, M. G., and Reimann, E. M.,** The specific inhibitor of calcium-dependent proteases. A heat-stable phosphoprotein with properties similar to the inhibitors of phosphoprotein phosphatases, *Adv. Prot. Phosphatases*, 1, 259, 1985.
30. **Mellgren, R. L., Lane, R. D., and Mericle, M. T.,** The binding of large calpastatin to biologic membranes is mediated in part by interaction of an amino terminal region with acidic phospholipids, *Biochim. Biophys. Acta*, 999, 71, 1989.
31. **Mellgren, R. L., Lane, R. D., and Kakar, S. S.,** Isolated bovine myocardial sarcolemma and sarcoplasmic reticulum vesicles contain tightly bound calcium-dependent protease inhibitor, *Biochem. Biophys. Res. Commun.*, 142, 1025, 1987.
32. **Mellgren, R. L., Lane, R. D., and Kakar, S. S.,** A sarcolemma-associated inhibitor is capable of modulating calcium-dependent proteinase activity, *Biochim. Biophys. Acta*, 930, 370, 1988.
33. **Mellgren, R. L.,** On the mechanism of binding of calpastatin, the protein inhibitor of calpains, to biologic membranes, *Biochem. Biophys. Res. Commun.*, 150, 170, 1988.
34. **Suzuki, K., Imajoh, S., Kawasaki, H., Minami, Y., and Ohno, S.,** Regulation of activity of calcium-activated neutral protease, *Adv. Enzyme Regul.*, 27, 153, 1988.
35. **Imajoh, S., Kawasaki, H., Emori, Y., and Suzuki, K.,** Calcium-activated neutral protease inhibitor from rabbit erythrocytes lacks the N-terminal region of the liver inhibitor but retains three inhibitory units, *Biochem. Biophys. Res. Commun.*, 146, 630, 1987.
36. **Coolican, S. A. and Hathaway, D. R.,** Effect of L-α-phosphatidylinositol on a vascular smooth muscle Ca^{2+}-dependent protease, *J. Biol. Chem.*, 259, 11627, 1984.
37. **Pontremoli, S., Salamino, F., Sparatore, B., Michetti, M., Sacco, O., and Melloni, E.,** Following association to the membrane, human erythrocyte procalpain is converted and released as fully active calpain, *Biochim. Biophys. Acta*, 831, 335, 1985.
38. **Suzuki, K., Ohno, S., Emori, Y., Imajoh, S., and Kawasaki, H.,** Calcium-activated neutral protease (CANP) and its biological and medical implications, *Prog. Clin. Biochem. Med.*, 5, 43, 1987.
39. **Rehno, W. M., Mellgren, R. L., and Lane, R. D.,** unpublished data, 1989.

Chapter 9

Ca^{2+}-DEPENDENT PROTEASES AND THEIR INHIBITORS IN *DROSOPHILA*

Peter Friedrich and Marianna Pintér

TABLE OF CONTENTS

I. INTRODUCTION

The Ca^{2+}-activated neutral thiol proteinase, calpain (EC 3.4.22.17), is a ubiquitous enzyme in vertebrates. It has been found in most mammalian tissues in various amounts.[1-5] In mammals, calpain occurs as two isoemzymes[6] and has a specific high M_r protein inhibitor[7,8] called calpastatin and a low M_r activator protein.[9,10] These components constitute a complex, Ca^{2+}-dependent proteolytic system.

Much is known about the mechanism of activation and regulation of calpains (for reviews see Reference 11 and this volume). As for physiological function, there is a plethora of observations that implicate calpain action in a variety of processes involving limited proteolysis (see Reference 12 and this volume). However, further investigations are needed to test the relevance *in vivo* of most of these phenomena.

In contrast to vertebrates, calpain-like activities have been detected only in a few invertebrates to date. These include the marine worm *Myxicola infundibulum,*[13] the Chilean squid *Dosidicus gigas,*[14] the crayfish *Procambarus clarkii,*[15] the Bermuda land crab *Gecarcinus lateralis,* and the lobster *Homarus americanus.*[16,17]

We have set out to examine whether the calpain system is present in another invertebrate, the fruit fly *Drosophila melanogaster,* in view of the approaches offered by this organism through genetic manipulations. Our report on the detection of calpains and calpastatin in *Drosophila* appeared recently.[18]

II. DETECTION OF CALPAIN ACTIVITY IN *DROSOPHILA* HEAD EXTRACT

Attempts to detect Ca^{2+}-dependent proteolytic activity in homogenates or homogenate supernatants of whole flies or heads were unsuccessful. However, activity became detectable after dialysis or fractionation at pH 4.5. The activity was measured with alkali-denatured casein as substrate according to Guroff.[19] The calpain activities in fly heads and rat kidney after pH 4.5 fractionation are compared in Table 1. Specific activities related to the original extracts show that rat kidney contains apparently about 80 times as much calpain activity as do *Drosophila* heads. (It should be mentioned that unfractionated rat kidney extracts exhibited no calpain activity either.) The low abundance of calpain activity in *Drosophila* samples necessitated unusually long (8 to 16 h) incubation times with Guroff's method.[19] In later stages of our work the use of [^{3}H]methyl-casein allowed us to reduce incubation times considerably.

An alternative way of detecting calpain activity from fly head extract was to fractionate by gel chromatography. As seen in Figure 1, a Ca^{2+}-dependent proteolytic activity peak emerged at apparent M_r = 280,000, which was unaffected by phenylmethyl-sulfonyl fluoride (PMSF), but was abolished by iodoacetate, as expected for a thiol proteinase.

III. CHARACTERIZATION OF CALPAINS I AND II FROM *DROSOPHILA* HEADS

A. SEPARATION OF CALPAIN ISOENZYMES BY DEAE-CELLULOSE CHROMATOGRAPHY

On DEAE-cellulose chromatography of the fly head extract three proteinase peaks appeared (Figure 2). The predominant peak was Ca^{2+} dependent, and it eluted at 0.26 *M* NaCl, which is the position of vertebrate calpain II.[19,20] The middle peak comprised both Ca^{2+}-activated and Ca^{2+}-inhibited fractions, and the third, minor peak was Ca^{2+} independent. Since vertebrate calpain I eluted under such conditions between 0.09 and 0.15 *M* NaCl,[6] which range overlaps with that of calpastatin,[21-23] these fractions were pooled and subjected

TABLE 1
Calpain activities in *Drosophila* Heads and Rat Kidney

Sample	Protein concn (mg/ml)	Specific calpain activity (nmol of tyrosine per hour per mg of protein)	
		In sample	Related to original extract[a]
Fly head extract after fractionation at pH 4.5	0.76	4.6	0.36
Crude rat kidney calpain	3.5	65	28

Note: The data are the means of three independent experiments: the S.D. was 12%. Wild type fly (Canton-S) head and rat kidney extracts (supernatants) were subjected to fractionation at pH 4.5 as described in Reference 18; the redissolved pH 4.5-pellets were analyzed.

[a] Calpain activity of sample related to the concentration of original extract, which was ~ 10 mg/ml for both fly heads and rat kidney.

From Pintér, M. and Friedrich, P., *Biochem. J.*, 253, 467, 1988. With permission.

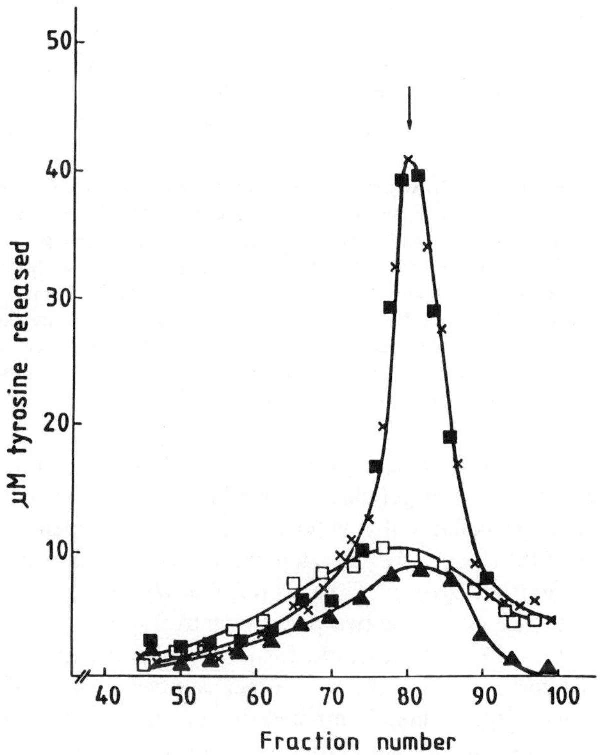

FIGURE 1. Gel-chromatographic analysis of Ca^{2+}-dependent-proteinase activity of fly head extract. A Sepharose CL-6B column (0.65 cm × 95.3 cm) was equilibrated with 10 m*M*-Hepes/KOH buffer, pH 7.5, containing 1 m*M*-EDTA, 135 m*M*-KCl and 0.5 m*M*-dithioerythritol, and 1.3 ml of fly head extract (13 mg of protein) was loaded on the column; 0.27 ml fractions were collected. The arrow indicates M_r = 280,000. Symbols: proteolytic activity in the presence (■) and absence (□) of Ca^{2+}, in the presence of Ca^{2+} and 0.88 m*M*-iodoacetate (▲), and in the presence of Ca^{2+} and 0.88 m*M*-PMSF (×). Reaction time was 13 h. (From Pintér, M. and Friedrich, P., *Biochem. J.*, 253, 467, 1988. With permission.)

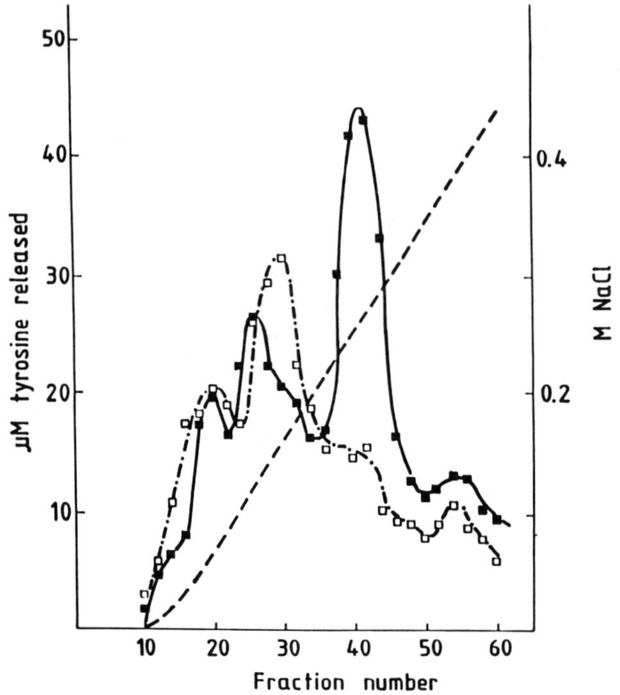

FIGURE 2. DEAE-cellulose chromatographic analysis of Ca^{2+}-dependent proteinase activity of fly head extract. The extract was dialyzed against 10 m*M*-Hepes/KOH buffer, pH 7.5, containing 1 m*M*-EDTA and 0.05 m*M*-dithioerythritol. The dialyzed extract (25 mg of protein) was applied to a DEAE-cellulose column (0.65 cm × 16 cm), and a linear NaCl gradient (----) was run. 1.14 ml fractions were collected. Symbols: proteolytic activity in the presence (■) and absence (□) of Ca^{2+}. (From Pintér, M. and Friedrich, P., *Biochem. J.*, 253, 467, 1988. With permission.)

to gel chromatography. As seen in Figure 3, a Ca^{2+}-dependent proteinase peak eluted at apparent M_r = 280,000. Similar gel chromatography of the major, putative calpain II peak from the DEAE-cellulose column also yielded a single Ca^{2+}-dependent proteolytic peak at apparent M_r = 280,000 (Figure 4). The activity in both peaks was inhibited by a crude calpastatin preparation from *Drosophila* heads (cf. below). On the basis of elution positions from DEAE-cellulose, we assign the two peaks tentatively to calpain I and II, respectively.

The apparent M_r = 280,000 observed for *Drosophila* calpains is considerably higher than those obtained for calpains from most other sources, which range from 110,000 to 150,000 (cf. Reference 20). It should be noted that the M_r = 280,000 value was found for the *Drosophila* enzymes whether gel chromatography was applied to the extract or to the pooled fractions from DEAE-cellulose. It is not clear as yet what the reason is for this strikingly high apparent M_r. One may hypothesize that *Drosophila* calpains associate with some other component in the head extract. Indeed, in mammalian systems, brain calpain I[24] and skeletal muscle calpains I and II[25] have been found to associate with protein kinase C through several steps of purification. On the other hand, crustacean calpains were found to range in M_r between 310,000 and 59,000.[17] Clearly, the true M_r of *Drosophila* calpains should become apparent after the purification and thorough characterization of these enzymes.

B. Ca^{2+}-SENSITIVITY OF CRUDE CALPAINS I AND II

Preliminary assessment of the Ca^{2+}-sensitivity of calpains I and II obtained by DEAE-

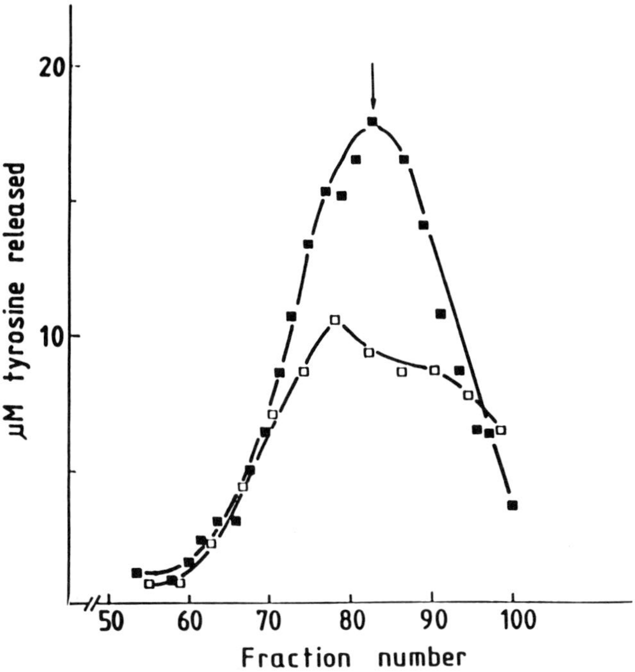

FIGURE 3. Gel-chromatographic detection of calpain I. Fractions 23—29 from Figure 2. were pooled and concentrated by ultrafiltration. A 1-ml sample (1.5 mg of protein) was applied to a Sepharose CL-6B column. The arrow indicates $M_r = 280{,}000$. For other details see Figure 1. Symbols: proteolytic activity in the presence (■) and absence (□) of Ca^{2+}. (From Pintér, M. and Friedrich, P., *Biochem. J.*, 253, 467, 1988. With permission.)

cellulose chromatography and Sepharose CL-6B gel chromatography yielded 50% activation at 1.0 and 1.4 m*M* Ca^{2+}, respectively.[18] However, the curve for calpain I was sigmoidal, exhibiting an apparent strong cooperativity in Ca^{2+}-activation. We deem that the anomalous Ca^{2+} curve of calpain I is due to other Ca^{2+}-binding species in the impure preparation.

C. PURIFICATION OF CALPAIN II CATALYTIC SUBUNIT

The calpain II fraction from the DEAE-cellulose column was further purified on Phenyl-Sepharose, Reactive-Red-Agarose, and Mono Q columns. In the Mono Q fractions, calpain activity coincided with an SDS-polyacrylamide gel electrophoretic band at $M_r \approx 90{,}000$ (Figure 5), which is presumed to be the catalytic subunit of the enzyme. The details of the purification procedure will be described elsewhere.

IV. DETECTION OF *DROSOPHILA* CALPASTATIN

Gel chromatography of fly head extract prepared in the presence of proteinase inhibitors revealed an inhibitory trough toward rat calpain at apparent $M_r = 310{,}000$ in the elution profile, followed by a broad range of inhibitors at lower M_r values (Figure 6). With extracts prepared without proteinase inhibitors, the $M_r = 310{,}000$ trough could not be observed. Mammalian calpastatins fall in the apparent M_r range of 270,000 to 300,000.[22] The inhibition toward lower M_r in the *Drosophila* profile is probably due to endogenous Ca^{2+} chelators and inhibitory peptides. On DEAE-cellulose chromatography of the pooled high-M_r fractions the inhibitor eluted at 0.10 NaCl (Figure 6 insert), which is within the elution range of

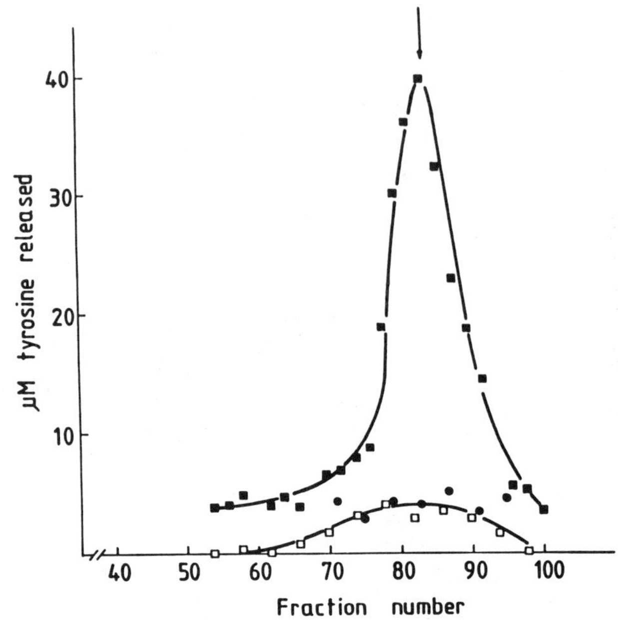

FIGURE 4. Sepharose CL-6B gel chromatography of calpain II separated by DEAE-cellulose chromatography. Fractions 38—46 from Figure 2 were pooled, concentrated (0.54 mg of protein in 0.8 ml), and applied to the column. The arrow indicates M_r = 280,000. For other details see Figure 1. Symbols: proteolytic activity in the presence (■) and absence (□) of Ca^{2+}, and in the presence of Ca^{2+} and crude *Drosophila* calpastatin (●). From Pintér, M. and Friedrich, P., *Biochem. J.*, 253, 467, 1988. With permission.)

calpastatin from mammals.[21-23] The high-M_r inhibitor was heat stable and sensitive to trypsin; it inhibited both *Drosophila* calpains I and II (cf. Figure 4. For more details, see Reference 18). It is concluded from these data that *Drosophila* heads contain the specific protein inhibitor of calpains, calpastatin.

V. CONVERSION OF PROTEIN KINASE C BY CALPAIN IN LARVAL BRAIN EXTRACT

In mammalian systems calpain activates protein kinase C irreversibly through limited proteolysis to yield protein kinase M, which no longer needs Ca^{2+} or phospholipids as activators.[24] Calpains tend to associate with protein kinase C.[24,25] It has been hypothesized that the release of protein kinase M into the cytosol is part of the physiological signal-transducing mechanism.[12]

We examined whether the above conversion of protein kinase C occurs in *Drosophila* nervous tissue. Larval brain extracts were incubated at 18°C with and without Ca^{2+} and the levels of protein kinase C and M activities were monitored. As seen in Figure 7, in the presence of Ca^{2+} protein kinase C activity rapidly decreased, while protein kinase M transiently increased. In the absence of Ca^{2+} there was only a slight decrease in both activities in 30 min. The Ca^{2+}-induced loss of protein kinase C activity could be prevented by leupeptin and by crude *Drosophila* calpastatin in a dose-dependent manner (Table 2). These data strongly suggest that the protein kinase C present in larval brain extract is cleaved by calpain(s).

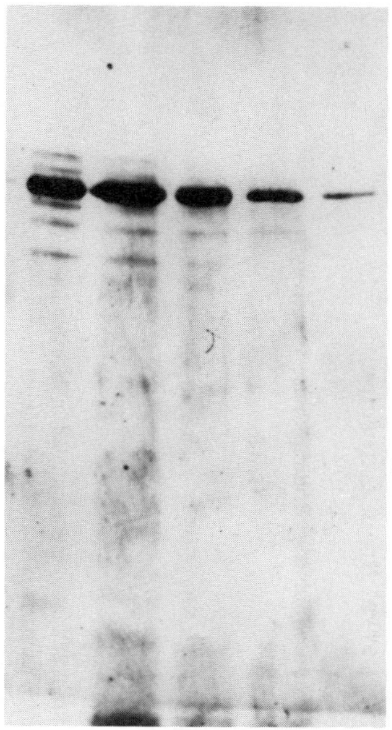

FIGURE 5. SDS-PAGE of Calpain II catalytic subunit. Consecutive fractions from a Mono Q column used in the purification of *Drosophila* caplain II were analyzed by SDS-PAGE. Protein bands were visualized by silver staining. The molecular weight markers are in Daltons.

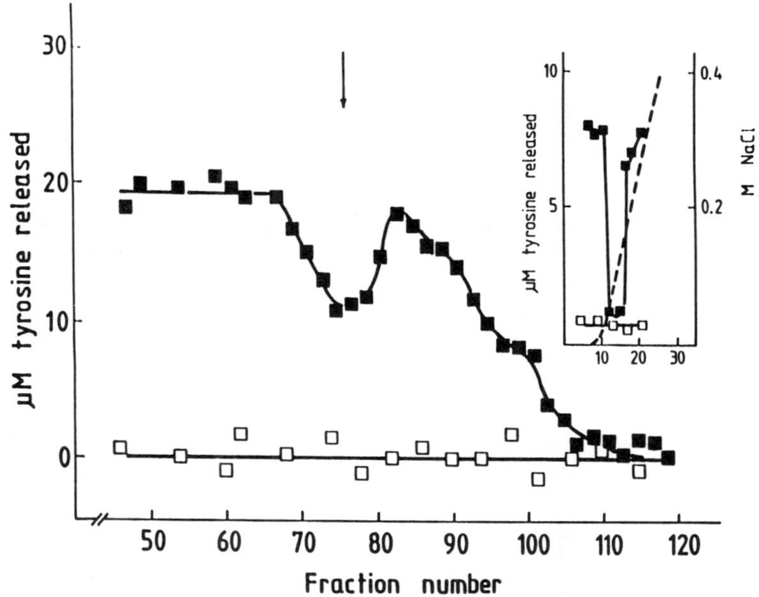

FIGURE 6. Detection of *Drosophila* calpastatin by gel chromatography. The extract was prepared in the presence of the following proteinase inhibitors: 1.9 m*M* PMSF. 4.8 m*M*benzamidine, 2.3 m*M*o-phenanthroline, 1.2 μ*M* trypsin inhibitor, 0.23 m*M* tosyl-L-phenylalanylchlormethyl ketone, 0.23 mg of leupeptin/ml and 0.1 mg of pepstatin A/ml. A 1-ml portion of extract was applied to the column. Calpastatin activity was detected with rat calpain stock solution. The arrow indicates M_r = 310,000. For other details, see Figure 1. Symbols: rat calpain activity in the presence of Sepharose CL-6B fractions with (■) and without (□) Ca^{2+}. Insert: DE-52 column (0.65 cm × 7 cm) chromatography of high -M_r calpastatin (fractions 70—81 in the main Figure). The pooled fractions were heat-treated, the supernatant was dialyzed and chromatographed (0.1 mg of protein) as in Figure 2; fractions were assayed for calpastatin activity. Rat calpain activity with (■) and without (□) Ca^{2+}. (From Pintér, M. and Friedrich, P., *Biochem. J.*, 253, 467, 1988. With permission.)

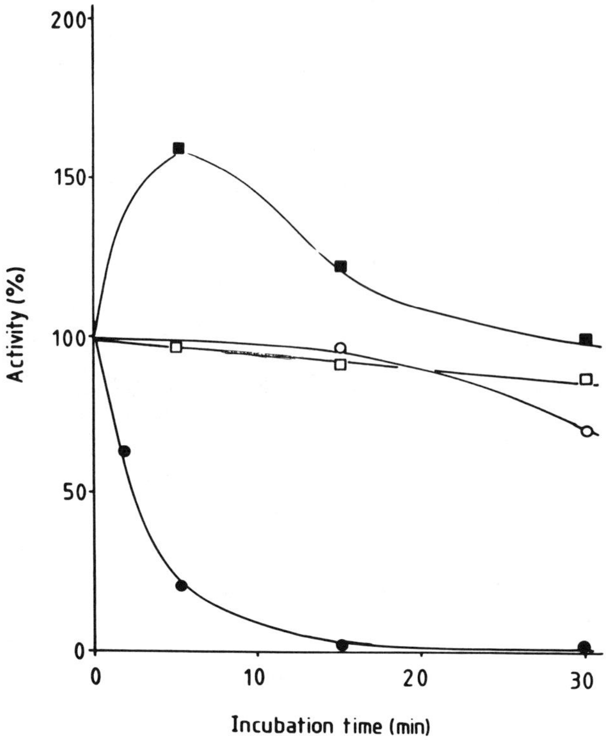

FIGURE 7. Effect of Ca^{2+} on protein kinase C and M activities in larval brain extract. The extract of Canton-S larval brains was incubated at 18 °C with 3 m*M* Ca^{2+} (●, ■) and without Ca^{2+} (○, □). The protein kinase C (●, ○) and protein kinase M (■, □) activities were determined with a synthetic nonapeptide substrate[26] and are expressed as percentage of the 0 time value. (Adapted from Dévay, P., Pintér, M., Kiss, I., Faragó, A., and Friedrich, P., *J. Neurogenetics,* 5, 119, 1989. With permission.

TABLE 2
Effect of Proteinase Inhibitors on the Ca^{2+}-Dependent Conversion of Protein Kinase C in Larval Brain Extract

Incubation time and addition	Protein kinase C activity (%)
No incubation	100
5 min	20
5 min. 0.1% leupeptin	100
30 min. 1 μg calpastatin	43
30 min. 2 μg calpastatin	92

Note: Larval brain extracts were incubated at 18 °C with 3 m*M* $[Ca^{2+}]$free. Calpastatin was a crude preparation from *Drosophila* heads.

Adapted from Dévay, P., Pintér, M., Kiss, I., Faragó, A., and Friedrich, P., *J. Neurogenetics,* 5, 119, 1989. With permission.

VI. SUMMARY AND PERSPECTIVES

The fruit fly *Drosophila melanogaster* possesses both calpain I- and II-like activities, as well as a calpastatin-like inhibitor. Although these proteins need more detailed characterization, they are likely to resemble their vertebrate counterparts, as suggested also by the calpain-induced conversion of protein kinase C in larval brain extract. *Drosophila* may prove an excellent organism to address questions on the function of calpain *in vivo,* by the isolation and study of mutant strains deficient in various components of this Ca^{2+}-dependent proteolytic system.

REFERENCES

1. **Murachi, T., Hatanaka, M., Yasumoto, Y., Nakayama, N., and Tanaka, K.,** Quantitative distribution study on calpain and calpastatin in rat tissues and cells, *Biochem. Int.,* 2, 651, 1981.
2. **Nelson, W. J. and Traub, P.,** Intermediate (10 nm) filament proteins and the Ca^{2+}-activated proteinase specific for vimentin and desmin in the cells from fish to man: an example of evolutionary conservation, *J. Cell Sci.,* 57, 25, 1982.
3. **Baudry, M., Simonson, L., Dubrin, R., and Lynch, G.,** Comparative study of soluble calcium-dependent proteolytic activity in brain, *J. Neurobiol,* 17, 15, 1986.
4. **Clark, A. F., DeMartino, G. N., and Croall, D. E.,** Fractionation and quantification of calcium-dependent proteinase activity from small tissue samples, *Biochem. J.,* 235, 279, 1986.
5. **Kawashima, S., Hayashi, M., Saito, Y., Kasai, Y., and Imahori, K.,** Tissue distribution of calcium-activated neutral proteinases in rat, *Biochim. Biophys. Acta,* 965, 130, 1988.
6. **Mellgren, R. L.,** Canine cardiac calcium-dependent proteases: Resolution of two forms with different requirements for calcium, *FEBS Lett.,* 109, 129, 1980.
7. **Waxman, L. and Krebs, E. G.,** Identification of two protease inhibitors from bovine cardiac muscle, *J. Biol. Chem,* 253, 5888, 1978.
8. **Nishiura, I., Tanaka, K., Yamato, S., and Murachi, T.,** The occurrence of an inhibitor of Ca^{2+}-dependent neutral protease in rat liver, *J. Biochem. (Tokyo),* 84, 1657, 1978.
9. **DeMartino, G. N. and Blumenthal, D. K.,** Identification and partial purification of a factor that stimulates calcium-dependent proteases, *Biochemistry,* 21, 4297, 1982.
10. **Takeyama, Y., Nakanishi, H., Uratsuji, Y., Kishimoto, A., and Nishizuka, Y.,** A calcium-protease activator associated with brain microsomal-insoluble elements, *FEBS Lett.,* 194, 110, 1986.
11. **Suzuki, K., Imajoh, S., Emori, Y., Kawasaki, H., Minami, Y., and Ohno, S.,** Calcium-activated neutral protease and its endogenous inhibitor, *FEBS Lett.,* 220, 271, 1987.
12. **Suzuki, K., Imajoh, S., Emori, Y., Kawasaki, H., Minami, Y., and Ohno, S.,** Regulation of activity of calcium activated neutral protease, *Adv. Enzyme Regul.,* 27, 153, 1988.
13. **Gilbert, D. S., Newby, B. J., and Anderton, B. H.,** Neurofilament disguise, destruction and discipline, *Nature,* 256, 586, 1975.
14. **Orrego, F.,** Protein degradation in squid giant axons, *J. Neurochem.,* 18, 2249, 1971.
15. **Nosek, T. M. and Crosland, M. B.,** Calcium-dependent proteolysis in neural tissue is activated at physiologic intracellular Ca^{2+} levels and inhibited by some anticonvulsants, *J. Neurobiol.,* 14, 9, 1983.
16. **Mykles, D. L. and Skinner, D. M.,** Ca^{2+}-dependent proteolytic activity in crab claw muscle. Effects of inhibitors and specificity for myofibrillar proteins, *J. Biol. Chem.,* 258, 10474, 1983.
17. **Mykles, D. L. and Skinner, D. M.,** Four Ca^{2+}-dependent proteinase activities isolated from crustacean muscle differ in size, net charge, and sensitivity to Ca^{2+} and inhibitors, *J. Biol. Chem.,* 261, 9865, 1986.
18. **Pintér, M. and Friedrich, P.,** The calcium-dependent proteolytic system calpain-calpastatin in *Drosophila melanogaster, Biochem. J.,* 253, 467, 1988.
19. **Guroff, G.,** A neutral, calcium-activated proteinase from the soluble fraction of rat brain, *J. Biol. Chem.,* 239, 149, 1964.
20. **Zimmerman, U.-J. P. and Schlaepfer, W. W.,** Calcium-activated neutral protease (CANP) in brain and other tissues, *Progr. Neurobiol.,* 23, 63, 1984.
21. **Waxman, L.,** Calcium-activated proteases in mammalian tissues, in *Methods in Enzymology,* Vol. 80, Lorand, L., Ed., Academic Press, New York, 1981, 664.

22. **Murachi, T., Tanaka, K., Hatanaka, M., and Murakami, T.,** Intracellular Ca^{2+}-dependent protease (calpain) and its high-molecular-weight endogenous inhibitor (calpastatin) *Adv. Enzyme Regul.*, 19, 407, 1981.
23. **Mellgren, R. L. and Carr, T. C.,** The protein inhibitor of calcium-dependent proteases: purification from bovine heart and possible mechanisms of regulation, *Arch. Biochem. Biophys.*, 225, 779, 1983.
24. **Kishimoto, A., Kajikawa, N., Shiota, M., and Nishizuka, Y.,** Proteolytic activation of calcium-activated, phospholipid-dependent protein kinase by calcium-dependent neutral protease, *J. Biol. Chem.*, 258, 1156, 1983.
25. **Savart, M., Belamri, M., Pallet, V., and Ducastaing, A.,** Association of calpains 1 and 2 with protein kinase C activities, *FEBS Lett.*, 216, 22, 1987.
26. **Romhányi, T., Seprödi, J., Antoni, F., Mészáros, Gy., Buday, L., and Faragó, A.,** The assay of the activity of protein kinase C with the synthetic oligopeptide substrate designed for histone kinase II, *Biochim. Biophys. Acta*, 888, 325, 1986.
27. **Dévay, P., Pintér, M., Kiss, I., Faragó, A., and Friedrich, P.,** Protein kinase C in larval brain of wild-type and dunce memory-mutant *Drosophila*, *J. Neurogenetics*, 5, 119, 1989.

Chapter 10

CALCIUM-DEPENDENT PROTEINASES IN CRUSTACEANS

Donald L. Mykles and Dorothy M. Skinner

TABLE OF CONTENTS

I. INTRODUCTION

Crustacean muscle constitutes a favorable model system for the study of intracellular proteolytic processes and their controls. As in vertebrate skeletal muscles, protein metabolsim in crustacean striated muscles is altered by growth, use, or injury. For example, atrophy is induced in claw muscles by denervation[1] or tenotomy[1] and in the thoracic musculature proximal to walking legs lost by limb autotomy.[2] Stretching of the new exoskeleton by the absorption of water[3] after shedding of the old exoskeleton stimulates the synthesis of protein and addition of sarcomeres to myofibrils.[4,5] One of the best understood systems at the biochemical level is the intermolt cycle-dependent atrophy and restoration that occurs in the claws of land crabs.[6-9] In the land crab, *Gecarcinus lateralis:* the claw closer muscle undergoes a profound atrophy, in which at least 40% of the protein is degraded, as the animal prepares to cast its exoskeleton at ecdysis.[6] The observations on atrophy of land crab claw muscle[6] were the first experimental demonstration to support the hypothesis proposed by Couch[10] in 1843 that reduction of muscle mass would facilitate withdrawal of appendages that have enlarged distal segments through their narrow basal joints of the exoskeleton.[11] After ecdysis, protein is resynthesized and the muscle restored at a mass slightly larger than in the previous intermolt period.[6]

Proteolysis during claw muscle atrophy in the land crab results in a fourfold reduction in myofibrillar cross-sectional area without affecting the general organization of the sarcomere.[12] However, a rearrangement of myofilament packing occurs during atrophy; both the thin:thick filament ratios and molar ratios of myosin heavy chain to actin decrease by approximately one third as the result of the preferential degradation of 11 thin filaments for each thick filament.[12,13] This atrophy is highly specific, occurring in claws with their large distal segments but not in walking legs, which have a relatively constant diameter throughout their lengths; fiber diameters and myofibrillar cross-sectional areas within fibers of walking legs remain unchanged regardless of intermolt stage.[7,14] Moreover, in the dimorphic claws of male fiddler crabs, a preferential atrophy occurs in the major claw; myofibrillar cross-sectional area decreases 4.5-fold in the major claw while the cross-sectional area is reduced only 32% in the minor claw.[15]

Calcium-dependent proteinases (CPs, or calpains in vertebrate systems) have been implicated in the nonlysosomal hydrolysis of myofibrillar proteins.[16] In vertebrate muscles maintained *in vitro,* elevated levels of calcium stimulate proteolysis, which is suppressed by inhibitors of cysteine proteinases.[17-23] A recent study, however, suggests that Ca^{2+} stimulates degradation of nonmyofibrillar, rather than myofibrillar, proteins.[24] Increased CP activity is measured in dystrophic muscle[25,26] and in muscle that is induced to atrophy by vitamin E deficiency[27] or corticosterone treatment.[28] Land crab CPs , which constitute more than 90% of the neutral proteolytic activity in the cytoplasm,[29] are partially active at micromolar levels of calcium, specifically degrade the major myofibrillar proteins (actin, myosin, paramyosin, troponin, and tropomyosin) to trichloroacetic acid-soluble material, and show more than a twofold increase in activity in atrophic muscle.[13,29] CPs are extracted from claw muscles of intermolt (anecdysial) and premolt (proecdysial) land crabs, regenerating limbs of land crabs, and claw and abdominal muscles of lobsters.[7,9,13,29,30] Thus, they are found in muscle that is not atrophying, such as that of regenerating limbs and abdominal muscle as well as that which is, such as claw. It is likely that CPs are not restricted to the degradation of contractile proteins during muscle atrophy, but are involved in protein catabolism during all phases of muscle development: growth and differentiation, steady-state turnover, and atrophy.[11]

The presence in Crustacea of calcium-dependent proteinases whose activity increases at a specific stage in the life cycle is not surprising given the marked fluctuations in the metabolism of calcium that occur in many crustacean species during the proecdysial period.[11] Ca^{2+} is released from the old exoskeleton as it is degraded.[31,32] In the land crab, and in

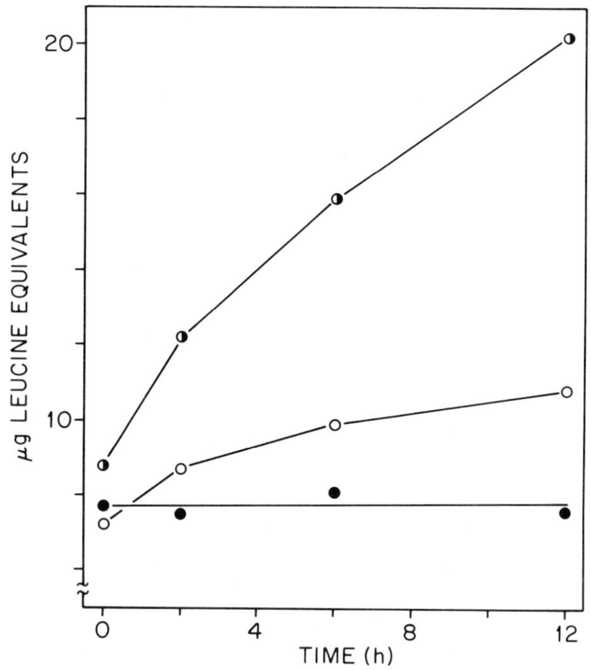

FIGURE 1. Reconstitution of calcium-dependent proteolytic activity in land crab claw muscle. Ca^{2+}-dependent release of trichloroacetic acid-soluble primary amines (μg leucine equivalents) from myofibrillar (●) and soluble (○) proteins incubated separately or together (◑) with 5 m*M* $CaCl_2$. (From Mykles, D. L. and Skinner, D. M., *J. Biol. Chem.* 258, 10474, 1983. With permission.)

other species with limited access to calcium,[33,34] and in some marine crustaceans as well,[3,35] Ca^{2+} makes its way to the anterior chamber of the stomach where it is stored in a set of two or four gastroliths, concretions of calcium carbonate with some phosphate that together can weigh as much as a gram in a 100-gram animal. Gastroliths are solubilized within 72 h following ecdysis,[3,31] releasing a load of calcium which is used primarily for calcification of the soft mouthparts. Ca^{2+} also accumulates in other organs, such as the digestive gland[35] and muscles, during proecdysis. In the land crab, for example, the content of calcium in the claw muscle doubles.[13]

II. ISOLATION AND CHARACTERIZATION OF FOUR CALCIUM-DEPENDENT PROTEINASE ACTIVITIES

Since crustacean CP activity is measured easily in crude preparations, we initially characterized the enzyme activity in homogenates and 90,000 × g supernatant fractions obtained from homogenates of land crab claw muscles.[13,29] The crab CPs are localized in the sarcoplasm and are not tightly bound to either membranes or myofibrils, since activity remains in the supernatant fractions, and washing myofibrils with a low-salt buffer removes all residual activity.[29] Reconstitution experiments using washed myofibrils and soluble extracts show that the myofibrillar proteins constitute the principle substrates for the enzymes.[29] In the presence of 5 m*M* Ca^{2+} there is no detectable release of trichloroacetic acid-soluble material from myofibrils incubated alone; soluble extracts incubated alone show only a small amount of proteolysis, which is associated with hydrolysis of small amounts of actin and a protein larger than myosin heavy chain but not other soluble proteins (Figure 1).[29] However, large amounts of primary amines are released when myofibrils and soluble extracts are incubated together. All the major myofibrillar proteins are completely hydrolyzed; myosin heavy chain is first cleaved to large fragments before subsequent Ca^{2+}-dependent degradation to acid-soluble material.[13,29]

The CPs differ from other proteolytic enzymes isolated from muscle as well as other crustacean tissues. We have recently purified and characterized a latent multicatalytic proteinase (MCP) from soluble extracts of muscle that is activated by SDS or heat and also degrades myofibrillar proteins.[36,37] The MCP is less effective than CPs in degrading actomyosin *in vitro;* it may therefore play a secondary role in the turnover of myofibrillar proteins.[36] Three cysteine proteinases, two with alkaline pH optima and one with a pH optimum of 4, and an aspartic proteinase with a pH optimum of 5, have been isolated from integumentary tissues of land crabs. These proteinases degrade exoskeletal proteins *in vitro.*[38,39] None of the four proteinases is inhibited by EDTA; therefore they are not CPs.

Four CP activities have been isolated from the claw and abdominal muscles of the lobster, *Homarus americanus,* that differ in molecular weight, net charge, and sensitivities to calcium and inhibitors.[30] Chromatography of active fractions from an organomercurial-agarose column on a HPLC preparative gel filtration column yields three peaks of CP activity: I, II, and III (Figure 2). Subsequent chromatography of each of the three peaks on a HPLC ion exchange column produces four discrete CP activities; peaks I and III each consist of a single enzyme while peak II consists of two enzymes (Figure 3). The four activities differ in net charge, as indicated by the concentration of NaCl required to elute them from the column: peak I, 0.22 *M*; peak IIa, 0.32 *M*; peak IIb, 0.39 *M*; and peak III, 0.31 *M*. When the four peaks of CP activity from ion exchange chromatography are chromatographed separately on a HPLC analytical gel filtration column, each enzyme elutes as a single, symmetrical peak.[30] The molecular weights of the lobster CPs are 310 kDa for peak I, 195 kDa for peak IIb, 125 kDa for peak IIa, and 59 kDa for peak III.

The four CPs do not arise from degradation of one or two native enzymes during purification or from contamination from tissues other than muscle as shown by the following experiment. Abdominal muscle, free of epidermis, nerve cord, and intestine, was homogenized and processed.[30] One aliquot was purified without delay. Other aliquots taken at several steps in the purification procedure were held on ice 24 h before chromatography on HPLC gel filtration and ion exchange columns. When compared to the sample processed immediately, the aged samples showed only minor differences in the relative amounts of peaks I, II, and III eluting from a preparative gel filtration column.[30] If degradation had occurred, a shift in molecular weight from larger (peak I) to smaller (peak III) forms would have been seen. Lack of degradation was confirmed by ion exchange chromatographic analysis, which resolved the same four CP activities eluting at the same NaCl concentrations shown in Figure 3.[30]

Specific inhibitors are used to classify proteolytic activities. E-64 and iodoacetamide inhibit cysteine proteinases; leupeptin inhibits cysteine proteinases and some serine proteinases; PMSF inhibits serine and some cysteine proteinases; and pepstatin inhibits aspartic proteinases.[40] Like other cysteine proteinases, both forms of the vertebrate calpains are strongly suppressed by leupeptin, E-64, and iodoacetamide while PMSF and pepstatin have little effect.[41] The same compounds also inhibit the CP activity in 90,000 $\times$ g supernatant fractions of crab claw muscle,[13,29] as well as the four lobster CPs partially purified by HPLC.[30] However, the four lobster enzymes show differential responses to pepstatin and PMSF. Peak I is inhibited by PMSF (26% inhibition) while peaks IIb and III are sensitive to pepstatin (26% and 90% inhibition, respectively). Peak IIa resembles vertebrate calpains in its insensitivity to both compounds.[30]

The distinct responses to the substrate analogs (leupeptin, E-64, and pepstatin) suggest that there are differences in substrate specificity among the four enzyme activities. Prelim-

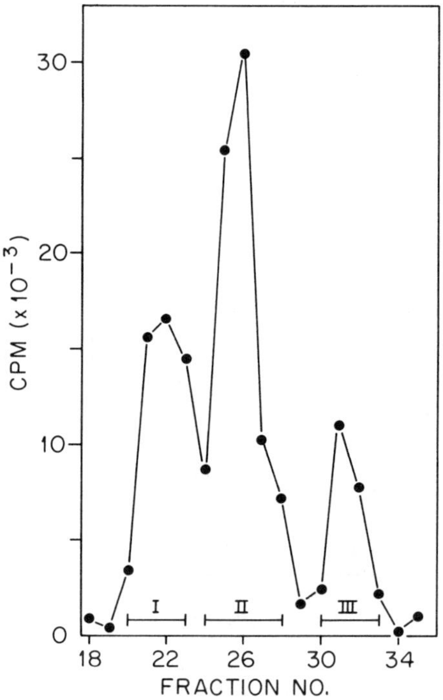

FIGURE 2. HPLC preparative gel filtration chromatography. Following homogenization and extraction of lobster muscles, a 90,000 × *g* supernatant fraction was precipitated with 65% ammonium sulfate, dissolved in buffer, and chromatographed on an organomercurial-agarose column. A 5-ml sample (100 mg protein) of the active fractions was injected into a Bio-Rad TSK-250 column (21.5 × 600 mm) and eluted at 5 ml/min. One-minute fractions were collected. Samples (40 μl) of each fraction were assayed for CP activity using ^{125}I-casein as substrate. (From Mykles, D. L. and Skinner, D. M., *J. Biol. Chem.*, 261, 9865, 1986. With permission.)

inary analysis of the hydrolytic characteristics of the four lobster CPs show that they constitute a family of closely related proteinases with distinct but overlapping substrate specificities. When incubated with purified myosin, all four enzymes generate identical fragments of myosin heavy chain.[42] In contrast, the CPs differ both qualitatively and quantitatively in hydrolysis of myofibrillar proteins in reaction mixtures containing actomyosin; degradation of myosin heavy and light chains, actin, tropomyosin, and troponin was quantified by densitometric scanning of SDS-polyacrylamide gels stained with Coomassie blue.[43] Of the four activities, peak IIb is the most effective in degrading actin, tropomyosin, and troponin-C_1; myosin heavy chain and α light chain are most susceptible to hydrolysis by peak IIa, while peak I preferentially degrades troponin-I_5 and peak III preferentially degrades troponin-I_1 (Table 1). Some of the apparent degradation of actin is, in fact, the hydrolysis of troponin-T, which comigrates with actin under the conditions used for SDS-PAGE.[44,45] CP activities from land crab claw muscle degrade both actin and troponin-T.[13,29] Peaks IIa and IIb have similar, but not identical, substrate specificities; both are effective in degrading actin, tropomyosin, myosin heavy chain and β light chain, and troponin-C_1 and -I_5. Peak IIa, however,

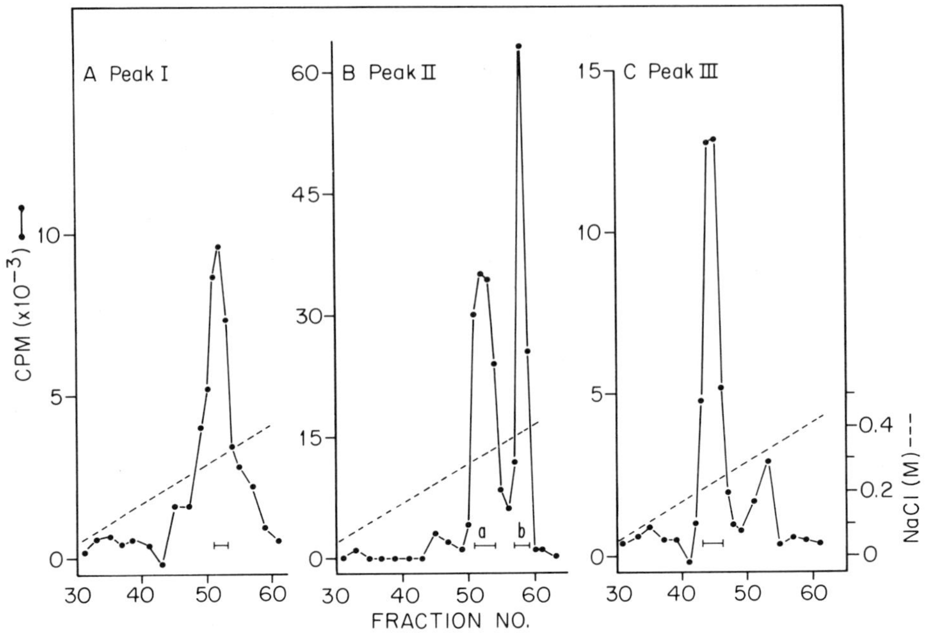

FIGURE 3. HPLC ion exchange chromatography (Bio-Rad TSK-DEAE-3-PW; 7.5 × 75 mm) of peaks I (A), II (B), and III (C) from HPLC preparative gel filtration (Figure 2). CP activities were eluted with a 30-min NaCl gradient from 0.05 to 0.40 *M* at a flow rate of 0.8 ml/min. One-minute fractions were collected. CP activity was assayed as in Figure 2. (From Mykles, D. L. and Skinner, D. M., *J. Biol. Chem.*, 261, 9865, 1986. With permission.)

TABLE 1
Degradation of Myofibrillar Proteins by Four Calcium-Dependent Proteinases (Peaks I, IIa, IIb, and III) Isolated from Lobster Muscle Expressed as Percent Decrease Per Hour[71]

Protein	I	IIa	IIb	III
Myosin heavy chain	0.2	7.6	3.0	4.8
Actin/Troponin-T	3.0	8.1	8.3	3.3
Tropomyosin	0.3	2.1	2.3	1.3
Troponin-I_1	0.1	ND	ND	0.4
Troponin-I_5	11.9	3.9	4.5	7.4
Myosin α light chain	ND	7.3	ND	ND
Troponin-C_1	1.0	NQ	15.6	2.5
Myosin β light chain	ND	7.1	10.6	2.3

Note: ND, no degradation; NQ, not quantifiable.

degrades myosin α light chain while peak IIb does not. Peak I differs from the other enzymes in that it preferentially degrades actin over other myofibrillar proteins except troponin-I_5; myosin heavy and light chains, in particular, constitute relatively poor substrates for this enzyme (Table 1). In summary, each enzyme degrades 6 to 8 of the 8 myofibrillar proteins analyzed; their specifities overlap so that all of the contractile and regulatory proteins are degraded by at least one of the CPs. The preferential hydrolysis of actin by peak I suggests that this enzyme may be differentially expressed or activated during molt-induced atrophy,

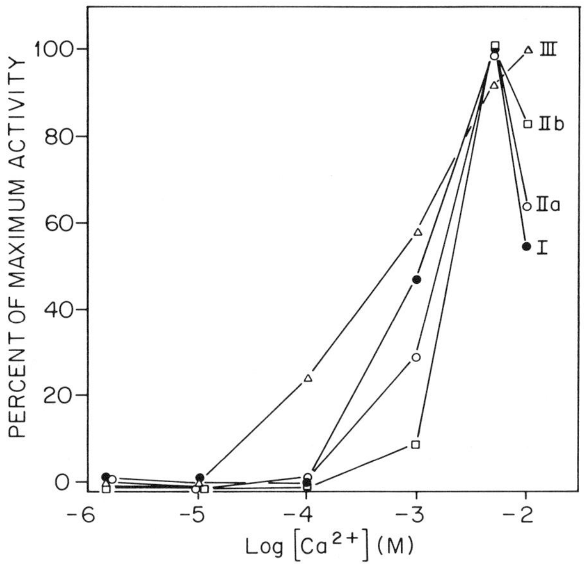

FIGURE 4. Calcium sensitivities of lobster CP activities. Activities of CP peaks from ion exchange chromatography (Figure 3) were assayed in reaction mixtures containing free calcium concentrations from 1.5 μ*M* to 10 m*M* and expressed as the percentage of maximum activity.[125] I-casein used as substrate. (From Mykles, D. L. and Skinner, D. M., *J. Biol. Chem.*, 261, 9865, 1986. With permission.)

since there is a reduction of thin myofilaments relative to thick myofilaments during proecdysis.[12,13]

Although none of the four partially purified lobster CPs is active at concentrations of calcium at 10 μ*M* or below, they show a differential sensitivity to higher calcium concentrations (Figure 4). In increasing order, the concentrations required for half-maximal activation are approximately 0.6 m*M* for peak III, 1 m*M* for peak I, 1.5 m*M* for peak IIa, and 2 m*M* for peak IIb. Of the four activities, only the peak III enzyme is active at 100 μ*M*. Since none of the enzymes is active at concentrations of calcium approaching physiological levels, they all can be classified as millimolar CPs. This contrasts with results of an earlier study of soluble extracts from crab claw muscle which evinced a broad sensitivity to calcium from 0.1 μ*M* to 10 m*M*.[29] Enzyme activity in such crude extracts exhibits a maximum at 1 m*M* with marked inhibition at 5 and 10 m*M* when ^{125}I-casein is used as subtrate. Maximum activity occurs at 5 m*M* Ca^{2+} in reaction mixtures containing myofibrils; this is probably due to the presence of myosin and other calcium-binding proteins as well as sarcoplasmic reticulum (SR).[9] A significant portion (28% of the maximum activity) of the enzyme(s) is active at 1.5 μ*M* calcium. CPs in soluble extracts of crayfish brain and nerve cord are also active at micromolar calcium concentrations.[46]

The lack of response of the lobster CPs to low (< 100 μ*M*) concentrations of Ca^{2+} suggests that a cofactor that alters the requirement for calcium is removed during purification. Such a cofactor might be similar to a calpain activator isolated from human neutrophils that increases the enzyme's affinity for Ca^{2+} by more than 100-fold[47] but would differ from an activator that has been purified from brain and has no effect on calcium sensitivity of the vertebrate enzyme.[48,49] Crustacean muscle contains a calcium-binding protein similar to parvalbumin and calmodulin that may function in activating CPs at physiological concentrations of calcium.[50] A second cofactor, which inhibits CP activity at high calcium concentrations (Figure 5), may be related to an endogenous CP inhibitor found in *Drosophila*

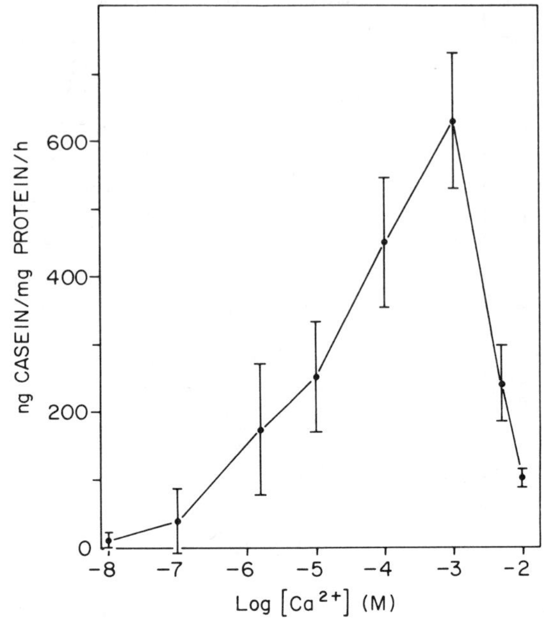

FIGURE 5. CP activity in 90,000 × g supernatant fractions of land crab claw muscle homogenates as a function of calcium concentration. Average of four experiments; standard deviation indicated by vertical bars. ^{125}I-casein used as substrate. (From Mykles, D. L. and Skinner, D. M., *J. Biol. Chem.*, 258, 10474, 1983. With permission.)

head extracts[51] and calpastatin purified from vertebrate tissues.[52] Preliminary experiments have shown that these cofactors in crustacean muscles are removed during organomercurial-agarose chromatography; when void fractions are recombined with active fractions, both activation at micromolar calcium and inhibition at m*M* calcium are restored.[8,53]

III. FUNCTIONS OF CRUSTACEAN CALCIUM-DEPENDENT PROTEINASES

Although crustacean CPs degrade casein commonly used as a substrate in bioassays,[29,30,46] their only known biological substrate is myofibrillar proteins.[13,29,36] The endogenous substrate of the CP activity in crude extracts from the brain and nerve cord of crayfish has not been established (casein yellow was used to assay activity);[46] the enzyme probably degrades neurofilament proteins, which appear to be the preferred substrates of CPs from nervous tissues of other invertebrates,[54-58] as well as vertebrates.[59] The Ca^{2+} sensitivity of the crayfish CPs is influenced by ionic strength. At an optimal concentration of KCl (0.15 *M*), enzyme activity at 0.1 μ*M* Ca^{2+} is 53% of the maximum.[46] The Ca^{2+} concentration required for maximal activation depended on the source of the animals. Preparations from Louisiana crayfish showed a maximal stimulation at 1 m*M* while those from California crayfish were maximally activated at 10 m*M* Ca^{2+}. The crayfish enzyme activity is inhibited by two anticonvulsant agents (valproic acid and pentobarbital) and caprylic acid (concentrations required for 50% inhibition at 1 m*M* Ca^{2+} are 1.7, 0.7, and 0.6 m*M* for valproic acid, caprylic acid, and pentobarbital, respectively).[46] It is likely that the effects of anticonvulsants on the electrophysiological properties of neurons are not mediated directly by CPs, since phenytoin, an antiepileptic drug, does not suppress CP activity and valproic and caprylic acids are both effective inhibitors of CPs that differ in anticonvulsant activity.

TABLE 2
Effects of Compounds that Elevate Intracellular Ca^{2+} on Release of Free Amino Acids and Tyrosine to the Culture Medium

Treatment	FAA Release[a] μmol/g/h	% Change	TYR Release[a] nmol/g/h	% Change
None	23.4 ± 12.6 (51)		76 ± 25 (8)	
A23187 (10μg/ul)	57.6 ± 21.6 (51)	+146***	221 ± 127 (8)	+192*
None	23.4 ± 3.6 (10)		ND[b]	
Caffeine (10 m*M*)	39.6 ± 11.4 (10)	+69***		
None	27.0 ± 7.2 (8)		ND	
Dibucaine (1 m*M*)	69.6 ± 30.6 (8)	+158**		

Note: *Significant at $p < 0.25$; **significant at $P < .005$; and ***significant at $p < .001$.

[a] Tissue preincubated with compound 30 min before measuring release over a 30-min interval. Mean ± 1 S.D.(n).
[b] Not determined.

From Mykles, D. L., *J. Exp. Zool.*, in press. With permission.

Recent studies of land crab claw muscle cultured *in vitro* have confirmed the importance of CPs in mediating muscle protein degradation.[60,61] Total proteolysis was measured by the release of primary amines and tyrosine from muscles into culture media containing cycloheximide to inhibit protein synthesis. Compounds that elevate intracellular Ca^{2+} concentrations, such as calcium ionophore A23187, caffeine, or dibucaine, stimulate proteolysis (Table 2). Inhibitors of CPs, such as leupeptin and lipophilic analogs of E-64 (Ep-475 and EST), suppress protein breakdown, while pepstatin has no effect (Table 3). In contrast, proteolysis is not altered by either chloroquine, monensin, or 3-methyladenine, compounds that disrupte lysosomal functions.[61] EST inhibits, by 30 to 35%, the increased proteolysis induced by A23187, while chloroquine does not.[61] The lack of effect of lysosomotropic reagents on degradation as well as the ultrastructural observations that lysosomes function primarily in the autolytic breakdown of membranous organelles during muscle atrophy,[12] suggest that lysosomal enzymes play a limited role in the turnover of muscle proteins under the culture conditions employed.

Measurements of the release of tyrosine and free amino acids do not discriminate between the hydrolysis of myofibrillar and nonmyofibrillar proteins. Since myofibrillar proteins constitute approximately 80% of the total protein and are preferentially hydrolyzed during molt-induced atrophy, experiments were performed to ascertain whether CPs degraded myosin and actin *in situ*. For these experiments, claw muscles were maintained in organ culture under three conditions: (1) no treatment (control), (2) treatment with A23187, and (3) preincubation with E-64 analogs before treatment with A23187. Following incubation, muscles were homogenized in a low-salt buffer and centrifuged to pellet myofibrils, which contained most (95 to 97%) of the myosin. Intact myosin and proteolytic fragments in soluble extracts were separated with SDS-PAGE and either stained with Coomassie blue or transferred to nitrocellulose and stained with a polyclonal antibody specific for crab claw muscle myosin heavy chain.[61] Elevated intracellular Ca^{2+} stimulates the degradation of actin and myosin; soluble extracts of claw muscles incubated 1 h with A23187 contain less actin[61] and myosin heavy chain than control extracts (Figure 6, compare lanes a,a′ with b,b′). In addition to a prominent heavy chain fragment (HC_f) present in myofibril and purified myosin preparations (Figure 6, lanes, i,i′), soluble extracts of control and A23187-treated muscles

TABLE 3
Effects of Proteinase Inhibitors on Release of Free Amino Acids and Tyrosine to the Culture Medium

Treatment	FAA Release[a] μmol/g/h	% Change	TYR Release[b] nmol/g/h	% Change
None	24.6 ± 10.2 (11)		74 ± 14 (8)	
Leupeptin (0.1 m*M*)	18.6 ± 10.2 (11)	−24*	49 ± 14 (8)	−34**
None	12.6 ± 7.2 (10)		ND[c]	
EST (1 m*M*)	9.0 ± 5.4 (10)	−33**		
None	ND		114 ± 22 (4)	
EST (0.1 m*M*)			53 ± 18 (4)	−53***
None	8.4 ± 3.0 (8)		ND	
Ep-475 (1 m*M*)	6.0 ± 1.2 (8)	−29**		
None	18.0 ± 4.2 (10)		ND	
Pepstatin A (0.1 m*M*)	19.8 ± 4.8 (10)	+10 (NS)[d]		

Note: *Significant at $p < 0.25$; **significant at $p < .01$; and ***significant at $p < .001$.

[a] Tissue preincubated 2 h with inhibitor before measuring release over a 30-min interval. Mean ± 1 S.D. (n)

[b] Tissue preincubated 1.5 h with inhibitor before measuring release over a 1-h interval. Mean ± 1 S.D. (n)

[c] Not determined

From Mykles, D. L., *J. Exp. Zool.*, in press. With permission.

contain 11 proteolytic fragments detected on immunoblots (Figure 6, lanes a′,b′, arrowheads). Since there is no increased accumulation of fragments in A23187-treated muscles, elevated Ca^{2+} must also stimulate the degradation of these fragments to peptides and amino acids. This is consistent with the hydrolytic characteristics of crustacean CPs, which degrade myofibrillar proteins to trichloroacetic acid-soluble material *in vitro.*[13,29] Preincubation with EST and Ep-475, which completely inhibits CP activity (Figure 6, lanes g,g′), partially inhibits the A23187-stimulated degradation of myosin heavy chain and reduces the number of proteolytic fragments produced (Figure 6, lanes c,c′). Densitometric quantification of the myosin heavy chain in lanes g and h of Figure 6 confirmed that the inhibition of CPs was complete.[61]

To compare the degradation of myosin *in situ* with the degradation of myosin by CPs *in vitro,* either purified myosin or myofibrils were incubated with a soluble extract containing CP activity in the presence or absence of 5 m*M* Ca^{2+} and analyzed as above. In general, the fragment pattern *in situ* is similar, but not identical, to the heavy chain fragments produced in reaction mixtures *in vitro* (Figure 6, compare lanes a′,b′,c′ with d′,e′). The major difference between *in situ* and *in vitro* preparations is the presence of a 145-kDa fragment in reaction mixtures containing purified myosin (lane d′, asterisk) that is absent in soluble extracts (lanes a′,b′,c′) and reaction mixtures containing myofibrils (lane e′). These results suggest that CPs degrade myofilaments directly; initial depolymerization of myosin filaments followed by subsequent hydrolysis apparently does not occur in the intact tissue.

In vertebrate skeletal muscle, elevated Ca^{2+} induces removal of the Z line[62] and increases the amount of easily releasable myofilaments.[18] Vertebrate calpains preferentially degrade Z lines in isolated myofibrils.[63,64] A23187 induces a rapid degradation of Z lines in crustacean muscles (Figure 7B) whereas control fibers maintained *in vitro* retain the fine structure of

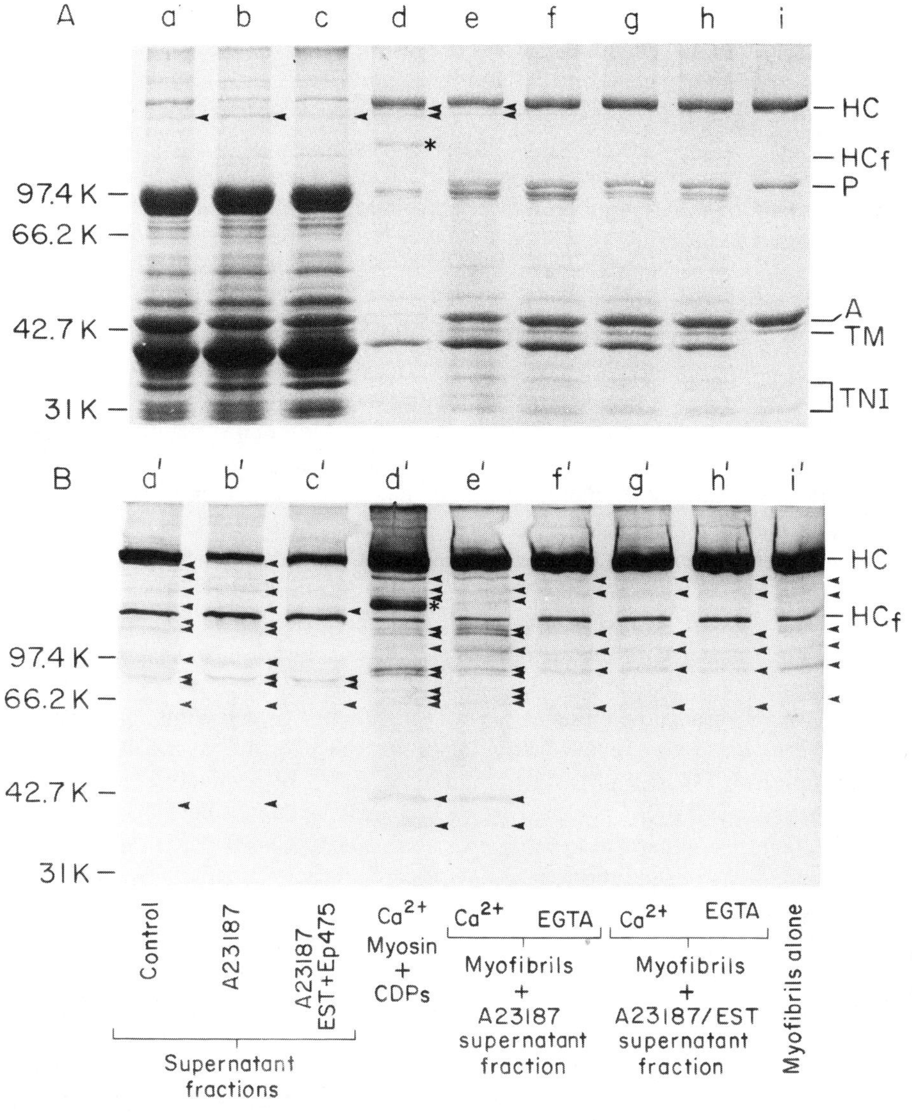

FIGURE 6. Effect of A23187 and cysteine proteinase inhibitors on degradation of myosin *in situ* and *in vitro*. For *in situ* studies (lanes a-c,a′-c′), claw muscles from land crab were preincubated 2 h with or without 1 m*M* EST and 1 m*M* Ep-475 before incubation for 1 h with or without A23187, homogenized, and soluble protein was extracted. For *in vitro* studies (lanes e-h,e′-h′), muscles were preincubated 1.5 h with or without 1 m*M* EST, incubated 1 h with A23187, and homogenized; soluble extracts were incubated with myofibrils and either 5 m*M* Ca^{2+} or 5 m*M* EGTA for 6 h at 37°C. Identical gels were either stained with Coomassie blue (A) or transferred to nitrocellulose and stained with a polyclonal antibody to myosin heavy chain (B). Lanes a,a′, soluble extract (60 μg) from control muscle incubated without A23187 and inhibitors; b,b′, soluble extract (60 μg) from muscle incubated with A23187 alone, c,c′, soluble extract (60 μg) from muscle incubated with A23187 and inhibitors; d,d′, reaction mixture containing monomeric myosin and extract incubated 2 h with Ca^{2+}, e,e′, reaction mixture containing myofibrils and A23187 extract incubated with Ca^{2+}; f,f′, reaction mixture containing myofibrils and A23187 extract incubated with EGTA; g,g′, reaction mixture containing myofibrils and A23187/EST extract incubated with Ca^{2+}, h.h′, reaction mixture containing myofibrils and A23187/EST extract incubated with EGTA; and i,i′, myofibrils (4 μg) alone. A volume equivalent to 2 μl of each of the *in vitro* reaction mixtures was analyzed (lanes e-h,e′-h′). Soluble extracts and reaction mixtures containing myofibrils lacked a 145-kDa fragment (*) present in reaction mixtures containing purified myosin (compare lanes a′-c′,e′ with d′). Proteolytic fragments of myosin heavy chain (HC) indicated by arrowheads. A, actin; HC_f, a prominent fragment of myosin heavy chain; P, paramyosin; TM, tropomyosin; and TNI, troponin-I isoforms. Positions of molecular weight standards indicated at left. (From Mykles, D. L., *J. Exp. Zool.*, in press. With permission.)

freshly dissected tissue.[12] Crustacean slow muscles characteristically are made up of myofibrils consisting of long (>4 μm) sarcomeres demarcated by thickened Z lines; a well-developed sarcoplasmic reticulum (SR) surrounds each myofibril (Figure 7A). In contrast, fibers incubated 1 h with A23187 lack the normal banding pattern found in control fibers. Although the SR appears normal, the Z lines are absent and the organization of the myofilament lattice is completely disrupted (Figure 7B). These observations suggest that crustacean CPs may be involved in initial proteolytic events that remove myofilaments from myofibrils for subsequent degradation.

IV. RELATIONSHIP OF INVERTEBRATE AND VERTEBRATE CALCIUM-DEPENDENT PROTEINASES

Calcium-dependent proteinases are distributed among a variety of invertebrate groups: annelids,[55-57] molluscs,[54,58,65] and arthropods.[13,29,30,46,51] Although vertebrate and invertebrate calcium-dependent proteolytic systems share some common features, such as activation by Ca^{2+}, neutral pH optima, degradation of cytoskeletal proteins, preferential hydrolysis of the Z-line, and sensitivity to cysteine proteinase inhibitors, they differ in many physical and catalytic properties. Most invertebrate CPs are larger than vertebrate calpains, both of which are about 110 kDa.[52,66] The two CPs from *Drosophila* are about 280 kDa, while three of the four lobster CPs are 125 kDa or larger. Crustacean muscle CPs completely hydrolyze myofibrillar proteins while vertebrate muscle calpains are more selective and cleave proteins to large fragments.[67,68] The enzymes also differ in stability when Ca^{2+} is present. Calpains undergo autoproteolysis in the presence of Ca^{2+} and are inactivated within minutes at 37°C.[66] In contrast, crustacean CPs remain active after 48 h at 37°C in 5m*M* $CaCl_2$.[13] Calpains I and II are active at μ*M* and m*M* concentrations of Ca^{2+}, respectively.[66] None of the four partially purified lobster CPs is active at 10 μ*M* Ca^{2+} or less, suggesting that crustacean muscles lack a form comparable to calpain I.[30] As noted in Chapter 9, both forms of fruitfly CP also require high concentrations of Ca^{2+} for activity (half-maximal activation at 1 to 1.4 m*M*).[51]

Both forms of the vertebrate calpains have been purified and their genes cloned and sequenced.[66] Since none of the invertebrate CPs has been purified, the relationship between vertebrate and invertebrate CPs remains to be elucidated. CPs appear to fulfill different functions in the turnover of myofibrillar proteins in the two animal groups. Since calpains degrade the Z-line but show only a limited cleavage of a few contractile proteins, it has been proposed that these enzymes are involved in the initial removal of myofilaments from the myofibrils for subsequent degradation by calcium-independent proteolytic processes.[16,24,69,70] Crustacean CPs, on the other hand, readily hydrolyze to acid-soluble material actin, myosin, paramyosin, troponin, and tropomyosin, which suggests that they are involved in the complete hydrolysis of myofibrillar proteins.

ACKNOWLEDGMENTS

This research was supported by grants from the National Science Foundation (DCB-8502543 to DMS and DCB-8617386 to DLM) and by the Office of Health and Environmental Research, United States Department of Energy, under contract DE-AC05-84OR21400 with Martin Marietta Energy Systems, Inc.

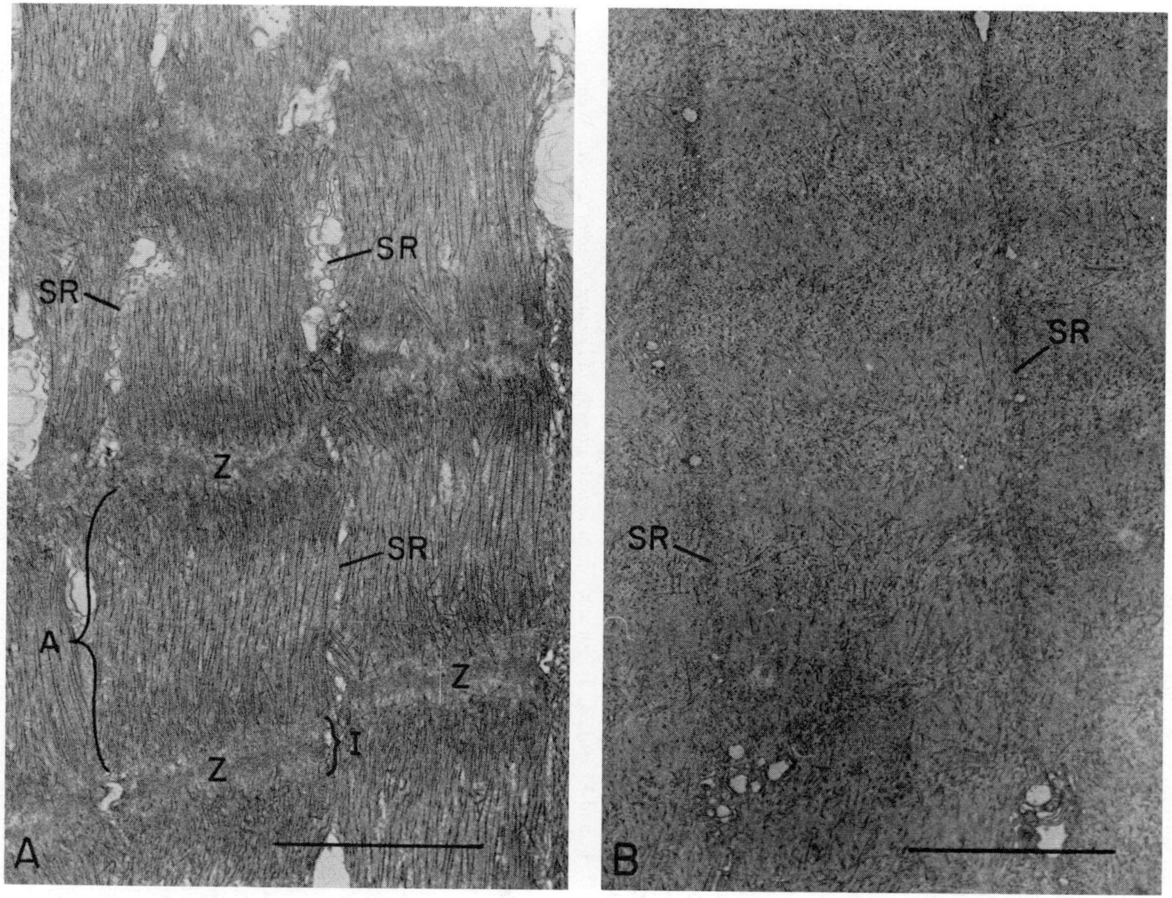

FIGURE 7. Calcium-dependent degradation of Z-lines *in situ*. Claw muscles from land crab were incubated without (A) or with (B) A23187 (10 μg/ml) for 1 h and processed for electron microscopy. In longitudinal sections of control fibers, sarcomeres have well-defined Z-lines (Z), I-bands (I), and A-bands (A). In fibers treated with A23187, sarcomeric organization is disrupted; Z-lines are absent and the arrangement of myofilaments is disorganized. SR, sarcoplasmic reticulum. Scale bar = 3 μm. (From Mykles, D. L., *J. Exp. Zool.*, in press. With permission.)

REFERENCES

1. **Boone, L. P. and Bittner, G. D.,** Morphological and physiological measures of trophic dependence in a crustacean muscle, *J. Comp. Physiol.,* 89, 123, 1974.
2. **Moffett, S.,** Muscles proximal to the fracture plane atrophy after limb autotomy in decapod crustaceans, *J. Exp. Zool.,* 244, 485, 1987.
3. **Mykles, D. L.,** The mechanism of fluid absorption at ecdysis in the American lobster, *Homarus americanus, J. Exp. Biol.,* 84, 89, 1980.
4. **El Haj, A. J., Govind, C. K., and Houlihan, D. F.,** Growth of lobster leg muscle fibers over intermolt and molt, *J. Crust. Biol.,* 4, 536, 1984.
5. **El Haj, A. J. and Houlihan, D. F.,** *In vitro* and *in vivo* protein synthesis rates in a crustacean muscle during the molt cycle, *J. Exp. Biol.,* 127, 413, 1987.
6. **Skinner, D. M.,** Breakdown and reformation of somatic muscle during the molt cycle of the land crab, *Gecarcinus lateralis, J. Exp. Zool.,* 163, 115, 1966.
7. **Mykles, D. L. and Skinner, D. M.,** Crustacean muscles: atrophy and regeneration during molting, in *Basic Biology of Muscles: A Comparative Approach,* Twarog, B. M., Levine, R. J. C., and Dewey, M. M., Eds., Raven Press, New York, 1982, 337.
8. **Mykles, D. L. and Skinner, D. M.,** The role of calcium-dependent proteinase in molt-induced claw muscle atrophy, in *Intracellular Protein Catabolism,* Khairallah, E. A., Bond, J. S., and Bird, J. W. C., Eds., Alan R. Liss, New York, 1985, 141.
9. **Mykles, D. L. and Skinner, D. M.,** Muscle atrophy and restoration during molting, in *Factors in Adult Growth,* Wenner, A. M., Ed., A. A. Balkema, Rotterdam, 1985, 31.
10. **Couch, J.,** On the process of exuviation and growth in crabs and lobsters, and other British species of stalk-eyed crustacean animals, *Annu. Rep. R. Cornwall Polytechnic Soc.,* 11, 1, 1843.
11. **Skinner, D. M.,** Molting and regeneration, in *The Biology of Crustacea,* Vol. 9, Bliss, D. E. and Mantel, L. H., Eds., Academic Press, New York, 1985, 43.
12. **Mykles, D. L. and Skinner, D. M.,** Preferential loss of thin filaments during molt-induced atrophy in crab claw muscle, *J. Ultrastruct. Res.,* 75, 314, 1981.
13. **Mykles, D. L. and Skinner, D. M.,** Molt-cycle associated changes in calcium-activated proteinase activity that degrades actin and myosin in crustacean muscle, *Dev. Biol.,* 92, 386, 1982.
14. **Houlihan, D. F. and El Haj, A. J.,** An analysis of muscle growth, in *Factors in Adult Growth,* Wenner, A. M., Ed., A. A. Balkema, Rotterdam, 1985, 15.
15. **Ismail, S. Z. M. and Mykles, D. L.,** Differential atrophy in the dimorphic claws of male *Uca, Am. Zool.,* 27, 81A, 1987.
16. **Goll, D. E., Edmonds, T., Kleese, W., Sathe, S. K., and Shannon, J. D.,** Some properties of the Ca^{2+}-dependent proteinase, in *Intracellular Protein Catabolism,* Khairallah, E. A., Bond, J. S., and Bird, J. W. C., Eds., Alan R. Liss, New York, 1985, 151.
17. **Sugden, P. H.,** The effects of calcium ions, ionophore A23187 and inhibition of energy metabolism on protein degradation in the rat diaphragm and epitrochlearis muscles *in vitro, Biochem. J.,* 190, 593, 1980.
18. **van der Westhuyzen, D. R., Matsumoto, K., and Etlinger, J. D.,** Easily releasable myofilaments from skeletal and cardiac muscles maintained *in vitro, J. Biol. Chem.,* 256, 11791, 1981.
19. **Lewis, S. E. M., Anderson, P., and Goldspink, D. F.,** The effects of calcium on protein turnover in skeletal muscles of the rat, *Biochem. J.,* 204, 257, 1982.
20. **Zeman, R. J., Kameyama, T., Matsumoto, K., Bernstein, P., and Etlinger, J. D.,** Regulation of protein degradation in muscle by calcium. Evidence for enhanced nonlysosomal proteolysis associated with elevated cytosolic calcium, *J. Biol. Chem.,* 260, 13619, 1985.
21. **Baracos, V., Greenberg, R. E., and Goldberg, A. L.,** Influence of calcium and other divalent cations on protein turnover in rat skeletal muscle, *Am. J. Physiol.,* 250, E702, 1986.
22. **Furuno, K. and Goldberg, A. L.,** The activation of protein degradation in muscle by Ca^{2+} or muscle injury does not involve a lysosomal mechanism, *Biochem. J.,* 237, 859, 1986.
23. **Turner, P. R., Westwood, T., Regen, C. M., and Steinhardt, R. A.,** Increased protein degradation results from elevated free calcium levels found in muscle from *mdx* mice, *Nature,* 335, 735, 1988.
24. **Goodman, M. N.,** Differential effects of acute changes in cell Ca^{2+} concentration on myofibrillar and nonmyofibrillar protein breakdown in the rat extensor digitorum longus muscle *in vitro, Biochem J.,* 241, 121, 1987.
25. **Neerunjun, J. S. and Dubowitz, V.,** Increased calcium-activated neutral protease activity in muscles of dystrophic hamsters and mice, *J. Neurol. Sci.,* 40, 105, 1979.
26. **Johnson, P. and Hammer, J. L.,** Calpain and calpastatin levels in dystrophic hamster skeletal muscles, *Int. J. Biochem.,* 20, 1227, 1988.
27. **Dayton, W. R., Schollmeyer, J. V., Chan, A. C., and Allen, C. E.,** Elevated levels of a calcium-activated muscle protease in rapidly atrophying muscles from vitamin E-deficient rabbits, *Biochim. Biophys. Acta,* 584, 216, 1979.

28. **Kayali, A. G., Young, V. R., and Goodman, M. N.,** Sensitivity of myofibrillar proteins to glucocorticoid-induced muscle proteolysis, *Am. J. Physiol.*, 252, E621, 1987.
29. **Mykles, D. L. and Skinner, D. M.,** Ca^{2+}-dependent proteolytic activity in crab claw muscle. Effects of inhibitors and specificity for myofibrillar proteins, *J. Biol. Chem.*, 258, 10474, 1983.
30. **Mykles, D. L. and Skinner, D. M.,** Four Ca^{2+}-dependent proteinase activities isolated from crustacean muscle differ in size, net charge, and sensitivity to Ca^{2+} and inhibitors, *J. Biol. Chem.*, 261, 9865, 1986.
31. **Skinner, D. M.,** The structure and metabolism of a crustacean integumentary tissue during a molt cycle, *Biol. Bull.*, 123, 635, 1962.
32. **Stringfellow, L. A. and Skinner, D. M.,** Molt-cycle correlated patterns of synthesis of integumentary proteins in the land crab *Gecarcinus lateralis, Dev. Biol.*, 128, 97, 1988.
33. **Travis, D. F.,** The deposition of skeletal structures in the Crustacea. I. The histology of gastrolith skeletal tissue complex and the gastrolith in the crayfish, *Orconectes (Cambarus) virilis,* Hagen — Decapoda, *Biol. Bull.*, 118, 137, 1960.
34. **McWhinnie, M. A.,** Gastrolith growth and calcium shifts in the freshwater crayfish, *Orconectes virilis, Comp. Biochem. Physiol.*, 7, 1, 1962.
35. **Travis, D. F.,** The molting cycle of the spiny lobster, *Panulirus argus* Latreille. II Preecdysial histological and histochemical changes in the hepatopancreas and integumental tissues, *Biol. Bull.*, 108, 88, 1955.
36. **Mykles, D. L.,** A high-molecular-weight serine proteinase from lobster muscle that degrades myofibrillar proteins, *J. Exp. Zool.*, 250, 244, 1989.
37. **Mykles, D. L.,** Purification and characterization of a multicatalytic proteinase from crustacean muscle. Comparison of latent and heat-activated forms, *Arch. Biochem. Biophys.*, 274, 216, 1989.
38. **O'Brien, J. J. and Skinner, D. M.,** Characterization of enzymes that degrade crab exoskeleton: I. Two alkaline cysteine proteinase activities, *J. Exp. Zool.*, 243, 389, 1987.
39. **O'Brien, J. J. and Skinner, D. M.,** Characterization of enzymes that degrade crab exoskeleton: II. Two acid proteinase activities, *J. Exp. Zool.*, 246, 124, 1988.
40. **Barrett, A. J.,** The many forms and functions of cellular proteinases, *Fed. Proc.*, 39, 9, 1980.
41. **Waxman, L.,** Calcium-activated proteases in mammalian tissues, *Meth. Enzymol.*, 80, 664, 1981.
42. **Mattson, J. M. and Mykles, D. L.,** Degradation of myofibrillar proteins by calcium-dependent proteinases in lobster, *Am. Zool.*, 27, 81A, 1987.
43. **Mattson, J. M. and Mykles, D. L.,** unpublished data.
44. **Mykles, D. L.,** Heterogeneity of myofibrillar proteins in lobster fast and slow muscles: variants of troponin, paramyosin, and myosin light chains comprise four distinct protein assemblages, *J. Exp. Zool.*, 234, 23, 1985.
45. **Mykles, D. L.,** Histochemical and biochemical characterization of two slow fiber types in decapod crustacean muscles, *J. Exp. Zool.*, 245, 232, 1988.
46. **Nosek, T. M. and Crosland, M. B.,** Calcium-dependent proteolysis in neural tissue is activated at physiologic intracellular Ca^{2+} levels and inhibited by some anticonvulsants, *J. Neurobiol.*, 14, 9, 1983.
47. **Pontremoli, S., Melloni, E., Michetti, M., Salamino, F., Sparatore, B., and Horecker, B. L.,** An endogenous activator of the Ca^{2+}-dependent proteinase of human neutrophils that increases its affinity for Ca^{2+}, *Proc. Natl. Acad. Sci. U.S.A.*, 85, 1740, 1988.
48. **DeMartino, G. N. and Croall, D. E.,** Calcium-dependent proteases from liver and heart, in *Intracellular Protein Catabolism,* Khairallah, E. A., Bond, J. S., and Bird, J. W. C., Eds., Alan R. Liss, New York, 1985, 117.
49. **Takeyama, Y., Nakamichi, H., Uzatsuji, Y., Kishimoto, A., and Nishizuka, Y.,** A calcium-protease activator associated with brain microsomal-insoluble elements, *FEBS Lett.*, 194, 110, 1986.
50. **Wnuk, W. , Cox, J. A., and Stein, E. A.,** Structural changes induced by calcium and magnesium in a high affinity calcium-binding protein from crayfish sarcoplasm, *J. Biol. Chem.*, 256, 11538, 1981.
51. **Pintér, M. and Friedrich, P.,** The calcium-dependent proteolytic system calpain-calpastatin in *Drosophila melanogaster, Biochem. J.*, 253, 467, 1988.
52. **Murachi, T.,** Intracellular regulatory system involving calpain and calpastatin, *Biochem. Int.*, 18, 263, 1989.
53. **Mykles, D. L. and Skinner, D. M.,** Four Ca^{2+}-dependent proteinases in crustacean muscles: possible regulation by activators and inhibitors, *J. Cell. Biol.*, 103, 401a, 1986.
54. **Orregó, F.,** Protein degradation in squid giant axons, *J. Neurochem.*, 18, 2249, 1971.
55. **Gilbert, D. S., Newby, B. J., and Anderton, B. H.,** Neurofilament disguise, destruction and discipline, *Nature,* 256, 586, 1975.
56. **Pant, H. C., Terakawa, S., and Gainer, H.,** A calcium-activated protease in squid axoplasm, *J. Neurochem.*, 32, 99, 1979.
57. **Eagles, P. A. M., Gilbert, D. S., and Maggs, A.,** The location of phosphorylation sites and Ca^{2+}-dependent proteolytic cleavage sites on the major neurofilament polypeptides from *Myxicola infundibulum, Biochem. J.*, 199, 101, 1981.

58. **Brown, A. and Eagles, P. A. M.,** Squid neurofilaments. Phosphorylation and Ca^{2+}-dependent proteolysis *in situ, Biochem. J.,* 239, 191, 1986.
59. **Zimmerman, U. P. and Schlaepfer, W. W.,** Characterization of a brain calcium-activated protease that degrades neurofilament proteins, *Biochemistry,* 21, 3977, 1982.
60. **Haire, M. F. and Mykles, D. L.,** Regulation of calcium-dependent proteolysis in crab claw muscle: effects of protease inhibitors and protein kinase C antagonists, *J. Cell Biol.,* 107, 262a, 1988.
61. **Mykles, D. L.,** Calcium-dependent proteolysis in crustacean claw closer muscle maintained *in vitro, J. Exp. Zool.,* in press.
62. **Ishiura, S., Nonaka, I., and Sugita, H.,** Suppression of calcium-induced removal of the Z-line by a thiol-protease inhibitor, E-64-c, *J. Biochem. (Tokyo),* 90, 283, 1981.
63. **Reddy, M. K., Etlinger, J. D., Rabinowitz, M., Fischman, D. A., and Zak, R.,** Removal of Z lines and alpha-actinin from isolated myofibrils by a calcium-activated neutral protease, *J. Biol. Chem.,* 250, 4278, 1975.
64. **Dayton, W. R., Goll, D. E., Zeece, M. G., Robson, R. M., and Reville, W. J.,** A Ca^{2+}-activated protease possibly involved in myofibrillar protein turnover. Purification from porcine muscle, *Biochemistry,* 15, 2150, 1976.
65. **Stafford, W. F. and Yphantis, D. A.,** Existence and inhibition of hydrolytic enzymes attacking paramyosin in myofibrillar extracts of *Mercenaria mercenaria, Biochem. Biophys. Res. Commun.,* 49, 848, 1972.
66. **Suzuki, K., Imajoh, S., Emori, Y., Kawasaki, H., Minami, Y., and Ohno, S.,** Regulation of activity of calcium activated neutral protease, *Adv. Enzyme Reg.,* 27, 153, 1988.
67. **Ishiura, S., Nonaka, I., and Sugita, H.,** Calcium-activated neutral protease: its degradative role in muscle cells, In *Muscular Dystrophy,* Ebashi, S., Ed., University of Tokyo Press, Tokyo, 1982, 265.
68. **Hara, K., Ichihara, Y., and Takahashi, K.,** Purification and characterization of a calcium-activated neutral protease from monkey cardiac muscle, *J. Biochem. (Tokyo),* 93, 1435, 1983.
69. **Sugita, H., Ishiura, S., Kamakura, K., Nakase, H., Hagiwara, K., Nonaka, I., and Tomomatsu, K.,** Ca-activated neutral protease in physiological and pathological conditions, in *Calcium Regulation in Biological Systems,* Ebashi, S., Ed., Academic Press, New York, 1984, 243.
70. **Lowell, B. B., Ruderman, N. B., and Goodman, M. N.,** Evidence that lysosomes are not involved in the degradation of myofibrillar proteins in rat skeletal muscle, *Biochem. J.,* 234, 237, 1986.
71. **Mattson, J. M. and Mykles, D. L.,** unpublished data.

II. Function of the Calcium-Dependent Proteolytic System in Normal Physiology and Pathophysiology

Chapter 11

INTRODUCTION: PHYSIOLOGICAL SIGNIFICANCE OF THE CALPAIN-CALPASTATIN SYSTEM

Takashi Murachi

TABLE OF CONTENTS

TABLE 1
Reported Endogenous Substrates of Calpain (Examples)[a]

Category	Susceptible protein
Enzymes	Phosphorylase b kinase, pyruvate kinase, protein kinase C, tyrosine hydroxylase, transglutaminase, etc.
Cytoskeletal and membrane proteins	Neurofilament, vimentin, myofibrillar proteins, MAP 2, fordin, crystallin, keratin, etc.
Receptors	Steroid hormone (estrogen and glucocorticoid) receptor, EGF receptor, α-adrenelin receptor

[a] For further information, see earlier (References 1, 2, 8, 13, 16, and 23 to 25) and more recent (Reference 26) reviews.

From Murachi, T., *Biochem. Int.*, 18, 263, 1989. With permission.

Ever since the concept that calpain and calpastatin constitute an intracellular regulatory system which can be modulated by changes in Ca^{2+} concentration within a cell was formed,[1,2] the major question has been, and still is, what the natural, or endogenous substrate proteins for calpain are.[2] A great number of cellular proteins have been reported to be susceptible to calpain hydrolysis, and they can be grouped into three categories: (1) enzyme proteins, (2) cytoskeletal and membrane proteins, and (3) receptor proteins (Table I).[3]

I. UBIQUITY AND DIVERSITY

Ubiquitous distribution of the calpain-calpastatin system in animal cells seems to lend support to the notion of a general, uniform function of calpain in all different cells. On the other hand, distinct unevenness of the absolute and relative amounts of calpains I and II and calpastatin in different cells[4] can be a clear indication of the individual, highly diversified significance of the calpain-calpastatin system in individual cells. Thus, ubiquity and diversity is the characteristic feature of the distribution of the system, and this must be the fundamental key to the understanding of its function or functions.

Proposed proteolytic down-regulation of protein kinase C by calpain[5] can be one of the strong candidates in the former category of general, uniform function, while the calpain-catalyzed cleavage of fodrin in synaptic membranes,[6] talin in platelets,[7] keratin in epidermal tissue,[8] etc. may belong to the latter category of more tissue/cell-specific functions.

II. FREE OR BOUND

Figure 1 is a schematic representation of the intracellular localization and interaction of proteases, substrate proteins and specific inhibitors. A protease can be localized (1) in the cytosol in a purely soluble state, or (2) in association with the membrane and cytoskeletal elements and (3) with receptor proteins or (4) inside an organelle.

All these four possibilities hold for the calpain-calpastatin system, although classical biochemical analysis had shown that calpain and calpastatin were localized mostly in the cytosol.[2] The home station for calpain and calpastatin must therefore be the cytosolic, soluble fraction of a cell, but they can be translocated from time to time to the membrane,[9,10] nuclei,[11] and mitrochondria.[12] The existence of membrane-associated calpastatin[13] and translocation of phosphorylated calpastatin to the membrane[3] have recently been demonstrated. The association of calpain with cytoskeletal proteins, which is loose enough to be broken simply by homogenization, is highly probable, since a number of cytoskeletal proteins such as tubulin, actin, microtubule-associated protein, myelin basic protein, etc. are known to be quite susceptible to calpain hydrolysis.[14]

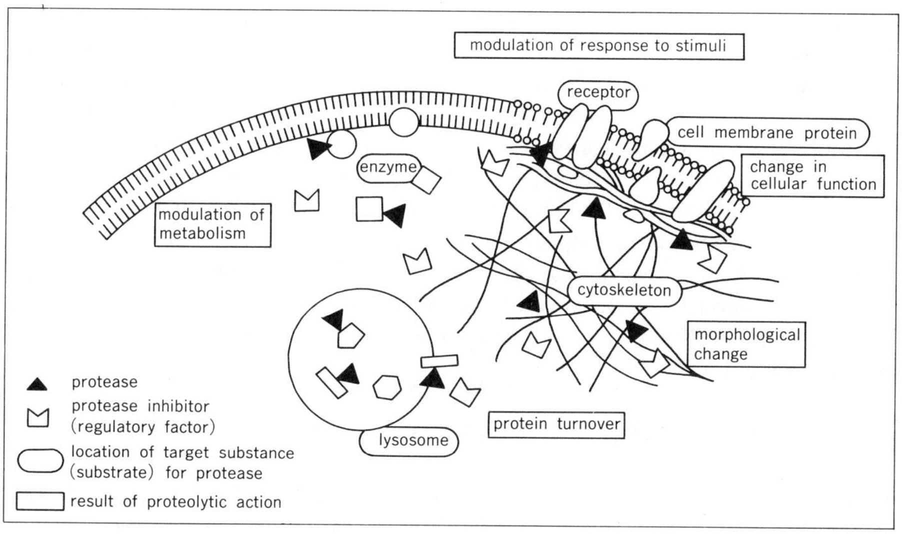

FIGURE 1. Location and function of intracellular proteases and their endogenous inhibitors. This is a simplified scheme showing only one pair consisting of protease and its specific inhibitor present in the system. In reality, a great number of proteases and inhibitors are located together in the cytosol or inside organelles such as lysosomes, nuclei, endoplasmic reticulum, etc., as well as in association with membranes and cytoskeletons. Possible substrate proteins and the expected results of proteolytic actions are categorically illustrated. Proteases and inhibitors may undergo translocation from one compartment to the other, e.g., from the cytosol to the membrane, under specified conditions, but such movements of the molecules are not shown in the figure. It is understood that calpain and calpastatin do not occur in lysosomes.

III. PRIMARY OR SECONDARY

There is no evidence for the presence of the calpain-calpastatin system in lysosomes. This would mean that the primary, physiological function of calpain is different from that of lysosomal cathepsins, which can be regarded as the agents responsible for general catabolism of intracellular proteins.

The latter statement, however, should not exclude the possibility that calpain exerts its strong proteolytic power with broad specificity[15] to result in general protein catabolism under specified circumstances, notably under pathologic conditions. This can be described as the secondary, pathological function of calpain, and should be clearly distinguished from the aforementioned primary, physiological function. The distinction lies with whether the intracellular Ca^{2+} concentration is being kept within the range of physiological fluctuation or whether it has become abnormally high. Two typical examples of the latter cases are calpain-catalyzed breakdown of myofibrillar proteins in several muscle-wasting diseases[16,17] and that of eye lens proteins in senile and experimental cataract.[18] In these cases, the function of calpain in such tissues should be regarded as secondary, because the primary factor must be an excessive increase in Ca^{2+} concentration within the cell.

IV. DISCRIMINATING OR NONDISCRIMINATING

The difference between primary and secondary functions of the calpain-calpastatin system may correspond to the distinction between "discriminating" and "nondiscriminating" cleavages of peptide bonds catalyzed by calpain. Thus, under physiological conditons, a calpain molecule in a given cell should attack a target substrate protein or proteins only, while under pathological circumstances calpain does not discriminate targets, so that it may cause general protein breakdown. It is likely that calpain was originally evolved so as to function in a discriminative fashion in each cell, but it can happen that it attacks surrounding proteins in a nondiscriminative way, if the normal control of the cell has been lost. Therefore, a given cellular component protein can be defined as a physiological substrate of calpain only when unequivocal evidence can be presented of discrimination and, most likely, when limited cleavage of that protein takes place *in situ* under physiological conditions. An *in vitro* experiment showing apparently selective and limited hydrolysis of a certain isolated protein by purified calpain preparations may not be sufficient basis to claim that that protein is one of the physiological targets of calpain. Similarly, demonstration of breakdown of several protein bands, as shown, for example, by polyacrylamide gel electrophoresis (PAGE) in the presence of sodium dodecyl sulfate (SDS) after the cell has been exposed to a Ca^{2+}-ionophore cannot be directly interpreted as indicating that such degradation actually occurs under physiological conditions.

V. MEMBRANES AND CYTOSKELETONS

For a calpain molecule to be activated, Ca^{2+} is obligatory, but a third factor seems to be also indispensable. This must be the factor that ensures that a Ca^{2+}-activated calpain molecule will attack its physiological target protein or proteins in a discriminative way, leaving other calpain-cleavable proteins intact. It is highly probable that such discrimination can be achieved by selective anchoring of a calpain molecule together with the target protein onto a certain structural element of the cell. The third factor can therefore be offered by structural elements of a cell, such as membranes and cytoskeletons. Membrane and cytoskeletal proteins by themselves can thus become specific targets of calpain, while numbers of cytosolic proteins can also be translocated to the surface of membranes and cytoskeletons so that they are juxtapositioned with similarly translocated calpain molecules. Specific proteolysis by calpain of protein kinase C is postulated to occur via such a mechanism.[9,10] The elucidation of the mechanism by which calpain as well as potential substrate protein molecules are jointly translocated to the structural elements must be of prime importance in identifying the true physiological substrate of calpain.

The involvement of the structural elements on which direct interaction of calpain and substrate molecules takes place can also be a clue as to how to avoid a triangular situation that might otherwise ensue among calpain, its substrate, and its inhibitor calpastatin molecules.[3,19]

VI. INTRACELLULAR OR EXTRACELLULAR

For a long time, calpain was thought to be exclusively an intracellular protease. The physiological significance of the calpain-calpastatin system has therefore been studied with tacit confinement of its involvement in intracellular mechanisms. A low, or μM Ca^{2+} requirement, of calpain I seems to be compatible with the notion that this molecular species of calpain has its physiological significance within the cell. In contrast, a high, or close-to-mM Ca^{2+} requirement of calpain II poses much more difficult problems in postulating the possible mechanism of its activation inside a cell, except for its secondary, pathological function as mentioned above.

All the puzzles pertaining to the existance of high-Ca^{2+}-requiring calpain II can readily be solved, if one simply assumes extracellular appearance of calpain II. The mM level of Ca^{2+} in the extracellular compartment would be ideal for a calpain II molecule to be fully activated. In fact, pieces of evidence are now accumulating for the extracellular appearance of calpain and calpastatin in the synovial fluid of the knee joint,[20,21] although it is premature yet to draw a general conclusion.

Enhanced expression of calpain II together with calpastatin in several cell-line T-lymphocytes infected with HTLV-I, a retrovirus responsible for adult T-cell leukemia, has been reported.[22] HTLV-I-infected cells strongly express the gene for interleukin-2 receptor protein which, after protein synthesis, is destined to be located in the cell membrane, half-externalized. It would be most tempting to postulate a similar destination of the calpain-calpastatin system which is expressed concomitantly with interleukin-2 receptor.

VII. SUMMARY

In order to elucidate the physiological significance of the calpain-calpastatin system, the following four lines should at least be taken into consideration.

1. Ubiquitous but diversified expression of the genes for calpain and calpastatin in different cells must be indicative of duality of their functions. Calpain may exert a general, uniform function in most, if not all, different cells, while it can also catalyze diversified functions different from one cell to the other. Both types of calpain functions must equally be considered important.
2. Involvement of calpain in general protein breakdown under such conditions that intracellular Ca^{2+} concentration is elevated beyond its normal limit should not be confused with the primary, physiological function of the calpain-calpastatin system in each cell. The former can cause nondiscriminating cleavages of various intracellular proteins and is to be regarded as the secondary, pathological phenomenon. The primary, physiological function must be of discriminative nature and focused on a few target substrate proteins that can be quite different depending on the sort of the cell and its physiologic state.
3. The discriminative nature of calpain toward a selected target substrate protein or proteins must be, most likely, dependent on its interaction with, or more precisely, on its anchoring on the structural elements of the cell. Completely free collision among Ca^{2+}, calpain, substrate, and calpastatin molecules with mutual competition may not lead to any ordered consequence desirable under physiological conditions. A whole variety of structural elements, such as membranes and cytoskeletons, must provide numerous selective sites for such strict discrimination.
4. The possibility of extracellular appearance or externalization, even transient, of calpain II should be more seriously considered in future studies.

REFERENCES

1. **Murachi, T.,** Intracellular Ca^{2+}-protease and its inhibitor protein: calpain and calpastatin, in *Calcium and Cell Function,* Cheung, W. Y., Ed., Academic Press, New York, 4, 377, 1983.
2. **Murachi, T.,** Calpain and calpastatin, *Trends Biochem. Sci.,* 8, 167, 1983.
3. **Murachi, T.** Intracellular regulatory system involving calpain and calpastatin, *Biochem. Int.,* 18, 263, 1989.
4. **Murachi, T., Hatanaka, M., Yasumoto, Y., Nakayama, N., and Tanaka, K.,** A quantitative distribution study on calpain and calpastatin in rat tissues and cells, *Biochem. Int.,* 2, 651, 1981.
5. **Kishimoto, A., Mikawa, K., Hashimoto, K., Yasuda, I., Tanaka, S., Tominaga, M., Kuroda, T., and Nishizuka, Y.,** Limited proteolysis of protein kinase C subspecies by calcium-dependent neutral protease (calpain), *J. Biol. Chem.,* 264, 4088, 1989.
6. **Siman, R., Baudry, H., and Lynch, G.,** Brain fodrin: substrate for calpain I, an endogenous calcium-activated protease, *Proc. Natl. Acad. Sci. U.S.A.,* 81, 3572, 1984.
7. **Fox, J. E., Goll, D. E., Reynolds, C. C., and Phyllips, D. R.,** Identification of two proteins (actin-binding protein and P235) that are hydrolyzed by endogenous Ca^{2+}-dependent protease during platelet activation, *J. Biol. Chem.,* 260, 1060, 1985.
8. **Ando, Y., Miyachi, Y., Imamura, S., Kannagi, R., and Murachi, T.,** Purification and characterization of calpains from pig epidermis and their action on epidermal keratin, *J. Invest. Dermatol.,* 90, 26, 1988.

9. **Pontremoli, S., and Melloni, E.,** Extralysosomal protein degradation, *Annu. Rev. Biochem.,* 55, 455, 1986.
10. **Suzuki, K., Imajoh, S., Emori, Y., Kawasaki, H., Minami, Y., and Ohno, S.,** Calcium-activated neutral protease and its endogenous inhibitor. Activation at the cell membrane and biological function, *FEBS Lett.,* 220, 271, 1987.
11. **Schollmeyer, J. E.,** Calpain II involvement in mitosis, *Science,* 240, 911, 1988.
12. **Beer, D. G., Hjelle, J. J., Peterson D. R., and Malkinson, A. M.,** Calcium-activated proteolytic activity in rat liver mitochondria, *Biochem. Biophys. Res. Commun.,* 109, 1276, 1982.
13. **Mellgren, R. L., Lane, R. D., and Kakar, S. S.,** Isolated bovine myocardial sarcolemma and sarcoplasmic reticulum vesicles contain tightly bound calcium-dependent protease inhibitor, *Biochem. Biophys. Res. Commun.,* 142, 1025, 1987.
14. **Murachi, T., Hatanaka, M., and Hamakubo, T.,** Calpain and neuropeptide metabolism, in *Neuropeptides and Their Peptidases,* Turner, A. J., Ed., Ellis-Horwood, Chichester, England, 1987, 202.
15. **Sasaki, T., Kikuchi, T., Yoshimura, N., and Murachi, T.,** Comparative specificity and kinetic studies on porcine calpain I and calpain II with naturally occurring peptides and synthetic fluorogenic substrates, *J. Biol. Chem.,* 259, 12489, 1984.
16. **Imajoh, K., Suzuki, K., and Sugita, H.,** Ca^{2+}-dependent neutral protease and the development of muscle dystrophy, in *Calcium and Cell Function,* Cheung, W. Y., Ed., Academic Press, New York, 3, 1982, 473.
17. **Kay, J.,** Ca^{2+}-activated proteinases, protein degradation and muscular dystophy, in *Proteases; Potential Role in Health and Disease,* Heiland, A., and Horl, W. H., Eds.,Plenum Press, New York, 519, 1983.
18. **David, L. L. and Shearer, T. R.,** Calcium-activated proteolysis in lens during selenite cataractogenesis, *Invest. Ophthalmol. Vis. Sci.,* 25, 1275, 1984.
19. **Mellgren, R. L.,** Calcium-dependent proteases: an enzyme system active at cellular membranes?, *FASEB J.,* 1, 110, 1987.
20. **Fukui, I., Tanaka, K., and Murachi, T.,** Extracellular appearance of calpain and calpastatin in the synovial fluid of the knee joint, *Biochem. Biophys. Res. Commun.,* 162, 559, 1989.
21. **Suzuki, K., Shimizu, K., Hamamoto, T., Nakagawa, Y., Hamakubo, T., and Yamamuro, T.,** Biochemical demonstration of calpains and calpastatin in osteoarthritic synovial fluid, *Arthritis Rheum.,* in press.
22. **Adachi, Y., Takano, E., Murachi, T., and Hatanaka, M.,** Distribution and expression of calpastatin in human hematopoietic system cells, *Biol. Chem. Hoppe-Seyler,* 369 (Suppl.), 223, 1988.
23. **Ishiura, S.,** Calcium-activated proteolysis in living cells, *Life Sci.,* 29, 1079, 1981.
24. **Murachi, T., Tanaka, K., Hatanaka, M., and Murakami, T.,** Intracellular Ca^{2+}-dependent protease (calpain) and its high-molecular-weight endogenous inhibitor (calpastatin), *Adv. Enzyme Regul.,* 19, 407, 1981.
25. **Murachi, T.,** Calcium-dependent proteinases and specific ihibitiors: calpain and calpastatin, *Biochem. Soc. Symp.,* 49, 149, 1984.
26. **Wang, K. K. W., Villalobo, A., and Roufogalis, B. D.,** Calmodulin-binding proteins as calpain substrates, *Biochem. J.,* 262, 693, 1989.

Chapter 12

THE CALPAIN-CALPASTATIN SYSTEM AND MEMBRANE FUSION

Nechama S. Kosower and Tova Glaser

TABLE OF CONTENTS

I. NATURAL AND EXPERIMENTALLY INDUCED MEMBRANE FUSION: GENERAL ASPECTS

Membrane fusion is a ubiquitous cellular event, important for a variety of biological processes. Some types of membrane fusion occur within the individual cell. Other fusion events involve cell-cell interaction. In the sperm acrosome reaction, required for subsequent fertilization, multiple fusions occur between the outer acrosomal membrane and the overlying sperm plasma membrane. Actual fertilization involves cell-cell fusion between the acrosome-reacted sperm cell and the egg cell membranes.[1] During development, myoblasts fuse to form multinucleated muscle cells.[2] Within cells, fusion of organelle membranes allows the directed movement of membrane proteins and lipids to specific sites. The exocytosis of hormones and neurotransmitters involves intracellular fusion between the secretory vesicles and the plasma membrane.[3] Certain viral infections involve endocytosis, followed by fusion of the virus membrane with the endosome membrane. Some viruses fuse directly with the cell membrane.[4]

Membrane fusion can be induced experimentally. Experimental membrane fusion is useful for investigating mechanisms of fusion, for defining factors, and for delineating stages involved. Cell and membrane fusion is used in applications such as the formation of hybridomas and introduction of drugs into cells. Various membrane systems have been studied, such as cells, erythrocyte ghosts, vesicles prepared from biological membranes, viruses, and viral envelopes, as well as different types of artificial lipid membranes, without or reconstituted with proteins. Experimental membrane fusion may be achieved by using some viruses, certain types of phospholipids, some chemical reagents (polyethylene glycol (PEG), fusogenic lipids and membrane mobility agents), and by electric pulse.[5-11]

An increase in intracellular, cytoplasmic calcium ions is required for the various natural membrane fusion events. Calcium ions are also necessary for most types of experimentally induced membrane fusion. The increase in cytosolic Ca^{2+} could result from an enhanced influx from the extracellular medium or be due to release of Ca^{2+} from intracellular stores.[1-3,7-10] In some cases micromolar Ca^{2+} concentrations are sufficient to trigger the fusion event, indicating the possible participation of a high-affinity Ca^{2+}-binding protein and enzyme activation in the process.

Some features are general to most fusion cases, with some special features pertaining to specific types of fusion. Naturally occurring fusion events involve specific recognition of membrane type and sites. Exocytotic vesicles fuse with the inner layer of the cell membrane, but not with lysosomal membrane. Virus-cell interaction involves receptors, and specific recognition is essential for gamete fusion. Myoblasts fuse with other myoblasts but not with other types of cells.

In most cases of experimentally induced fusion, the recognition step is bypassed and the steps and factors directly involved in membrane fusion can be studied. As a first step, membrane fusion requires a direct, tight apposition between the two original membranes. The initial sizes of the areas actually in direct contact may be quite small relative to the size of the final fused product. The actual fusion involves a merging of the tightly apposed sites of the two membranes, with mixing of the aqueous compartments. This direct fusion is to be distinguished from an ''indirect'' process in the cases in which components dissociate from one membrane before being transferred into the other membrane, thereby causing the growth of the latter at the expense of the former.[12]

It is generally agreed that the actual fusion between the tightly apposed membranes takes place in protein-free lipid areas and that membrane protein components have to move and redistribute to create such lipid, fusion-potent membrane regions. In addition, reorganization of membrane phospholipid domains, with exposure of certain classes of phospholipids, may facilitate fusion.[1-3,12-14] Phosphatidylethanolamine and phosphatidylserine are important for vesicle and cell fusion.[13,14] It has been proposed that the actual fusion between the lipid bilayers proceeds via inverted micelle mechanism.[12,15] Permeation of water into a region of tight apposition between the perturbed membranes would lead to the formation of an inter-bilayer inverted micelle, the resolution of which will result in communication between the two aqueous compartments.[12]

Overall fusion can be assessed by light microscopy with certain details revealed by electron microscopy, using transmission thin sections and freeze-fracture techniques.[3] The tight apposition of the membranes can be seen in transmission electron microscopy as a pentalaminar structure. Freeze-fracture studies have shown membrane particle-free areas in association with fusion processes. These areas are presumed to represent the fusion-potent membrane regions.

II. MEMBRANE FUSION INDUCED BY THE MEMBRANE MOBILITY AGENT, A_2C

Membrane-mobility agents are long-chain esters designed to promote disorder within bilayers and enhance motion of membrane components in cell membranes. The membrane-mobility agent A_2C (2-(methoxyethoxy)ethyl *cis*-8-(2-octylcyclopropyl)octanoate) forms small, oil-like particles (0.5 to 1.0 μm in diameter) when dispersed in aqueous medium by a short sonication.

$$A_2C \qquad CH_3(CH_2)_7\overset{\overset{CH_2}{/\;\;\backslash}}{CH - CH}(CH_2)_7COOCH_2OCH_2OCH_3$$

After addition of an A_2C particle suspension to cells, the particles stick to and then fuse with the cell membrane; the particle content is dispersed within the membrane. A_2C has been found to promote the lateral mobility of certain ligand-membrane receptor complexes and inhibit cytokinesis in certain types of cells.[16-19] A_2C has been found to promote membrane fusion.[8,12,20-23] The results obtained using A_2C-induced erythrocyte membrane fusion have allowed the identification of stages in the overall fusion process, formulation of molecular mechanism for the process and identification of some cellular factors involved.

Among erythrocytes of various species, we have found differences in the occurrence of A_2C-induced fusion. Nucleated erythrocytes of different avian sources as well as non-nucleated erythrocytes of different mammals vary in their fusibility. Erythrocytes of some species fuse easily, whereas those of others do not.[20,21] Mixed fusion, that between fusible erythrocytes from different species, is also induced by A_2C. Moreover, the potential for fusion is a transferable characteristic, fusion being induced by A_2C in otherwise nonfusible cells if mixed with fusible erythrocytes.[21]

In the case of fusible cells, the addition of A_2C and Ca^{2+} to erythrocyte suspension in buffer containing dextran (to facilitate particle-cell and cell-cell adhesion[6]), results initially in the fusion of A_2C particles with the cell membrane, followed by the diffusion of particle content throughout the cell membrane. A morphological change of the erythrocyte to a spherical, round cell, then occurs, especially noted in the case of nucleated oval avian erythrocytes.[8,12,21] Two or more such cells stick to each other and fuse, the whole sequence taking about 30 to 60 min with most of the cells fusing to form large bodies (Figure 1).

Freeze-fracture replicas of A_2C-treated cells have shown an altered distribution of intramembrane particles, with some membrane regions devoid of particles.[19,21] Transmission electron micrographs of cells undergoing fusion show apposition of the membranes at multiple points, with areas of apposed membranes in tight contact (Figures 2A,2B). Within these areas, regions which correspond to ''channels'' of communication between the cells are noted, and these contain disordered membrane remnants (Figure 2B). Between the apposed

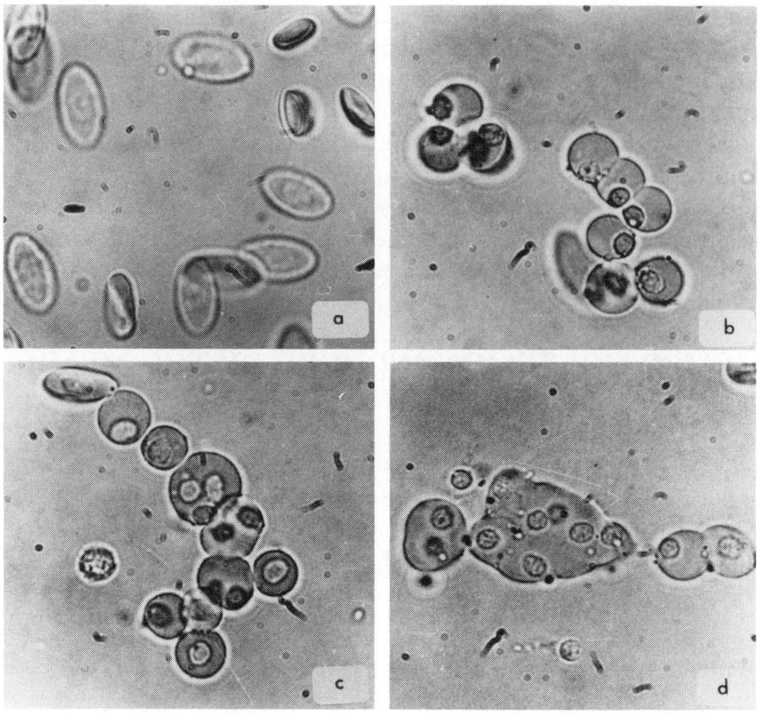

FIGURE 1. Fusion of leghorn hen erythrocytes by the membrane mobility agent A_2C. Cells were suspended in buffer containing 0.5 m*M* Ca^{2+} and 100 mg/ml dextran, pH5.6. One volume of 8% cell suspension was mixed with one volume of A_2C suspension (0.5 μl/ml 150 m*M* NaCl[8]). Cell suspensions were then incubated at 37 °C. (a) 5 min, cells are similar in shape and size to control cells. (b) 25 min, rounding and contact between cells may be seen. (c) 40 min, fusion to binucleated and trinucleated cells. (d) 60 min, multinucleated cells. (From Kosower, E. M., Kosower, N. S., and Wegman, P., *Biochim. Biophys. Acta,* 401, 530, 1975. With permission.)

membranes of the fusing cells, extracellular delimited spaces (''vesicles'') are seen. At a late stage of fusion, the membrane ''vesicles'' disappear, with some remnants left (Figure 2C). The sequence of steps and the morphological aspects of A_2C-induced fusion are thus similar to those of other membrane fusion cases studied.[1-6]

III. CALPAIN AND CALPASTATIN ACTIVITIES IN FUSIBLE AND NONFUSIBLE ERYTHROCYTES

As pointed out above, erythrocytes of some species have been found to fuse easily by A_2C, whereas others do not. Rat erythrocytes, which fuse easily, and human erythrocytes, which do not fuse under the same conditions (Figure 3), have then been studied in an attempt to clarify the molecular basis for the differences observed. A_2C enters and diffuses throughout the membrane in both rat and human cells.[21,24] It alters phospholipid organization, with phosphatidylethanolamine domains becoming exposed[25] and promotes the entry of Ca^{2+} into both types of cells.[26] Since A_2C-induced fusion requires Ca^{2+}, we investigated the response of the cells to Ca^{2+}, introduced into the cells by the ionophore A23187. No fusion was seen in rat or human erythrocytes treated with Ca^{2+} and ionophore.[26] In the rat cells thus treated, significant membrane protein degradation occurs, and little high molecular weight (HMW) protein is observed after treatment. In contrast, in human erythrocytes treated similarly, little degradation of membrane proteins is seen, and significant amounts of HMW proteins are

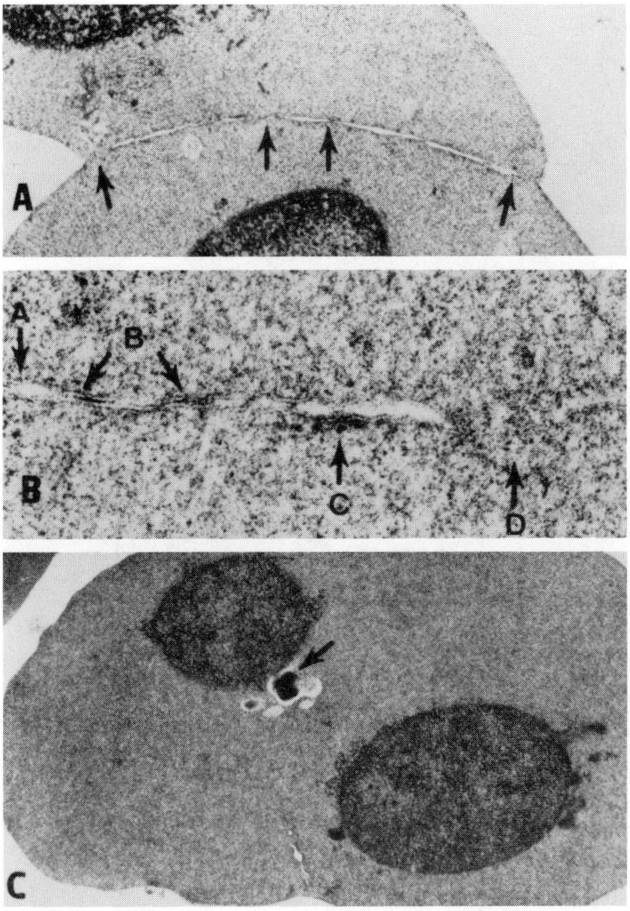

FIGURE 2. Transmission electron micrographs of Leghorn hen red cells during fusion. (A) Apposition of membranes at multiple points (arrows) is illustrated. (B) A fusing region is shown under high magnification. Arrow A indicates an apposition between the two membranes. Arrows labeled B show a tight apposition between the two bilayers and the beginning of channel formation between the cells. Note the "pentalaminar" structure in which the tightly apposed outer layers of the two membranes give the appearance of a dense middle "layer". Arrow C points at a delimited space ("vesicle") between two fusing areas. Arrow D indicates a communicating channel in the process of formation, with membrane remnants still present. (C) The end stage of cell-cell fusion between two Leghorn hen red cells in which membrane remnants (arrow), some heavily stained, are seen with a vesicle along the line of fusion. (From Tavassoli, M. T., Kosower, N. S., Halverson, C., Aoki, M., and Kosower, E. M., *Biochim. Biophys. Acta,* 601, 544, 1980. With permission.)

detected (Figure 4). As Lorand et al. have shown, HMW proteins are produced by Ca^{2+}-activated cytoplasmic transglutaminase.[27]

The Ca^{2+}-induced protein degradation in the rat erythrocyte membrane was found to require the presence of cytoplasm.[26] Marked protein degradation was found in "red" ghosts (ghosts containing hemolysate), with little effect shown by Ca^{2+} in "white" ghosts (ghosts devoid of cytosol).

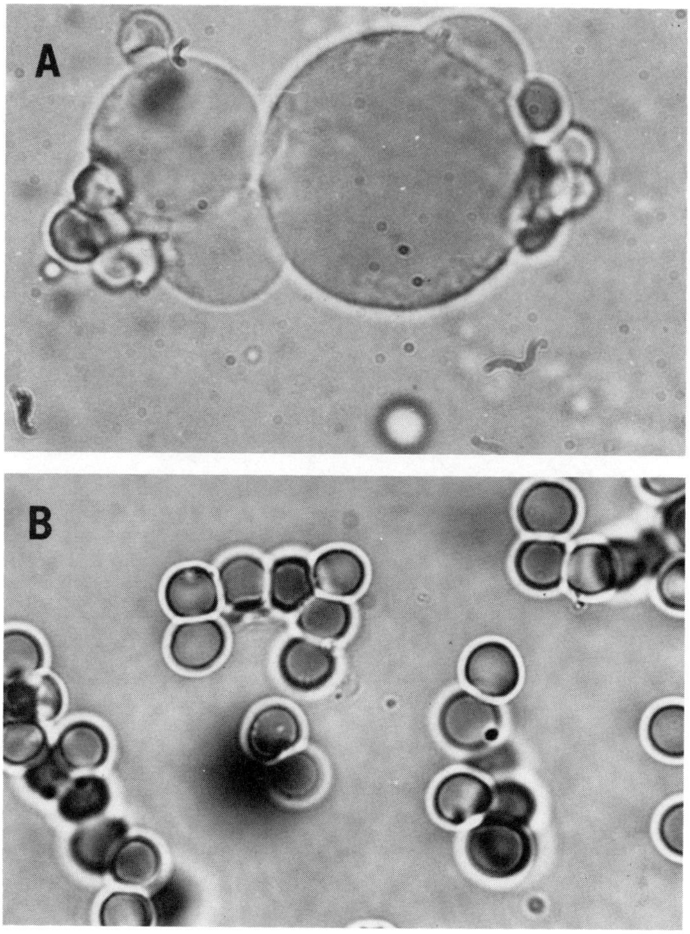

FIGURE 3. Micrographs showing fusing rat erythrocytes and nonfusing human erythrocytes. Cells were suspended in buffer, as described in Figure 1. Cell suspensions were incubated for 45 min. (A) Rat erythrocytes, (B) Human erythrocytes. (From Kosower, N. S., Glaser, T., and Kosower, E. M., *Proc. Natl. Acad. Sci. U.S.A.*, 80, 7542, 1983. With permission.)

Rat erythrocyte ghosts containing hemolysate and Ca^{2+} fused when treated with A_2C, similar to the fusion induced in the intact rat cells.[26] Proteolysis but no fusion was seen when "red" ghosts were treated with Ca^{2+} alone. No degradation nor fusion occurred in ghosts treated with A_2C in the absence of Ca^{2+}.[26] Thus, Ca^{2+}-induced protein degradation is necessary but not sufficient for the cell membranes to undergo fusion. Both membrane proteolysis and the membrane mobility agent are required for the rat erythrocyte membranes to fuse.

Membrane protein degradation and fusion were inhibited in intact rat erythrocytes or "red" ghosts by EDTA, leupeptin, or alkylation of cytoplasmic thiol groups, but not by pepstatin, phenylmethylsulfonyl fluoride, soybean trypsin inhibitor, or by 1,10-phenanthroline.[28] In further experiments, ghosts and hemolysate were prepared from rat erythrocytes that had been treated with fusion inhibitors. The ghosts isolated from inhibitor-treated cells fused when loaded with control hemolysate. In contrast, control ghosts did not fuse when reconstituted with hemolysate from inhibitor-treated cells. These results indicated that an active cytoplasmic Ca^{2+}-dependent protease was necessary for fusion, whereas the fusing membranes did not contain any inhibitor-sensitive factor or substrate.

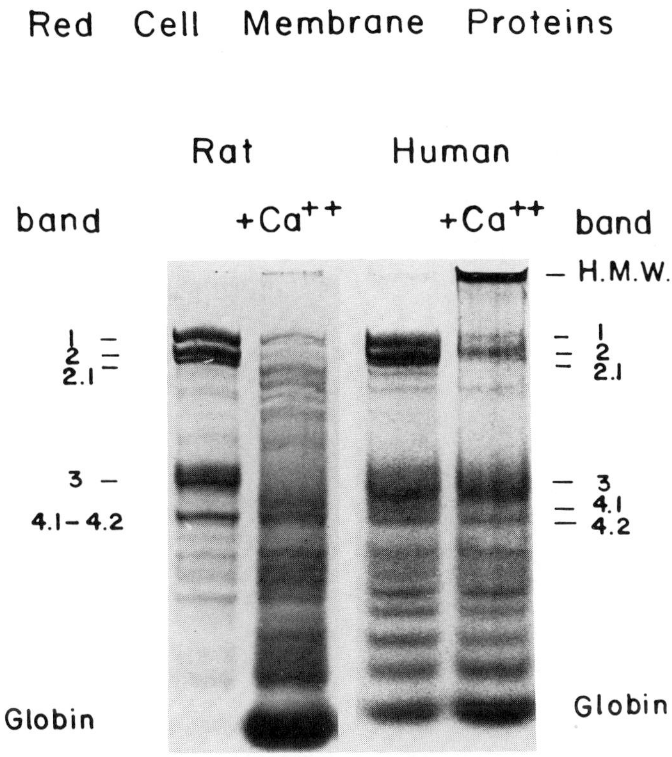

FIGURE 4. SDS-polyacrylamide gel electrophoretic patterns of erythrocyte membrane proteins isolated from rat and from human cells treated with Ca^{2+} and ionophore. (Controls: no additions.) Cell suspentions (8%) in buffer were incubated with 0.5 m*M* Ca^{2+} and 10 *M* ionophore A23187 at 37°C for 20 min (rat cells) or 90 min (human cells). Cell suspensions were then mixed with 2 m*M* EDTA, cooled to 4°C, and cetrifuged; ghosts were prepared and solubilized. Lanes: A and C, controls; B and D, with Ca^{2+}. HMW, high M_r. (From Kosower, N. S., Glaser, T., and Kosower, E. M., *Proc. Natl. Acad. Sci. U.S.A.*, 80, 7542, 1983. With permission.)

We then purified the cytoplasmic enzyme required for the membrane proteolysis and fusion and identified it as being the μ*M* Ca^{2+}-requiring protease, calpain I.[28] The μ*M* and m*M* Ca^{2+}-proteases,[29,30] calpain I and II, have been identified in various tissues.[29-32] Calpain I was identified in rat erythrocytes by Murachi et al.[31,32] Erythrocytes, among other cells, also contain an endogenous inhibitor specific to calpain, calpastatin.[31,32] Calpastatin activity has been determined by the inhibition of calpain-induced degradation of casein.[31,32] Based on the assay of casein degradation, calpastatin is present to a different extent in different cells. The rat erythrocyte contains an excess of calpain activity over that of calpastatin, whereas the human cell contains an excess of calpastatin.[31]

Based on the studies of Murachi et al.,[31,32] we proceeded to investigate whether calpain *vs.* calpastatin activities could be responsible for the different fusion behavior of the rat and human erythrocytes, or whether fusibility was due to a specific membrane structure. Calpain was isolated from rat or from human erythrocytes and calpastatin isolated from human cells. We found that calpain, trapped in the presence of Ca^{2+} within erythrocyte ghosts, led to membrane protein degradation. These ghosts fused when they were further treated with A_2C.[33] No degradation nor fusion were seen when Mg^{2+} was used instead of Ca^{2+}.[33] Ca^{2+} seems to be necessary only during the first step of limited proteolysis, since removal of Ca^{2+} by washing and the additon of EGTA at that time did not prevent the subsequent fusion. Both rat and human cell ghosts underwent degradation by calpain isolated from either cell (Figures 5 and 6). Both rat and human calpain-treated cell ghosts underwent

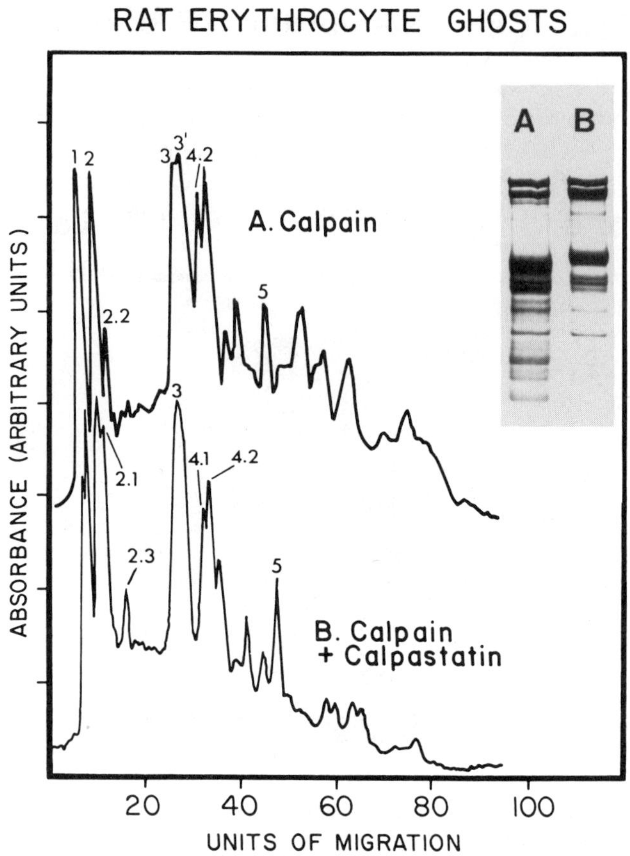

FIGURE 5. Densitometric scans of SDS-polyacrylamide gel electrophoretic patterns of membrane proteins from rat erythrocyte ghosts. Ghost suspensions (4% in buffer containing 135 m*M* NaCl, 10 m*M* Tris-HCl (pH 7.4), 0.1 m*M* EGTA, 0.5 m*M* DTT, 0.5 m*M* $CaCl_2$ and 20 mg/ml Dextran (T-70, Pharmacia)) were mixed at 4°C with a solution of calpain or calpain and calpastatin, kept at 4°C for 20 min, then incubated at 37°C for 40 min. Ghost suspensions were then mixed with 5 m*M* EDTA, centrifuged, and solubilized. Ghosts incubated with (A) 4 units calpain/ml ghosts suspension, (B) 4 units calpain and 6 units of calpastatin (about 30 to 50 units of calpain and about 60 to 80 units of calpastatin are obtained /ml erythrocytes).

fusion when further treated with A_2C (Figure 7). Calpastatin, isolated from the human cell, inhibited proteolysis and fusion in both types of cell membranes (Figures 5 to 7).

Thus, both types of cell membranes are capable of fusion, provided membrane protein degradation occurs as a precondition for fusion. The difference in fusibility could be ascribed to different calpain/calpastatin activities in the two types of erythrocytes.

As discussed in Section I, motion and redistribution of membrane components are generally considered to be necessary for fusion. Partial degradation of certain proteins could allow the motion of membrane components. The actual mobility must be enhanced with a reagent such as A_2C.

The steps in the overall A_2C-induced erythrocyte membrane fusion is summarized in Scheme 1.

IV. SENSITIVITY OF ERYTHROCYTE MEMBRANE PROTEINS TO CALPAIN

The erythrocyte membrane has been extensively used as a convenient system for the study of the structure and behavior of eukaryotic cell membranes. Like other membranes,

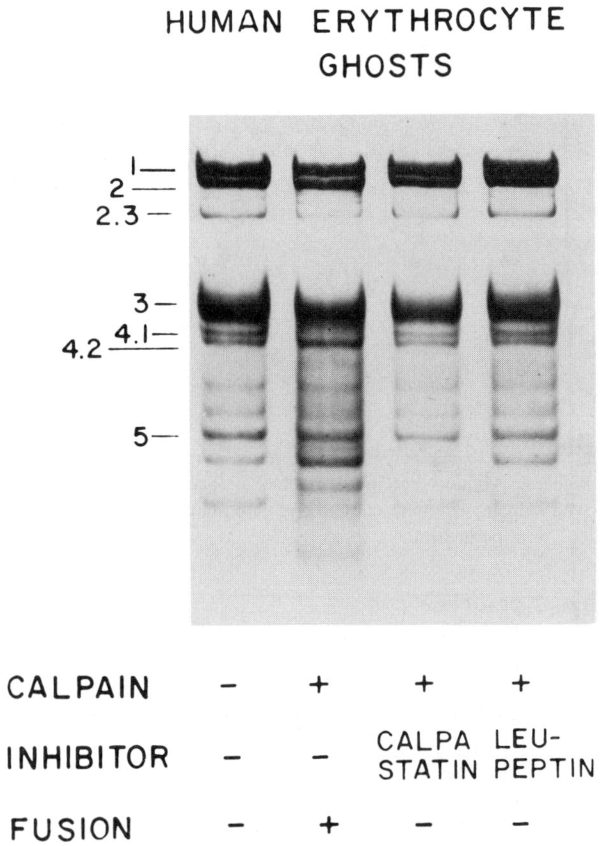

FIGURE 6. SDS-polyacrylamide gel electrophoretic patterns of membrane proteins isolated from fusing and nonfusing human erythrocyte ghosts. Ghosts were suspended in buffer and incubated, as described in Figure 5. Ghost suspensions were then centrifuged, resuspended in buffer, and mixed with half a volume of A_2C (0.5 l A_2C/ml 150 m*M* NaCl). Suspensions were further incubated at 37 °C for 60 min. Ghosts incubated (A) without calpain, (B) with 8 units calpain/ml ghost suspension, (C) with 8 units calpain and 15 units calpastatin, and (D) with 8 units calpain and 1.0 m*M* leupeptin. (From Glaser, T. and Kosower, N. S., *FEBS Lett.*, 206, 115, 1986. With permission.)

the erythrocyte is composed of a lipid bilayer, with some proteins traversing it (integral proteins), while other proteins are associated with the cytoplasmic surface (peripheral proteins). The main integral, transmembrane proteins are glycophorin, which traverses the membrane once, and band 3, with multiple membrane-spanning segments. The peripheral proteins include the cytoskeleton proteins and proteins which bind the cytoskeleton to the integral proteins and lipids. The main cytoskeletal proteins are spectrin, actin, band 4.1 and band 4.9, with several other less well-characterized proteins.[34,35]

The principal linking proteins that connect the cytoskeleton to the overlying membrane are band 2.1 (ankyrin) and band 4.1. Ankyrin anchors spectrin molecules to the membrane via band 3, by binding to a specific segment of β spectrin and to the cytoplasmic segment of band 3.[34-36] Band 4.1 promotes the interaction of spectrin with actin. In addition, it links the cytoplasmic assembly to the membrane. Proposed membrane-binding sites for band 4.1 include band 3, glycophorins A and C. Band 4.1 may also associate directly with phosphatidylserine.[34-36] Band 3, being the major integral membrane component, may be the important site for 4.1 binding to the membrane.

The cytoskeleton structural proteins and binding proteins are responsible for the stability

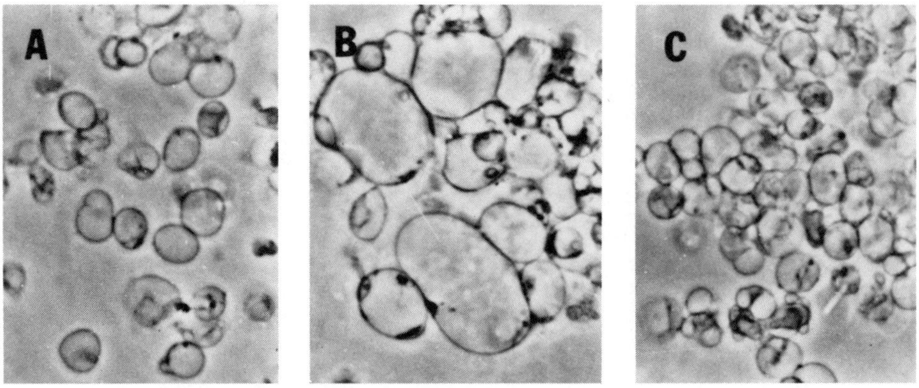

FIGURE 7. Micrographs showing fusing and nonfusing human erythrocyte ghosts. Ghost suspensions were incubated as described in Figure 6. Aliquots were photographed after 60 min of incubation. Ghosts were incubated (A) without calpain, (B) with calpain, (C) with calpain and calpastatin. (From Glaser, T. and Kosower, N. S., *FEBS Lett.*, 206, 115, 1986. With permission.)

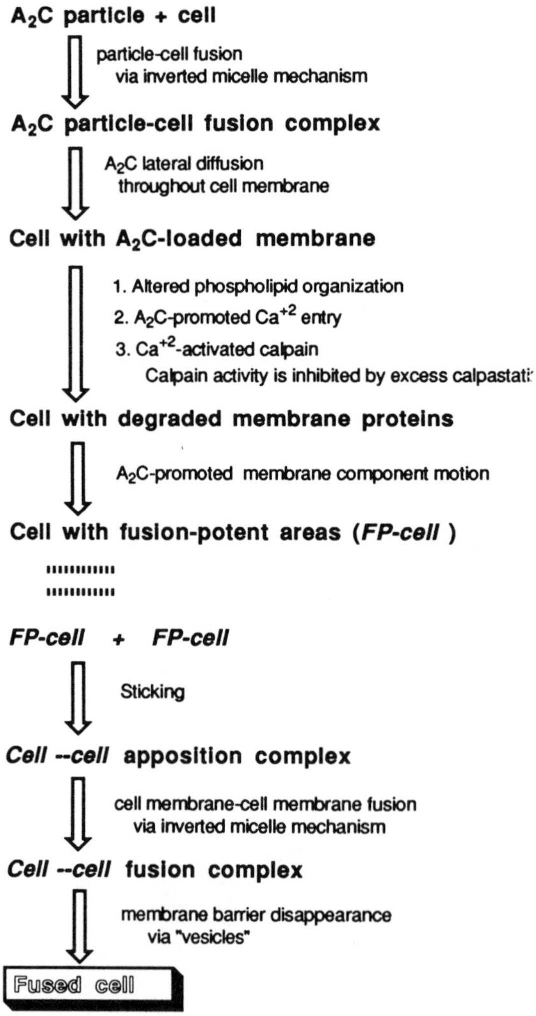

Scheme 1

and shape of the erythrocyte and play a role in maintaining the bilayer phospholipid asymmetry.[37] The binding of band 3 to the cytoskeleton via ankyrin and band 4.1 would restrict the lateral mobility of band 3. Glycophorin mobility would also be restricted because of its interaction with band 4.1. Partial degradation of the cytoskeleton, of the binding proteins or of intrinsic membrane proteins could release the restraints and allow the motion of membrane components, necessary for membrane fusion.

We found that a limited and selective calpain-induced proteolysis was a prerequisite for A_2C-induced fusion (Figures 5 to 7). A degradation of band 4.1 and a partial degradation of band 2.1 occurred in both rat and human erythrocytes. Thus, proteolysis of the binding proteins may be the necessary step in this membrane fusion process. Alterations in membrane proteins, however, were not identical in the rat and human erythrocytes. In the rat cell membrane, calpain also induced a partial degradation of band 3, along with the degradation of 2.1 and 4.1. The addition of leupeptin prior to calpain treatment prevented proteolysis and subsequent A_2C-induced fusion. The addition of the inhibitor following the limited protein degradation prevented further proteolysis of band 3, but did not prevent the subsequent A_2C-induced fusion. Under the conditions used for promoting membrane fusion, spectrin in the rat cell membrane was much less sensitive to calpain than band 3 (Figure 5). In contrast, in the human erythrocyte cell membrane, band 3 was not noticeably degraded by calpain (Figure 6). Some degradation of spectrin in the treated human erythrocyte membranes occurred, though it was not noticeable in the beginning of fusion. Glycophorins were not degraded by calpain in either of the cell membranes. Actin also did not appear to be degraded in either of the cell membranes.

The different sensitivity to calpain may be due to differences in the structural arrangements of membrane lipids and proteins and/or to differences in target proteins and in substrate specificity. We found that the differential sensitivity among the membrane proteins and the differences between the two types of cells were retained in partially purified membrane vesicles or proteins. Proteins band 2.1 and band 4.1 were degraded in both rat and human cell membrane vesicles freed of spectrin (Figure 8), whereas band 3 was visibly degraded only in rat cell vesicles (not shown).

Band 3 preparation (membranes stripped of peripheral proteins at a high pH) and spectrin, isolated from rat and human cell membranes, behaved similarly to the proteins in the intact membrane. Rat band 3 was more sensitive to calpain than human band 3 (Figure 9). Human spectrin was degraded more than rat spectrin (Figure 10). In all of these preparations, proteolysis was inhibitied by calpastatin, isolated from the human erythrocyte. Glycophorins, isolated from both types of membranes were not degraded, as was the case for glycophorins in the intact membranes.

Degradation of bands 2.1, 4.1, spectrin, and band 3 by calpain has been observed in various membrane preparations,[38-42] with differences noted among the studies in the degree of sensitivity expressed by the individual membrane proteins.[38-41] Our results confirm and extend these observations by finding that substrate specificity across species may not be as well conserved as previously assumed. The fact that both rat and human cell membranes were rendered fusible by calpain indicates, however, that the same functional consequence may be achieved in dissimilar membranes.

V. RELATION TO OTHER MEMBRANE FUSION PROCESSES

A number of natural and experimental membrane fusion processes may be understood in terms of the mechanism and factors discussed here. These include processes such as virus-cell fusion, acrosome reaction, neurotransmitter release and myoblast fusion. Okada et al.[43,44] have shown that the virus-induced fusion may be divided into two stages, the virus-cell fusion and cell-cell fusion. The virus adheres to the cell membrane and fuses with it. Viral components are dispersed within the cell membrane. Mobility of membrane components and

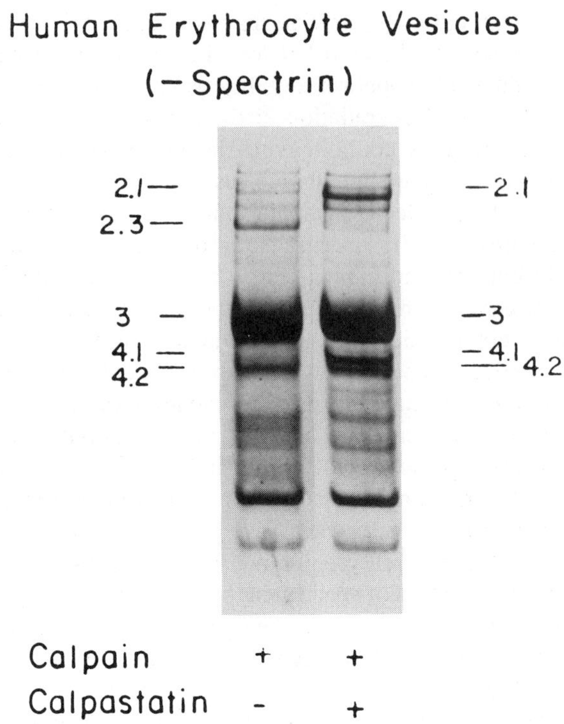

FIGURE 8. SDS-polyacrylamide gel electrophoretic patterns of human erythrocyte membrane proteins from spectrin-free vesicles. Ghosts were suspended in spectrin extracting buffer (0.1 m*M* EDTA, 0.3 m*M* Na phosphate, pH8.0) at 37°C for 30 min, then centrifuged at 40,000 × g for 60 min and resuspended in buffer as described in Figure 5. Vesicles were incubated with calpain or with calpain and calpastatin. For details, see Figure 5.

the redistribution of intramembraneous particles in the erythrocyte membrane induced by Sendai virus is a prerequisite for cell-cell fusion, which proceeds at multiple points along the contact area.[45-47] Thus, the overall stages of virus-induced fusion are similar to A_2C-induced fusion.

A redistribution of membrane components and the formation of particle-depleted areas have been reported in a variety of other natural and experimental fusion processes such as exocytosis of synaptic vesicles, degranulation of mast cells, sperm acrosome reaction, myoblast fusion, endocytosis and chemically-induced cell fusion.[48-51]

In all of these cases protease activity may facilitate the formation of fusion-potent regions, by releasing membrane components from restraints. The actual redistribution may be promoted by certain reagents or by endogenous molecules, such as endogenous phospholipids, unsaturated fatty acids or hydrophobic proteins.[50-52]

Several types of proteases such as serine proteases, metalloprotease and Ca^{2+}-activated proteases have been implicated in fusion events.[1,2,51,53-55] Direct evidence for a requirement for any of the enzymes during these fusion processes is still lacking. The substrates involved have not been identified and the relation to the overall fusion processes not clarified.

In the case of sperm acrosome reaction, trypsin-like enzyme and metalloenzyme have been suggested as playing a role in the process.[53] However, all cases of acrosome reaction require Ca^{2+} and it is of interest that calpain has been found in the acrosome region, as identified by antibodies.[56] Acrosome reaction, induced by lysophosphatidylcholine and by A_2C, is inhibited by leupeptin but not by PMSF (Yanagimachi et al.,[74] unpublished results). Thus, calpain may play a role in sperm acrosome reaction and such a possibility should be further studied.

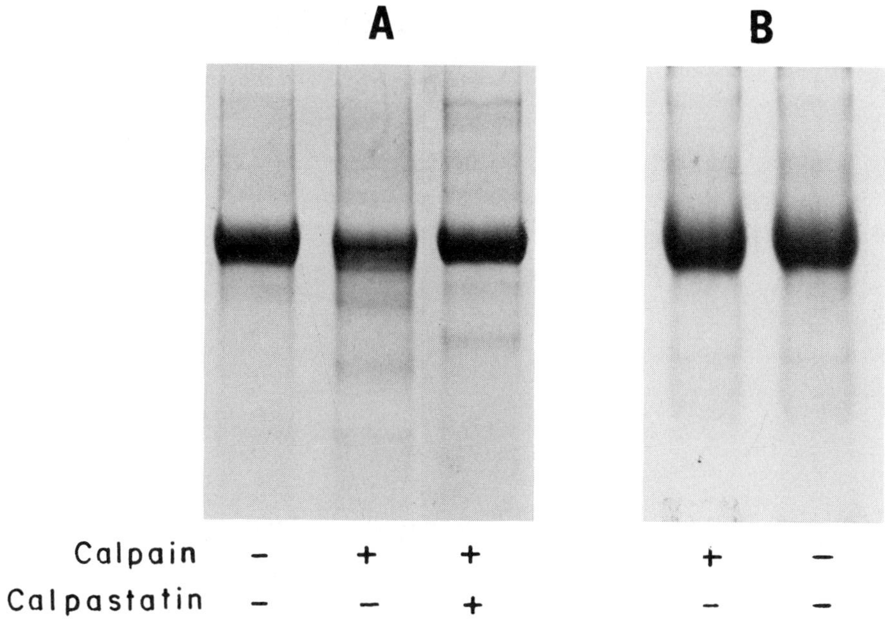

FIGURE 9. SDS-polyacrylamide gel electrophoretic patterns of band 3 from rat and human erythrocyte membranes. Ghosts were stripped of peripheral proteins by treatment with 0.1 *N* NaOH in ice for a few seconds, centrifuged, washed, and resuspended in buffer pH7.4, as described in Figure 5. Aliquots (0.2 mg protein/ml) were incubated at 37°C without or with calpain, or with calpain and calpastatin. For details see Figure 5. (A) Rat erythrocyte band 3; (B) human erythrocyte band 3.

Metalloenzymes may be involved in exocytosis and in myoblast fusion.[54] Their relation to Ca^{2+} requirement for this fusion processes is, however, not clear. Calpain activity has been shown to appear in differentiating myoblasts, concomitant with myotube formation.[57] The undifferentiated myoblasts have been shown to contain a calpain inhibitor (protein in nature but unidentified) which is lost during myoblast differentiation.[57] Calpain II has been identified by antibodies in myoblasts. It exhibits a peripheral, membrane-associated distribution in fusing myoblasts, whereas it is diffuse in the cytoplasm in nonfusing cells.[58] Based on these findings, it has been proposed that calpain II may facilitate myoblast fusion by promoting the disassembly of the cytoskeleton-membrane linkage.[58] It would be important to further study the involvement of calpain in myoblast fusion.

No clear information is available on other types of membrane fusion events. Heterogeneity in fusibility has been observed in certain cells.[59,60] The involvement of the calpain-calpastatin system in some of these cases remains to be studied. Also, the possibility of calpain-induced protein degradation in other fusion processes (e.g., induced by virus,[4] PEG,[5,60] electro-fusion[11]) should be investigated.

The specific site proteolyzed on a protein may determine the consequent fusion behavior. Thus, ankyrin has two regulatory domains present at opposite ends of the molecule. A deletion of one end of the molecule (by a noncalpain enzyme or by an alternate splicing of mRNA) gives rise to a fragment with enhanced binding to band 3. In contrast, calpain induces degradation of ankyrin at the opposite end of the molecule, resulting in a fragment of a similar size, but one which has lost the ability to bind to band 3.[42]

Proteins analogous to the erythrocyte membrane proteins have been identified in a wide variety of cell types.[34-36] These include analogues to spectrin, band 3, ankyrin, and band 4.1. As has been pointed out,[40] the high susceptibility of band 4.1 to calpain may be of great importance to membrane function. A specific cleavage of ankyrin[42] and of band 4.1

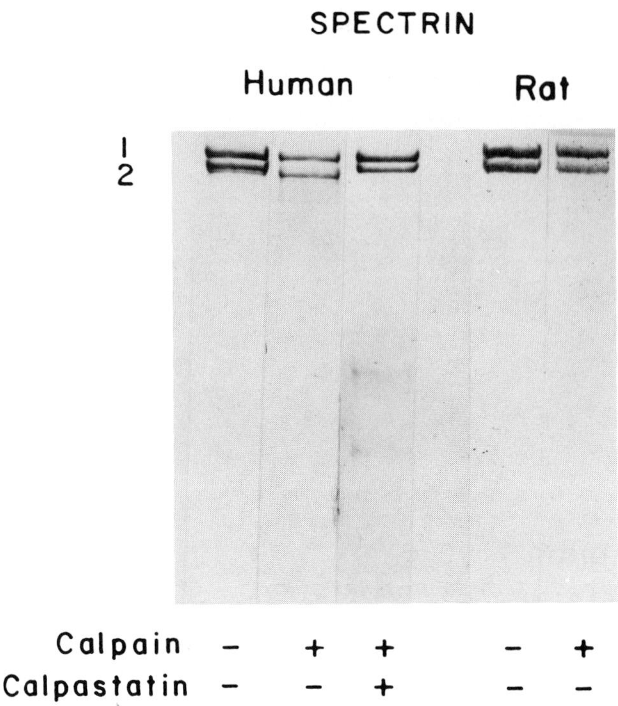

FIGURE 10. SDS-polyacrylamide gel electrophoretic patterns of spectrin from rat and human erythrocyte membranes. Ghosts were suspended in spectrin extracting buffer and treated as described in Figure 8. Supernatant (extracted spectrin) was incubated with calpain or with calpain and calpastatin solution in buffer containing 135 m*M* NaCl, 10 m*M* Tris-HCl, pH 7.4 in the presence of 0.2 m*M* PMSF, 0.2 m *M* Ca^{2+} and 0.5 m*M* DTT, then analyzed by gel electrophoresis. (A) Human spectrin solution (0.1 mg/ml) incubated with 4 units of calpain with or without 6 units of calpastatin; (B) rat spectrin solution incubated with 6 units of calpain.

could thus help in the separation of cytoskeleton elements from other intracellular or plasma membrane components. The degradation of spectrin-like or of band 3 analogues may also occur in certain cells and help in fusion-associated events.

Questions have been raised in regards to the activation of calpain under physiological conditions.[29-31,61-67] The regulation of calpain activity involves the availability of Ca^{2+}, the interaction with calpastatin and localization of the enzyme near the substrates. As has been well established, calpain is a cytoplasmic enzyme, present ubiquitously in a variety of cells, with two isozymes identified. These are calpain I and calpain II, requiring micromolar and millimolar concentrations of Ca^{2+}, respectively, for activation.[29-31,61-67] For membrane fusion processes, Ca^{2+} enters the cell or is mobilized from intracellular stores.[3,68] For some fusion events, such as sperm acrosome reaction and myoblast fusion, external millimolar Ca^{2+} is necessary.[1,2] The influx of Ca^{2+} may activate calpain II, which has been identified in the sperm acrosome and in association with the membrane in fusing myoblasts.[56,58] In other cases, such as some exocytotic fusion events, intracellular micromolar Ca^{2+} seems to suffice.[3] In these cases, calpain I may be involved. In additon, calpain II may be activated by autoproteolysis.[62-67] Thus, the transient influx of Ca^{2+} in certain fusion processes may convert the m*M* form of calpain to a form requiring less Ca^{2+} for activity. The translocation of calpain I or II to specific membrane locations would ensure calpain activation near the substrates,[63-67] provided calpain is not rapidly inactivated by further autoproteolysis[62,69] and provided calpastatin is not present there.[65-67]

Calpastatin is also widely present and is found in different amounts and molecular forms in various cells.[31,66] Calpastatin binds to calpain in solution in Ca^{2+}-dependent manner, with one calpastatin molecule inactivating multiple molecules of calpain.[66,70-72] Under usual experimental conditions (i.e., using calpastatin samples after heating of solutions) calpain is inactivated by calpastatin at Ca^{2+} concentrations lower than those necessary for calpain activation.[71] However, the nature of the interaction of native calpastatin with calpain in the cell is not yet clear and the factors regulating this interaction are not known. The demonstrated activity of calpastatin in modulating membrane fusibility provides an example for calpastatin activity in cells and cell fractions.

Little is known about the cellular localization of calpastatin and about its interactions with cellular and membrane factors. Some types of calpastatin may be associated with membranes, whereas other types are not.[73] The binding of calpastatin to a membrane may require an amino terminal region. Mellgren has suggested that a limited proteolysis of calpastatin would release it from the membrane and allow calpain activity at that site.[66,73] In addition, the intracellular location of calpastatin may be more restricted than the distribution of calpain, so that some sites may have calpain but not calpastatin.[71,73] Further studies are required for clarifying the regulation of calpain-calpastatin interaction within cells and relation to processes such as membrane fusion. Roles of cellular and membrane components such as adducin, calmodulin, and calpactins in these processes also remain to be studied.[34]

REFERENCES

1. **Yanagimachi, R.**, Mammalian fertilization, in *The Physiology of Reproduction,* Knobil E. and Neill, J., Eds., Raven Press, New York, 1988, 135.
2. **Wakelam, M. S. O.**, The fusion of myoblasts, *Biochem. J.*, 228, 1, 1985.
3. **Blumenthal, R.**, Membrane fusion, *Curr. Top. Membr. Transp.*, 29, 203, 1987.
4. **Okada, Y.**, Analysis of giant polynuclear cell formation caused by HVJ virus in Ehrlich ascites tumor cells. I. Microscopic observation of giant polynuclear cell formation, *Exp. Cell Res.*, 26, 98, 1962.
5. **Lucy, J. A.**, Biochemical and ultrastructural studies on chemically induced cell fusion, in *Membrane-membrane Interactions,* Gilula, W. B., Ed., Raven Press, New York, 1980, 81.
6. **Kao, K. and Michayluk, M. R.**, A method for high-frequency intergeneric fusion of plant protoplasts, *Planta,* 115, 355, 1973.
7. **Kosower, E. M., Kosower, N. S., Faltin, Z., Diver, A., Saltoun, G., and Fransdorf, A.**, Membrane mobility agents. A new class of biologically active molecules, *Biochim. Biophys. Acta,* 363, 261, 1974.
8. **Kosower, E. M., Kosower, N. S., and Wegman, P.**, Membrane mobility agents. II. Active promoters of cell fusion, *Biochim. Biophys. Acta,* 401, 530, 1975.
9. **Fleming, A. D. and Yanagimachi, R.**, Effects of various lipids on the acrosome reaction and fertilizing capacity of guinea pig spermatozoa, with special reference to the possible involvement of lysophospholipids in the acrosome reaction, *Gamete Res.*, 4, 253, 1981.
10. **Saunders, J. A. and Bates, G. W.**, Chemically induced fusion of plant protoplasts, in *Cell Fusion,* Sowers, A. E., Ed., Plenum Press, New York, 1987, 397.
11. **Sowers, A. E.**, Fusion events and nonfusion contents mixing events induced in erythrocyte ghosts by an electric pulse, *Biophys. J.*, 54, 619, 1988.
12. **Kosower, E. M., Kosower, N. S., and Wegman, P.**, Membrane mobility agents. IV. The mechanism of particle-cell and cell-cell fusion, *Biochim. Biophys. Acta,* 471, 311, 1977.
13. **Eytan, D. E. and Eytan, E.**, Fusion of proteoliposomes and cells, *J. Biol. Chem.*, 255, 4992, 1980.
14. **Session, A. and Horwitz, A. F.**, Differentiation-related differences in the plasma membrane phospholipid asymmetry of myogenic and fibrogenic cells, *Biochim. Biophys. Acta,* 728, 103, 1983.
15. **Neher, E.**, Asymmetric membranes resulting from the fusion of two black lipid bilayers, *Biochim. Biophys. Acta,* 373, 327, 1974.

16. **Lustig, S., Pluznik, D. H., Kosower, N. S., and Kosower, E. M.,** Membrane mobility agent alters the consequences of lectin-cell interaction in a malignant cell membrane, *Biochim. Biophys. Acta,* 401, 458, 1975.
17. **Lustig, S., Kosower, N. S., Pluznik, D. H., and Kosower, E. M.,** Inhibition of cytokinesis in Friend leukemia cells by membrane mobility agents, *Proc. Natl. Acad. Sci. U.S.A.,* 74, 2884, 1977.
18. **Kosower, N. S., Faltin, Z., and Kosower, E. M.,** Cell membrane receptor classes delimited through cap formation either with diamide or with membrane mobility agent, A_2C, *J. Immunol. Methods,* 41, 215, 1981.
19. **Douglas, S. D., Zuckerman, S. H., and Cody, C. S.,** Alterations in intramembrane particle distribution during interaction of erythrocyte-bound ligands with immunoprotein receptors: effect of the membrane mobility agent A_2C on immunologic and nonimmunologic ligand-binding, *J. Reticuloendoth. Soc.,* 28, 91, 1980.
20. **Kosower, N. S., Wegman, P. O., Neiman, T., and Kosower, E. M.,** Membrane mobility agents. V. Genetic variability in the fusibility of hen red cells, *Exp. Cell Res.,* 116, 454, 1978.
21. **Tavassoli, M. T., Kosower, N. S., Halverson, C., Aoki, M., and Kosower, E. M.,** Membrane fusion induced by the membrane mobility agent, A_2C. Differentiation between fusible and nonfusible cells. Transfer of fusibility, *Biochim. Biophys. Acta,* 601, 544, 1980.
22. **Fleming, A. D., Kosower, N. S., and Yanagimachi, R.,** Promotion of capacitation of guinea pig spermatozoa by the membrane motility agent, A_2C and inhibition by the disulfide reducing agent, DTT, *Gamete Res.,* 5, 19, 1982.
23. **Kuhn, H. and Parish, R. W.,** Fusion of dictyostelium cells with one another and with erythrocyte ghosts, *Exp. Cell Res.,* 131, 89, 1981.
24. **Kosower, N. S., Zipser, Y., and Kosower, E. M.,** Membrane mobility agents: alteration of human red blood cell membrane properties, *Arch. Biochem. Biophys.,* 203, 325, 1980.
25. **Zipser, Y., Shachar, R., Goldfarb, A., Rachmilewitz, E., and Kosower, N. S.,** Organization of membrane phospholipids (PL) in β-thalassemia red cells (Abstr.), 4th Int. Congr. Cell Biol., Montreal, August 14—19, 1988, 218.
26. **Kosower, N. S., Glaser, T., and Kosower, E. M.,** Membrane-mobility agent-promoted fusion of erythrocytes: fusibility is correlated with attack by calcium-activated cytoplasmic proteases on membrane proteins, *Proc. Natl. Acad. Sci. U.S.A.,* 80, 7542, 1983.
27. **Siefring, G. E., Jr., Apostol, A. B., Velasco, P. T., and Lorand, L.,** Enzymatic basis for the Ca^{2+}-induced cross-linking of membrane proteins in intact human erythrocytes, *Biochemistry,* 17, 2598, 1978.
28. **Glaser, T. and Kosower, N. S.,** Fusion of rat erythrocytes by membrane mobility agent A_2C depends on membrane proteolysis by a cytoplasmic calpain, *Eur. J. Biochem.,* 159, 387, 1986.
29. **Dayton, W. R., Reville, W. J., Goll, D. E., and Stromer, M. H.,** A Ca^{2+}-activated protease possibly involved in myofibrillar protein turnover. Partial characterization of the purified enzyme. *Biochemistry,* 15, 2159, 1976.
30. **Mellgren, R. L.,** Canine cardiac calcium-dependent proteases: resolution of two forms with different requirements for calcium. *FEBS Lett.,* 109, 129, 1980.
31. **Murachi, T., Tanaka, K., Hatanaka, M., and Murakami, T.,** Intracellular Ca^{2+}-dependent protease (calpain) and its high-molecular-weight endogenous inhibitor (calpastatin), *Adv. Enzyme Regul.,* 19, 407, 1981.
32. **Murachi, T.,** Calcium-dependent proteinases and specific inhibitors: calpain and calpastatin, *Biochem. Soc. Symp.,* 49, 149, 1984.
33. **Glaser, T. and Kosower, N. S.,** Calpain-calpastatin and fusion: fusibility of erythrocytes is determined by a protease-protease inhibitor (calpain-calpastatin) balance, *FEBS Lett.,* 206, 115, 1986.
34. **Mangeat, P. H.,** Interaction of biological membranes with the cytoskeletal framework of living cells, *Biol. Cell,* 64, 261, 1988.
35. **Bennett, V.,** The spectrin-actin junction of erythrocyte membrane skeletons, *Biochim. Biophys. Acta,* 988, 107, 1989.
36. **Low, P. S.,** Structure and function of the cytoplasmic domain of band 3: center of erythrocyte membrane peripheral protein interactions, *Biochim. Biophys. Acta,* 864, 145, 1986.
37. **Sheetz, M. P.,** Membrane skeletal dynamics: role in modulation of red cell deformability, mobility of transmembrane proteins and shape, *Semin. Hematol.,* 20, 175, 1983.
38. **Murakami, T., Hatanaka, M., and Murachi, T.,** The cytosol of human erythrocytes contains a highly Ca^{2+}-sensitive thiol protease (calpain I) and its specific inhibitor protein (calpastatin), *J. Biochem.,* 90, 809, 1981.
39. **Siman, R., Baudry, M., and Lynch, G.,** Brain fodrin: substrate for calpain I, an endogenous calcium-activated protease. *Proc. Natl. Acad. Sci. U.S.A.,* 81, 3572, 1984.
40. **Croall, D. E., Morrow, J. S., and DeMartino, G. N.,** Limited proteolysis of the erythrocyte membrane skeleton by calcium-dependent proteinases, *Biochim. Biophys. Acta,* 882, 287, 1986.

41. **Pontremoli, S., Melloni, E., Sparatore, B., Michetti, M., and Horecker, B. L.,** A dual role for the Ca^{2+}-requiring proteinase in the degradation of hemoglobin by erythrocyte membrane proteinases, *Proc. Natl. Acad. Sci. U.S.A.,* 81, 6714, 1984.
42. **Hall, T. G. and Bennett, V.,** Regulatory domains of erythrocyte ankyrin, *J. Biol. Chem.,* 262, 10537, 1987.
43. **Okada, Y., Koseki, I., Kim, I., Maeda, Y., Hashimoto, T., Kanno, Y., and Matsui, Y.,** Modification of cell membranes with viral envelopes during fusion of cells with HVJ (Sendai virus), *Exp. Cell Res.,* 93, 368, 1975.
44. **Okada, Y.,** Sendai virus-mediated cell fusion, in *Curr. Top. Membr. Transp.,* 22, 297, 1988.
45. **Bachi, T., Aguet, M., and Howe, C.,** Fusion of erythrocytes by Sendai Virus studied by immuno-freeze-etching, *J. of Virol.,* 2, 1004, 1973.
46. **Sekiguchi, K. and Asano, A.,** Participation of spectrin in Sendai virus-induced fusion of human erythrocyte ghosts, *Proc. Natl. Acad. Sci. U.S.A.,* 75, 1740, 1978.
47. **Knutton, S.,** The mechanism of virus-induced cell fusion, *Micron,* 9, 133, 1978.
48. **Chi, E. Y., Lagunoff, D., and Koehler, J. K.,** Freeze-fracture study of mast cell secretion, *Proc. Natl. Acad. Sci. U.S.A.,* 73, 2823, 1976.
49. **Friend, D. S., Orci, L., Perrelet, A., and Yanagimachi, R.,** Membrane particle changes attending the acrosome reaction in guinea pig spermatozoa, *J. Cell Biol.,* 74, 561, 1977.
50. **Kalderon, N. and Gilula, N. B.,** Membrane events involved in myoblast fusion, *J. Cell Biol.,* 81, 411, 1979.
51. **Lucy, J. A.,** Fusogenic mechanisms, in *Cell Fusion,* CIBA Foundation Symposium, 1984, 28.
52. **Scott, J. H., Creutz, C. E., Pollard, H. B., and Ornberg, R.,** Synexin binds in a calcium-dependent fashion to oriented chromaffin cell plasma membranes, *FEBS Lett.,* 180, 17, 1985.
53. **Meizel, S.,** The importance of hydrolytic enzymes to an exocytotic event, the mammalian sperm acrosome reaction, *Biol. Rev.,* 59, 125, 1984.
54. **Strittmatter, W. J., Couch, C. B., and Mundy, D. I.,** Role of metalloendoprotease in the fusion of biological membranes, in *Cell Fusion,* Sowers, A. E., Ed., Plenum Press, New York, 1987, 99.
55. **Thomas, P., Limbrick A. R., and Allan, D.,** Limited breakdown of cytoskeletal proteins by an endogenous protease controls Ca^{2+}-induced membrane fusion events in chicken erythrocytes, *Biochim. Biophys. Acta,* 730, 351, 1983.
56. **Schollmeyer, J. E.,** Identification of Calpain II in porcine sperm, *Biol. Reprod.,* 34, 721, 1986.
57. **Kaur, H. and Sanwal, B. D.,** Regulation of the activity of a calcium-activated neutral protease during differentiation of skeletal myoblasts, *Can. J. Biochem.,* 59, 743, 1981.
58. **Schollmeyer, J. E.,** Possible role of calpain I and calpain II in differentiating muscle, *Exp. Cell Res.,* 163, 413, 1986.
59. **Koyama, A. H., Maeda, T., Toyama, S., Ohnishi, S., Uetake, H., and Toyama, S.,** Interaction between Sendai virus and KB cell lines with altered cell fusion ability: a study employing electron spin resonance, *Biochim. Biophys. Acta,* 508, 130, 1978.
60. **Roos, D. S., Davidson, R. L., and Choppin, P. W.,** Control of membrane fusion in polyethylene glycol-resistant cell mutants. Applications to fusion technology, in *Cell Fusion,* Sowers, A. E., Ed., Plenum Press, New York, 1987, 123.
61. **Goll, D. E., Edmunds, T., Kleese, W. C., Sathe, S. K., and Shannon, J. D.,** Some properties of the Ca^{2+}-dependent proteinase, *Prog. Clin. Biol. Res.,* 180, 151, 1985.
62. **Mellgren, R. L., Repetti, A., Muck, T. C., and Easly, J.,** Rabbit skeletal muscle calcium-dependent protease requiring millimolar Ca^{2+}, *J. Biol. Chem.,* 257, 7203, 1982.
63. **Suzuki, K., Imajoh, S., Emori, Y., Kawasaki, H., Minami, Y., and Ohno, S.,** Calcium-activated neutral protease and its endogenous inhibitor, *FEBS Lett.,* 220, 271, 1987.
64. **Fukui, I., Toyohara, H., Ito, K., Hamakubo, T., and Murachi, T.,** Molecular and catalytic characterization of intact heterodimeric and derived monomeric calpains isolated under different conditions from pig polymorphonuclear leukocytes, *Biochemistry,* 27, 3260, 1988.
65. **Pontremoli, S. and Melloni, E.,** Extralysosomal protein degradation, *Annu. Rev. Biochem.,* 55, 455, 1986.
66. **Mellgren, R. L.,** Calcium-dependent proteases: an enzyme system active at cellular membranes?, *FASEB J.,* 1, 110, 1987.
67. **Kuboki, M., Ishii, H., and Kazama, M.,** Procalpain is activated on the plasma membrane and the calpain acts on the membrane, *Biochim. Biophys. Acta,* 929, 164, 1987.
68. **Boss, W. F.,** Fusion-permissive protoplasts, a plant system for studying cell fusion, in *Cell Fusion,* Sowers, A. E., Ed., Plenum Press, New York, 1987, 145.
69. **Crawford, C., Willis, A. C., and Gagnon, J.,** The effects of autolysis on the structure of chicken calpain II, *Biochem. J.,* 248, 579, 1987.

70. **Maki, M., Takano, E., Mori, H., Sato, A., Murachi, T., and Hatanaka, M.,** All four internally repetitive domains of pig calpastatin possess inhibitory activities against calpains I and II, *FEBS Lett.*, 223, 174, 1987.
71. **Goll, D. E., Kleese, W. C., and Szpacenko, A.,** Skeletal muscle proteases and protein turnover, in *Animal Growth Regulation,* Campion, D. R., Hausman, G. J., and Martin, R. J., Eds., Plenum Press, New York, 1989, 141.
72. **Nakamura, M., Inomata, M., Imajoh, S., Suzuki, K., and Kawashima, S.,** Fragmentation of an endogenous inhibitor upon complex formation with high- and low-Ca^{2+}-requiring forms of calcium-activated neutral proteases, *Biochemistry,* 28, 449, 1989.
73. **Mellgren, R. L.,** On the mechanism of binding of calpastatin, the protein inhibitor of calpains, to biologic membranes, *Biochem. Biophys. Res. Commun.*, 150, 170, 1988.
74. **Yanagimachi, R. et al.,** unpublished results.

Chapter 13

CALPAINS AND REGULATION OF PROTEIN KINASE C

Keisuke Hashimoto, Katsuya Mikawa, Tsukasa Kuroda, Katsuhiko Ase, and Akira Kishimoto

TABLE OF CONTENTS

I. INTRODUCTION

Since its discovery in 1964,[1,2] the enzyme Ca^{2+}-dependent neutral protease (calpain) has been proposed to play a role in a variety of Ca^{2+}-mediated processes. Although this protease is found in various mammalian tissues, one of the major drawbacks in assessing the physiological significance of such proteolytic reaction is that the enzyme (calpain II) requires much higher concentrations of Ca^{2+} than are normally present within the cell. Another drawback is probably the irreversible nature of the process such as proteolytic activation of muscle glycogen phosphorylase kinase.[2] Nevertheless, one of these possible drawbacks appears to have been overcome by the finding of a new species of the enzyme (calpain I) which is active at much lower concentrations of this divalent cation.[3,4] Subsequently, nascent calpains have been proposed to be proenzymes, that may be converted to active enzymes by an autocatalytic reaction.[5-7] This kind of reaction is stimulated by the additon of either erythrocyte membrane or phospholipids. Thus, some efforts have been made to clarify the function of calpain in the modification of signal transduction across the membrane. Among them, growth factor receptor, transglutaminase, and cytoskeletal proteins are proposed to be regulated by proteolysis by calpain.[8-10]

Earlier reports from this laboratory[11,12] have shown that protein kinase C (PKC) is cleaved by calpain to produce a catalytically active fragment, and that calpain I, which is active at the micromolar range of Ca^{2+}, may react preferentially with the active form of PKC that is associated with membranes. Recent molecular cloning analysis indicates that PKC exists as a family of multiple subspecies; seven PKC clones named α-, βI-, βII-, γ-, δ-, ε-, and ζ-PKC have been identified from the rat brain.[13] The rat brain PKC has been resolved into three fractions, type I, II, and III, which correspond, respectively, to γ-, β- (βI and βII), and α-PKC, upon hydroxyapatite column chromatography.[14,15] The three types of PKC, α-, β-, and γ-PKC, are cleaved at one or two specific sites in the variable region with significantly different susceptibilities.[16] This difference of susceptibilities coincides with the rates of down-regulation of PKC subspecies within cells treated with the mitogenic phorbol ester TPA.[17] Thus, available evidence suggests that, upon stimulation with extracellular signals, the PKC subspecies become tightly associated with the membrane to be activated, and are subsequently cleaved by calpains due to their susceptibilities, resulting in the modification of this signal transduction pathway. This chapter will briefly summarize the specificity and possible role of the cleavage of PKC with calpain.

II. PKC SUBSPECIES AND EXPRESSION

An increasing body of evidence has indicated that PKC plays a pivotal role in the mechanism by which external signals are transmitted to the cell interior and in tumorigenesis.[18] When a ligand binds to its receptor on the cell surface, inositol phospholipids are hydrolyzed, producing diacylglycerol (DG) and inositol phosphates (IP_3), both being thought to act as "second messengers".[19,20] The primary effects of IP_3 and DG are to mobilize intracellular Ca^{2+} and to activate PKC, respectively, which in turn phosphorylates a range of cellular proteins. PKC also serves as the receptor for phorbol ester, a class of tumor promotors including 12-*O*-tetradecanoylphorbol-13-acetate (TPA); it is probably the inappropriate signals given through PKC which result in the tumor-promoting characteristics of these compounds.[21-23]

Four clones (α, βI, βII, and γ) of the PKC subspecies have been shown to correspond to the subtypes (Type I, II, and III) of PKC which have been chromatographically separated from rat brain extract and well characterized.[15] The structures of these subspecies are schematically shown in Figure 1. These subspecies have a common structure consisting of four conserved (C_1 to C_4) and five variable (V_1 to V_5) regions. The third variable region (V_3)

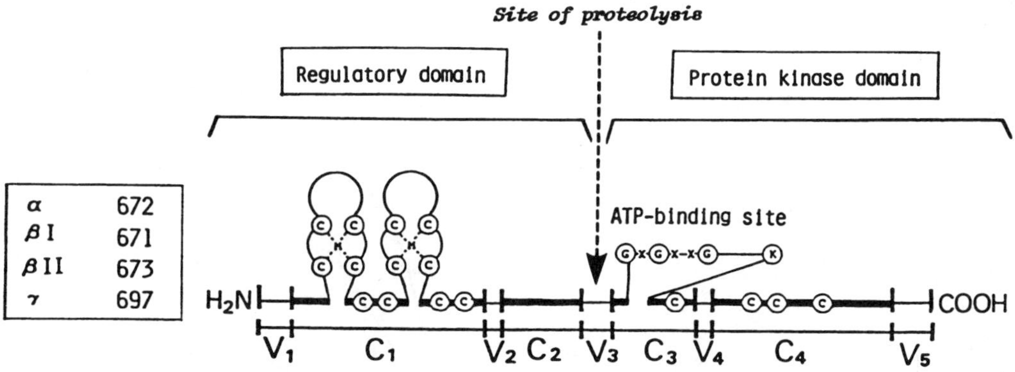

FIGURE 1. Common structure of α-, βI-, βII-, and γ-PKC subspecies. Four conserved (C_1—C_4) and five variable (V_1—V_5) regions are indicated. The numbers in this figure indicate the number of amino acid residues composing each subspecies.[24] C, G, K, X, and M represent cysteine, glycine, lysine, any amino acid, and metal, respectively.

connects the regulatory (N-terminal half) and protein kinase (C-terminal half) domains.[24] The conserved region C_1 contains a tandem repeat of a cysteine-rich sequence resembling the consensus sequence of a ''cysteine-zinc DNA-binding finger''. Although this sequence is found in many metalloproteins and DNA-binding proteins that are related to transcriptional regulation,[25,26] there is no evidence presently indicating that PKC will bind to DNA. The conserved region C_3 has an ATP-binding sequence, Gly-X-Gly-X-X-Gly.....Lys, which shows homology with many other protein kinases.[27] Although the conserved region C_4 contains a similar sequence, Gly-X-Gly-X-X-Gly.....(X), its significance is unknown. The sites involved in binding to calcium, DG, and phospholipid are presumably contained in C_1 and C_2 but have not yet been indentified with any certainty. The hydrolytic cleavage with calpain occurs at one or two specific sites in the region V_3, resulting in the release of the fully catalytically active fragment. For reasons discussed later, it may be related to down-regulation of the PKC molecule itself.

Using biochemical, immunochemical and cytochemical procedures with subspecies-specific antibodies as well as Northern blot and *in situ* hybridization analysis, the relative activity and individual pattern of expression of multiple PKC subspecies in several tissues and cultured cell lines have recently been extensively examined.[13,28-37] Among PKC subspecies, the γ-subspecies appears to be expressed solely in the brain and spinal cord and is particularly concentrated in the hippocampus, cerebellar cortex, and amygdaloid complex.[32] More specially, in cerebellar cortex, the γ-subspecies seems to be localized in the cell body and nerve endings of the Purkinje, whereas βI- and βII-subspecies are stained strongly in granular and molecular layers, respectively.[33-36] The γ-subspecies is associated with membranous structures of the cell in general, except for the nucleus and perhaps mitochondria, which express little PKC.[37] In the brain, the quantity of the γ-subspecies increases postnatally, reaching a maximum in the rat at around 3 weeks after birth.[38] Although it seems premature to discuss the function of this PKC subspecies, it may be important in specialized neuronal processes, such as long-term potentiation in the hippocampus.[39]

The tissue distribution of the βI- and βII-subspecies shows that there is much more βII than βI in the brain and many other tissues.[40,41] Cytochemical analysis with polyclonal antibodies against βI and βII that distinguish the two subspecies reveals a distinct pattern of cellular expression in certain tissues.[33-36] Many tissues including liver, kidney, spleen, lung, heart, and testis contain the β-subspecies (particularly βII) in variable quantities, but the α-subspecies is most widely distributed.[42] Some tissues such as heart, lung, adrenal cortex, and platelets, appear to contain several undefined subspecies,[42,43] Most cell types contain more than one subspecies. For instance, T-lymphocytes express the α-, βI, and βII-

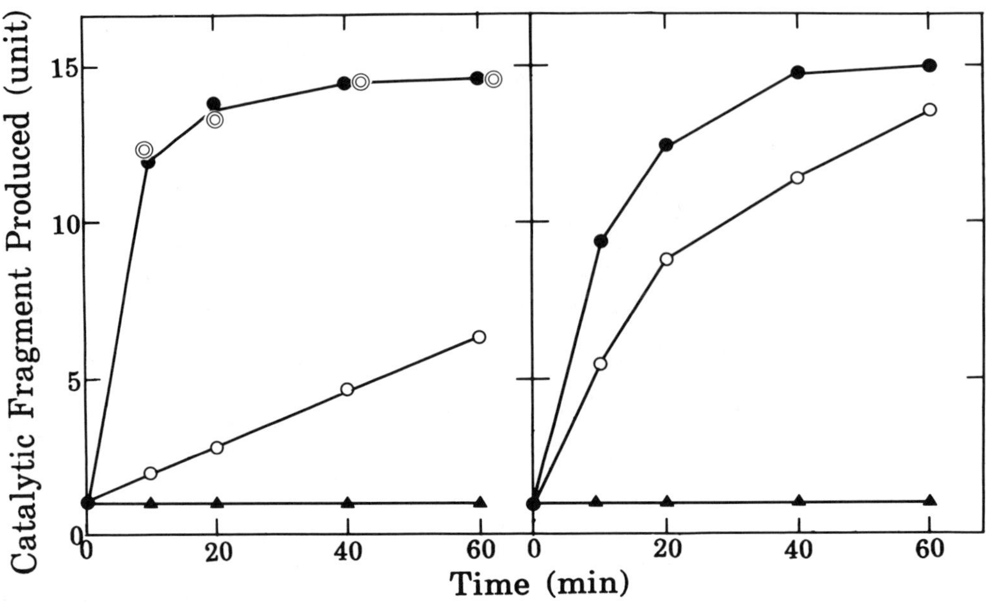

FIGURE 2. Proteolysis of PKC by calpain. Type II PKC was cleaved with either calpain I or II in the presence of $CaCl_2$. Where indicated, phospholipid (100 μg/ml) plus diolein (1 μg/ml), or phospholipid (100 μg/ml) plus TPA (50 n*M*) were added to the reaction mixture. At the various times indicated, an aliquot of the mixture was taken and immediately assayed for the active fragment with H1 histone as substrate in the presence of EGTA. In control experiments, calpain was omitted. (Left) and (right) with calpain I and II, respectively; ●, phospholipid plus diolein; ◎, phospholipid plus TPA; ○, minus phospholipid, diolein and TPA; ▲, minus calpain. The detailed experimental conditions have been described elsewhere.

subspecies in different ratios.[44] Their intracellular distribution may be partly dependent on the state of differentiation of the cell.

III. PROTEOLYSIS OF PKC SUBSPECIES BY CALPAIN

PKC alone is inactive, but exhibits enzymatic activity with the simultaneous addition of Ca^{2+}, phospholipid, and DG.[19] However, PKC subspecies, type I (γ), II (βI and βII), and III (α) each produce a fragment which is fully active without added Ca^{2+}, phospholipid, and DG, when incubated with either calpain I or II. A typical experimental result given in Figure 2 shows the time course of the formation of the active fragment from type II PKC. It is noted that this activation process, particulary with calpain I, is markedly enhanced by the simultaneous addition of phospholipid and DG, and almost full enzymatic activity is obtained after prolonged incubation. Phorbol esters such as TPA can substitute for DG and show a similar effect. With casein as substrate, neither calpain I nor II requires phospholipid and DG for the maximum enzymatic activity. Phospholipid, DG, or TPA alone show practically no effect. It has been postulated that DG or phorbol ester remain associated with the cellular membrane and recruit PKC to the membrane, where a quarternary complex consisting of PKC, Ca^{2+}, phospholipid, and DG or phorbol ester is formed. The conformation of the PKC in this quaternary complex probably differs from that of the inactive free form, and the active form may be preferentially cleaved by calpain, particularly calpain I. Essentially similar results are obtained for type I and III PKC.

Quantitative analysis with a fixed amount of calpain under comparable conditions given in Figure 3 shows that the susceptibilities of the PKC subspecies are distinctly different. Namely, the relative rates of cleavage of type I, II, and III PKC are approximately 100:16:2. This is probably due to the difference of the sequences around the sites of the cleavage (see

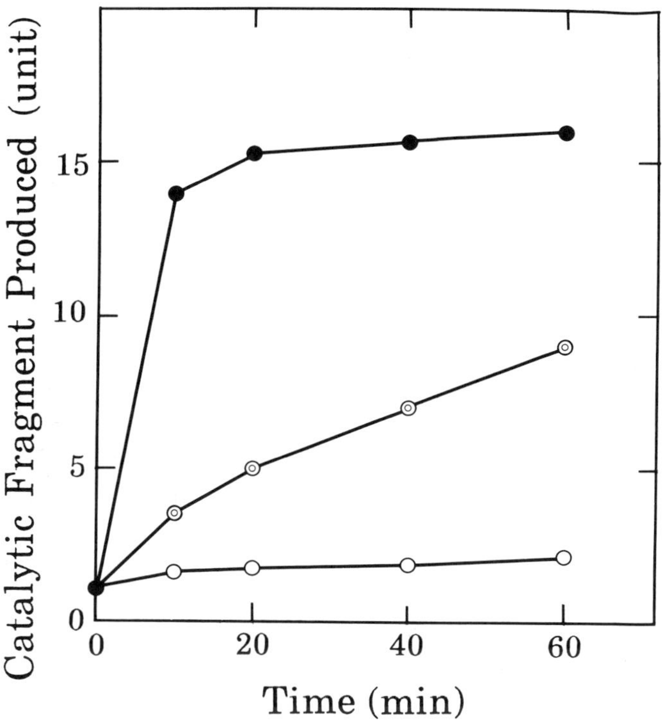

FIGURE 3. Relative susceptibility of type I, II, and III PKC to calpain I. Type I, II, and III PKC were separately subjected to proteolysis by a fixed amount of calpain I in the presence of Ca^{2+}, phospholipid, and diolein under conditions similar to those given in Figure 2. At the various times indicated, an aliquot of the mixture was taken and immediately assayed for the active fragment with H1 histone as substrate. In control experiments calpain I was omitted; ●, ◎, and ○ indicate cleavage of type I, II, and III PKC, respectively. The detailed experimental conditions have been described elsewhere.[16]

below). The results imply that within the cell various PKC subspecies may be cleaved at different rates under various physiological conditions, as discussed below.

IV. FRAGMENTS OF PKC GENERATED BY CALPAIN ACTION

The catalytic and regulatory fragments of PKC generated by calpain action are able to be separated and isolated on gel filtration after completion of the proteolysis. Estimation of the molecular weight of these fragments on SDS-PAGE is summarized in Table 1. Some combinations of the PKC subspecies and calpain produce two bands of catalytic fragments of the PKC molecule, due to proteolysis at two sites. This is confirmed by amino acid sequence analysis of the N-terminal region of each catalytic fragment (see below). The catalytic fragment is reactive with an antibody that has been raised against the C-terminal region V_5 of the respective type of PKC. Another fragment, the regulatory fragment, reacts with an antibody that has been raised against the N-terminal region C_1 of PKC. The regulatory fragment can also be obtained by trypsin treatment of the PKC molecule, and is shown to possess phorbol ester-binding activity.[45-47]

The sites in the PKC molecule which are cleaved with calpain I and II have been determined by analyzing the N-terminal sequence of the samples of catalytic fragments. Figure 4 summarizes the sites determined in various combinations of PKC subspecies and

TABLE 1
Catalytic and Regulatory Fragments of PKC Subspecies Generated by Calpain I and II.

PKC	Cal-pain	Molecular weight	
		Catalytic fragment	Regulatory fragment
Type I	I	49 kDa, 47 kDa	36 kDa
Type I	II	49 kDa, 47 kDa	36 kDa
Type II	I	46 kDa, 45 kDa	36 kDa
Type II	II	46 kDa,	36 kDa
Type III	I	46 kDa, 45 kDa	36 kDa
Type III	II	47 kDa	36 kDa

Note: Various PKC subspecies were fully cleaved with calpain I and II in the presence of Ca^{2+}, phospholipid, and diolein under conditions similar to those described in Figure 2. Then, catalytic and regulatory fragments were analyzed by SDS-PAGE. The detailed experimental conditions have been described elsewhere.[16]

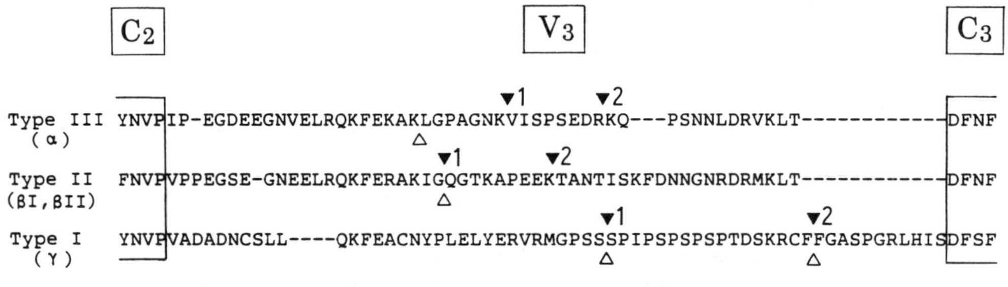

FIGURE 4. Sites of cleavage of PKC molecules with calpain I and II. The third variable region of the PKC molecules is referred to as V_3; (▼) and (▲), represent the bonds cleaved with calpain I and II, respectively. The numbers 1 and 2 attached to the arrows represent site 1 and site 2, respectively. The detailed experimental conditions have been described elsewhere.[16]

calpain I or II. In all cases, the cleaved sites are located at the third variable region (V_3), which connects the regulatory and catalytic domain. For type III PKC, calpain I and II cleave different sites. A single homogenous 36-kDa regulatory fragment always appears, whereas two catalytic fragments are often produced (Table 1). Appearance of the larger catalytic fragment is faster than that of the smaller one, and the amount of the larger one is decreased after prolonged proteolysis. The evidence suggests that of the two sites cleaved (site 1 and site 2, as indicated in Figure 4) site 1 is cleaved first, and then the smaller catalytic fragment is generated by subsequent cleavage at site 2. There is no obvious common sequence around the cleavage sites of various PKC subspecies. This seems to be a major reason for the difference of susceptibilities of PKC subspecies for proteolysis by calpain.

V. POSSIBLE INVOLVEMENT OF CALPAIN IN PKC REGULATION

Two possibilities for the role of PKC proteolysis by calpain may be considered. First, this proteolysis may be a process to activate PKC, and the resulting active fragment may be released from membranes and have its own function in the control of cellular activity.[48,49] Second, in contrast, this may be a process to initiate the degradation of PKC, eventually depleting the enzyme from the cell. Recent studies in several laboratories have shown that,

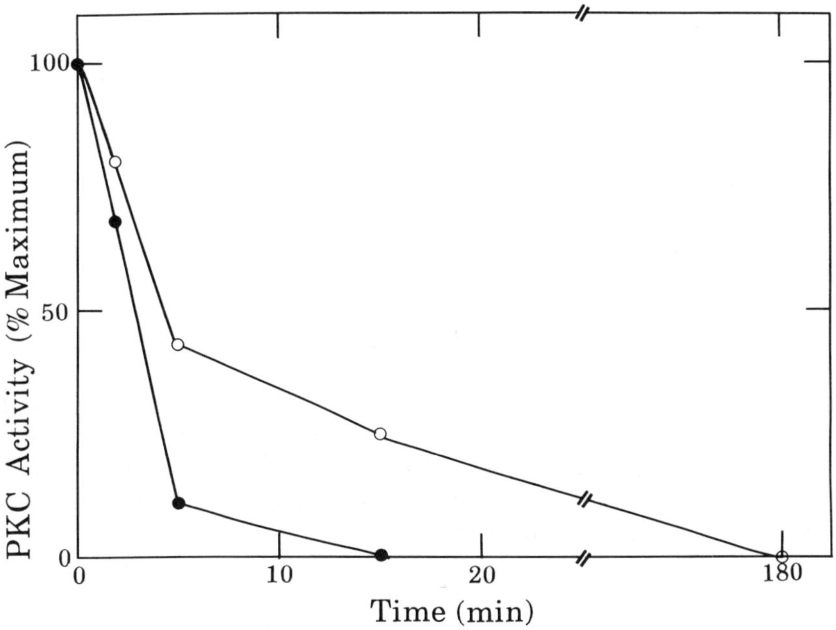

FIGURE 5. Time course of down-regulation of PKC subspecies in KM3 cells. KM3 cells were treated with 50 n*M* TPA. At the various times indicated, cells were washed and immediately homogenized. Then, PKC subspecies were separated and quantified on a hydroxyapatite column. Type II and III PKC activities are normalized to the activity of the respective enzyme in cells without TPA treatment: ● and ○, activities of type II and III PKC, respectively. The detailed experimental conditions have been described elsewhere.[17]

in a variety of tissues and cell types, TPA, the phorbol ester which induces persistent activation of PKC, elicits its translocation from the cytosol to the membrane, and subsequent depletion, termed down-regulation of PKC.[50-54] Since PKC forms a quaternary complex with TPA when translocated to the membrane, calpain may initiate the down-regulation of PKC, forming the catalytic fragment which is probably degraded further by other proteases. In fact, upon treatment with TPA, various subspecies of PKC coexpressed in a single cell type, such as KM3 and HL-60, disappear quickly at different rates, which coincide with the rate of the *in vitro* proteolysis as shown in Figures 3 and 5. It it more likely, then, that the limited proteolysis of the PKC molecules by calpain, particularly by calpain I, is directly related to the initiation of degradation of the enzyme, and that various subspecies of PKC are depleted from the cell at different rates due to the proteolysis at their variable region V_3.

VI. CODA

Recent experimental results obtained in this laboratory and in others have indicated that PKC has a dual action, providing positive forward control as well as negative feedback control, over various steps of signaling processes.[18] It is now well recognized that the synergistic interaction between the PKC and Ca^{2+} pathways, proposed earlier, underlies a variety of cellular responses to external stimuli, including gene expression. One of the major roles of PKC appears to be to decrease Ca^{2+} concentrations in the cell. PKC is proposed to decrease Ca^{2+} by activating the Ca^{2+}-transport ATPase and Na^{+}/Ca^{2+} exchange protein, and by inhibiting the receptor-mediated hydrolysis of inositol phospholipids. The negative feedback role of PKC may be extended to long-term responses such as cell proliferation.

The receptor for EGF has repeatedly been shown to be phosphorylated by PKC, resulting in a rapid decrease in high affinity binding of EGF as well as inhibition of ligand-induced tyrosine phosphorylation.[55] Such a dual effect of PKC is also seen in T-lymphocyte proliferation, where the synergistic action of TPA and Ca^{2+} ionophore has been demonstrated. Here, the activated PKC also has a negative feedback action in these cells as it phosphorylates and thereby induces down-regulation of the T-cell receptor, preventing any further stimulation by antigen to proliferation.[56] It will be important to investigate further the possible roles of limited proteolysis and subsequent down-regulation of the PKC molecule by calpain in terms of these events.

ACKNOWLEDGMENTS

The authors are grateful for the critical reading of the manuscript by, and the discussion with, Dr. Yasutomi Nishizuka. The skillful secretarial assistance of Mrs. S. Nishiyama and Mrs. Y. Bignon is appreciated. This investigation was supported in part by research grants from the Scientific Research Fund of Ministry of Education, Science, and Culture, Japan; and the Sankyo Foundation of Life Science.

REFERENCES

1. **Guroff, G.,** A neutral, calcium-activated proteinase from the soluble fraction of rat brain, *J. Biol. Chem.,* 239, 149, 1964.
2. **Meyer, W. L., Fischer, E. H., and Krebs, E. G.,** Activation of skeletal muscle phosphorylase b kinase by Ca^{2+}, *Biochemistry,* 3, 1033, 1964.
3. **Mellgren, R. L.,** Canine cardiac calcium-dependent proteases: resolution of two forms with different requirements for calcium, *FEBS Lett.,* 109, 129, 1980.
4. **Kishimoto, A., Kajikawa, N., Tabuchi, H., Shiota, M., and Nishizuka, Y.,** Calcium-dependent neutral proteases, widespread occurrence of a species of protease active at lower concentrations of calcium, *J. Biochem.,* 90, 889, 1981.
5. **Coolican, S. A. and Hathaway, D. R.,** Effect of L-α-phosphatidylinositol on a vascular smooth muscle Ca^{2+}-dependent protease, *J. Biol. Chem.,* 259, 11627, 1984.
6. **Pontremoli, S., Melloni, E., Sparatore, B., Salamino, F., Michetti, M., Sacco, O., and Horecker, B. L.,** Role of phospholipids in the activation of the Ca^{2+}-dependent neutral proteinase of human erythrocytes, *Biochem. Biophys. Res. Commun.,* 129, 389, 1985.
7. **Imajoh, S., Kawasaki, H., and Suzuki, K.,** The amino-terminal hydrophobic region of the small subunit of calcium-activated neutral protease (CANP) is essential for its activation by phosphatidylinositol, *J. Biochem.,* 99, 1281, 1986.
8. **Gates, R. E. and King, L. E.,** Different forms of the epidermal growth factor receptor kinase have different autophosphorylation sites, *Biochemistry,* 24, 5209, 1985.
9. **Ando, Y., Imamura, S., Yamagata, Y., Kitahara, A., Saji, H., Murachi, T., and Kannagi, R.,** Platelet factor XIII is activated by calpain, *Biochem. Biophys. Res. Commun.,* 144, 484, 1987.
10. **Phillips, D. R. and Jakábová, M.,** Ca^{2+}-dependent protease in human platelets, *J. Biol. Chem.,* 252, 5602, 1977.
11. **Inoue, M., Kishimoto, A., Takai, Y., and Nishizuka, Y.,** Studies on a cyclic nucleotide-independent protein kinase and its proenzyme in mammalian tissues. II. Proenzyme and its activation by calcium-dependent protease from rat brain, *J. Biol. Chem.,* 252, 7610, 1977.
12. **Kishimoto, A., Kajikawa, N., Shiota, M., and Nishizuka, Y.,** Proteolytic activation of calcium-activated, phospholipid-dependent protein kinase by calcium-dependent neutral protease, *J. Biol. Chem.,* 258, 1156, 1983.
13. **Nishizuka, Y.,** The molecular heterogeneity of protein kinase C and its implications for cellular regulation, *Nature,* 334, 661, 1988.
14. **Huang, K.-P., Nakabayashi, H., and Huang, F. L.,** Isozymic forms of rat brain Ca^{2+}-activated and phospholipid-dependent protein kinase, *Proc. Natl. Acad. Sci. U.S.A.,* 83, 8535, 1986.
15. **Sekiguchi, K., Tsukuda, M., Ase, K., Kikkawa, U., and Nishizuka, Y.,** Mode of activation and kinetic properties of three distinct forms of protein kinase C from rat brain, *J. Biochem.,* 103, 759, 1988.

16. **Kishimoto, A., Mikawa, K., Hashimoto, K., Yasuda, I., Tanaka, S., Tominaga, M., Kuroda, T., and Nishizuka, Y.,** Limited proteolysis of protein kinase C subspecies by calcium-dependent neutral protease (calpain), *J. Biol. Chem.*, 264, p. 4088, 1989.
17. **Ase, K., Berry, N., Kikkawa, U., Kishimoto, A., and Nishizuka, Y.,** Differential down-regulation of protein kinase C subspecies in KM3 cells, *FEBS Lett.*, 236, 396, 1988.
18. **Nishizuka, Y.,** Studies and perspectives of protein kinase C, *Science,* 233, 305, 1986.
19. **Kishimoto, A., Takai, Y., Mori, T., Kikkawa, U., and Nishizuka, Y.,** Activation of calcium and phospholipid-dependent protein kinase by diacylglycerol, its possible relation to phosphatidylinositol turnover, *J. Biol. Chem.*, 255, 2273, 1980.
20. **Berridge, M. J. and Irvine, R. F.,** Inositol triphosphate, a novel second messenger in cellular signal transduction, *Nature,* 312, 22, 1984.
21. **Castagna, M., Takai, Y., Kaibuchi, K., Sano, K., Kikkawa, U., and Nishizuka, Y.,** Direct activation of calcium-activated, phospholipid-dependent protein kinase by tumor-promoting phorbol esters, *J. Biol. Chem.*, 257, 7847, 1982.
22. **Kikkawa, U., Takai, Y., Tanaka, Y., Miyake, R., and Nishizuka, Y.,** Protein kinase C as a potential receptor protein of tumor-promoting phorbol esters, *J. Biol. Chem.*, 258, 11442, 1983.
23. **Uratsuji, Y., Nakanishi, H., Takeyama, Y., Kishimoto, A., and Nishizuka, Y.,** Activation of cellular protein kinase C and mode of inhibitory action of phospholipid-interacting compounds, *Biochem. Biophys. Res. Commun.*, 130, 654, 1985.
24. **Kikkawa, U., Ogita, K., Ono, Y., Asaoka, Y., Shearman, M. S., Fujii, T., Ase, K., Sekiguchi, K., Igarashi, K., and Nishizuka, Y.,** The common structure and activities of four subspecies of rat brain protein kinase C, *FEBS Lett.*, 223, 212, 1987.
25. **Berg, J. M.,** Potential metal-binding domains in nucleic acid binding proteins, *Science,* 232, 485, 1986.
26. **Johnston, M.,** Genetic evidence that zinc is an essential cofactor in the DNA binding domain of GAL4 protein, *Nature,* 328, 353, 1987.
27. **Hanks, S. K., Quinn, A. M., and Hunter, T.,** The protein kinase family: conserved features and deduced phylogeny of the catalytic domains, *Science,* 241, 42, 1988.
28. **Parker, P. J., Coussens, L., Totty, N., Thee, L., Young, S., Chen, E., Stabel, S., Waterfield, M. D., and Ullrich, A.,** The complete primary structure of protein kinase C-the major phorbol ester receptors, *Science,* 233, 853, 1986.
29. **Knopf, J. L., Lee, M.-H., Sultzman, L. A., Kriz, R. W., Loomis, C. R., Hewick, R. M., and Bell, R. M.,** Cloning and expression of multiple protein kinase C cDNAs, *Cell,* 46, 491, 1986.
30. **Ohno, S., Kawasaki, H., Imajoh, S., Suzuki, K., Inagaki, M., Yokokura, H., Satoh, T., and Hidaka, H.,** Tissue-specific expression of three distinct types of rabbit protein kinase C, *Nature,* 325, 161, 1987.
31. **Brandt, S. J., Niedel, J. E., Bell, R. M., and Young, W. S.,** Distinct patterns of expression of different protein kinase C mRNAs in rat tissues, *Cell,* 49, 57, 1987.
32. **Saito, N., Kikkawa, U., Nishizuka, Y., and Tanaka, C.,** Distribution of protein kinase C-like immunoreactive neurons in rat brain, *J. Neurosci.*, 8, 369, 1988.
33. **Huang, F. L., Yoshida, Y., Nakabayashi, H., Young, W. S., III, and Huang, K. P.,** Immunochemical identification of protein kinase C isozymes as products of discrete genes, *Biochem. Biophys. Res. Commun.*, 149, 946, 1987.
34. **Hidaka, H., Tanaka, T., Onoda, K., Hagiwara, M., Watanabe, M., Ohta, H., Ito, Y., Tsurudome, M., and Yoshida, T.,** Cell type-specific expression of protein kinase C isozymes in the rabbit cerebellum, *J. Biol. Chem.*, 263, 4523, 1988.
35. **Ase, K., Saito, N., Shearman, M. S., Kikkawa, U., Ono, Y., Igarashi, K., Tanaka, C., and Nishizuka, Y.,** Distinct cellular expression of βI- and βII-subspecies of protein kinase C in rat cerebellum, *J. Neurosci.*, 8, 3850, 1988.
36. **Huang, F. L., Yoshida, Y., Nakabayashi, H., Young, W. S., III, and Huang, K.-P.,** Immunocytochemical localization of protein kinase C isozymes in rat brain, *J. Neurosci.*, 8, 4734, 1988.
37. **Kose, A., Saito, N., Ito, H., Kikkawa, U., Nishizuka, Y., and Tanaka, C.,** Electron microscopic localization of type I protein kinase C in rat Purkinje cells, *J. Neurosci.*, 8, 4262, 1988.
38. **Hashimoto, T., Ase, K., Sawamura, S., Kikkawa, U., Saito, N., Tanaka, C., and Nishizuka, Y.,** *J. Neurosci.*, 8, 1678, 1988.
39. **Akers, R. F., Lovinger, D. M., Colley, P. A., Linden, D. J., and Routtenberg, A.,** Translocation of protein kinase C activity may mediate hippocampal long-term potentiation, *Science,* 7, 587, 1986.
40. **Shearman, M. S., Naor, Z., Kikkawa, U., and Nishizuka, Y.,** Differential expression of multiple protein kinase C subspecies in rat central nervous tissue, *Biochem. Biophys. Res. Commun.*, 147, 911, 1987.
41. **Naor, Z., Shearman, M. S., Kishimoto, A., and Nishizuka, Y.,** Calcium-independent activation of hypothalamic type I protein kinase C by unsaturated fatty acids, *Mol. Endocrinol.*, 88, 1043, 1988.
42. **Kosaka, Y., Ogita, K., Ase, K., Nomura, H., Kikkawa, U., and Nishizuka, Y.,** The heterogeneity of protein kinase C in various rat tissues, *Biochem. Biophys. Res. Commun.*, 151, 973, 1988.

43. **Tsukuda, M., Asaoka, Y., Sekiguchi, K., Kikkawa, U., and Nishizuka, Y.**, Properties of protein kinase C subspecies in human platelets, *Biochem. Biophys. Res. Commun.*, 155, 1387, 1988.
44. **Shearman, M. S., Berry, N., Oda, T., Ase, K., Kikkawa, U., and Nishizuka, Y.**, Isolation of protein kinase C subspecies from a preparation of human T lymphocytes, *FEBS Lett.*, 234, 387, 1988.
45. **Huang, K.-P. and Huang, F. L.**, Conversion of protein kinase C from a Ca^{2+}-dependent to an independent form of phorbol ester-binding protein by digestion with trypsin, *Biochem. Biophys. Res. Commun.*, 139, 320, 1986.
46. **Lee, M.-H. and Bell, R. M.**, The lipid binding, regulatory domain of protein kinase C, *J. Biol. Chem.*, 261, 14867, 1986.
47. **Hoshijima, M., Kikuchi, A., Tanimoto, T., Kaibuchi, K., and Takai, Y.**, Formation of a phorbol ester-binding fragment from protein kinase C by proteolytic digestion, *Cancer Res.*, 46, 3000, 1986.
48. **Tapley, P. M. and Murray, A. W.**, Modulation of Ca^{2+}-activated, phospholipid-dependent protein kinase in platelets treated with a tumor-promoting phorbol ester, *Biochem. Biophys. Res. Commun.*, 122, 158, 1984.
49. **Melloni, E., Pontremoli, S., Michetti, M., Sacco, O., Sparatore, B., and Horecker, B. L.**, The involvement of calpain in the activation of protein kinase C in neutrophils stimulated by phorbol myristic acid, *J. Biol. Chem.*, 261, 4101, 1986.
50. **Rodriguez-Pena, A. and Rozengurt, E.**, Disappearance of Ca^{2+}-sensitive, phospholipid-dependent protein kinase activity in phorbol ester-treated 3T3 cells, *Biochem. Biophys. Res. Commun.*, 120, 1053, 1984.
51. **Blackshear, P. J., Witters, L. A., Girard, P. R., Kuo, J. F., and Quamo, S. N.**, Growth factor-stimulated protein phosphorylation in 3T3-L1 cells, *J. Biol. Chem.*, 260, 13304, 1985.
52. **Ballester, R. and Rosen, O. M.**, Fate of immunoprecipitable protein kinase C in GH_3 cells treated with phorbol 12-myristate 13-acetate, *J. Biol. Chem.*, 260, 15194, 1985.
53. **Stabel, S., Rodriguez-Pena, A., Young, S., Rozengurt, E., and Parker, P. J.**, Quantitation of protein kinase C by immunoblot-expression in different cell lines and response to phorbol esters, *J. Cell. Physiol.*, 130, 111, 1987.
54. **Young, S., Parker, P. J., Ullrich, A., and Stabel, S.**, Down-regulation of protein kinase C is due to an increased rate of degradation, *Biochem. J.*, 244, 775, 1987.
55. **Schlessinger, J.**, Allosteric regulation of the epidermal growth factor receptor kinase, *J. Cell Biol.*, 103, 2067, 1986.
56. **Cantrell, D. A., Davies, A. A., and Crumpton, M. J.**, Activators of protein kinase C down-regulate and phosphorylate the T3/T-cell antigen receptor complex of human T-lymphocytes, *Proc. Natl. Acad. Sci. U.S.A.*, 82, 8158, 1985.
57. **Mikawa, K., et al.**, in preparation.

Chapter 14

CALPAINS AND THE CYTOSKELETON

Robert L. Shoeman and Peter Traub

TABLE OF CONTENTS

I. INTRODUCTION

The cytoskeleton of a typical mammalian cell consists of three different filament systems (microtubules, microfilaments and intermediate filaments), various associated proteins that crosslink these filament systems to each other and to other cell organelles, and finally the elements that form the membrane skeletons of the plasma membrane and nuclear envelope. Functional elements of the cytoskeleton permeate the entire cell and are involved in cell attachment, cell movement, endocytotic processes, exocytosis, and intracellular transport of macromolecules and organelles, as well as localization and stabilization of macromolecules and cell organelles. The boundaries or limits of the cytoskeleton are not abrupt or clearly defined since it connects or interacts with many cell organelles. Calpains have been shown to be capable of degrading many of the diverse protein subunits of the cytoskeleton (reviewed in References 1 and 2). This contribution will focus primarily on the ability of calpains to degrade components of the three-filament systems and their associated proteins. The reader should refer to other chapters in this book for discussions of the occurrence and activity of endogenous calpain inhibitors, the calpastatins, and for cellular mechanisms of regulation of intracellular Ca^{2+} concentration, both of which are responsible for regulating calpains. In addition, a great body of basic research has been summarized in the opening paragraphs of each of the following sections without citing the basic research articles; an interested reader can find additional information in books by Bershadsky and Vasiliev,[3] Schliwa,[4] Rousset[5] and Traub.[2]

Some of the best examples of the degradation of cytoskeletal elements by calpains include studies on the membrane skeleton[6-13] and neurofilaments[14-25] (a class of intermediate filaments); these topics are the subject of separate chapters in this work and will be dealt with only briefly in this contribution. Most of the data that will be discussed were obtained *in vitro,* thus ensuring that the degradation of the various cytoskeletal proteins observed was actually due to calpain and not other cellular proteinases. Where possible, these *in vitro* results will be supplemented with *in vivo* observations.

The major protein subunits of the cytoskeleton that have been observed to be substrates for calpains are listed in Table 1. Detailed descriptions of the various cytoskeleton elements, their function (where known), and the effect of calpain treatment will be presented in the following sections. For the sake of brevity, we have chosen to use the generic term calpain to describe the various calcium-activated neutral thiol proteinases utilized in these studies or found in the different tissues and will present more details (such as high Ca^{2+}- or low Ca^{2+}-requiring form) only when germane. A complete discussion of the properties of the various enzymes comprising the ''calpains'' can be found elsewhere in this book (Chapters 1 and 2). Our primary goal in this contribution is not to describe the multitude of calpain isolation and characterization studies but rather to present an overview of the cytoskeleton and the sites and consequences of action of calpain on it.

II. CYTOSKELETON COMPONENTS

A. MICROTUBULES

1. Protein Subunits, Polymerization, Structure and Function

Microtubules are found in essentially all eukaryotic cells. They are found throughout the cytoplasm as long filaments, in ordered arrays within cilia and flagella, and form the spindle structure responsible for the mitotic segregation of chromosomes. During mitosis, cytoplasmic microtubules are disassembled and the subunit proteins are utilized to build the microtubules of the spindle apparatus. After mitosis, the spindle apparatus disassembles and the subunit proteins are recycled to reform the cytoplasmic microtubules.

The major protein of microtubules is tubulin, which exists in two forms, α and β, of identical molecular weight. Both forms bind GTP and associate via specific protein inter-

TABLE 1
Cytoskeletal Protein Substrates of Calpain

Protein	Molecular weight	Localization	Cell type	Refs.
Microtubule-associated protein I (MAP1)	350,000	Microtubules	Neurons	34—39
Microtubule-associated protein II (MAP2)	340,000 (300,000)	Microtubules	Neurons	34, 35, 37—39
Tubulin	α = 55,000 β = 55,000	Microtubules	Platelets, neurons	34, 38
Neurofilament proteins	NF-H ≈ 200,000 NF-M ≈ 150,000 NF-L ≈ 70,000	Intermediate filaments	Neurons	14—25
Cytokeratins	40—70,000 (>20 different subunits in man)	Intermediate filaments	Epithelial cells	24, 61
Vimentin	54,000	Intermediate filaments	Mesenchymal cells; tissue culture cells	24, 41, 45, 49, 52, 54, 60, 62
Desmin	53,000	Intermediate filaments	Muscle cells	24, 45, 49, 50, 60
Glial fibrillary acidic protein (GFAP)	51,000	Intermediate filaments	Astroglial cells	24, 45, 48, 50, 53, 55
Nuclear lamins	A = 70,000 B = 66,000 C = 60,000	Nuclear lamina (meshwork in nuclear periphery)	Nucleated cells	62
Filamin (actin-binding protein)	250,000	Microfilaments	Platelets, smooth muscle cells	80—83
Talin (P 235)	235,000	Membrane skeleton	Platelets	85—90
Actin	43,000	Microfilaments Z-disks	Eye lens Muscle cells	57 76
α-Actinin	≈100,000	Membrane skeleton	Platelets	73—79
Vinculin	130,000	Microfilaments (at cell focal contact points)	Smooth muscle cells	72
Ankyrin	≈200,000	Membrane skeleton	Erythrocytes	6, 9, 12
Band 3 protein	93,000			6—9, 11, 13
Band 4 protein	≈ 80,000			7, 11—13
Spectrin	α = 240,000 β = 220,000	Membrane Skeleton	Erythrocytes, platelets	9, 10, 13, 68
Fodrin (brain spectrin)	α = 240,000 β = 230—235,000	Membrane skeleton	Neurons	69, 70
Glycophorin	31,000	Integral membrane protein	Erythrocytes	8
Myosin light chain	20,000	Cytoskeleton	Neutrophils	92, 93
Titin	>1,000,000	Cytoskeleton/myosin filaments	Muscle cells	94, 95
Nebulin	700,000	Cytoskeleton/actin filaments	Muscle cells	94, 95
Dystrophin	427,000	Sarcolemma	Muscle cells	96

Note: The molecular weight column lists the reported molecular weights (generally derived from apparent relative mobility in denaturing gel electrophoresis) and is included only as a guide to the relative size of these proteins. The cell type column lists the cells utilized in the individual studies described in the text and is not meant to represent a definitive survey of all cell types.

actions to form α-β dimers. The α- and β-tubulin dimers (protofilaments) polymerize in a helical array that folds up into hollow tubes with a diameter of approximately 25 nm. The protofilaments associate in a polar fashion (αβ → αβ → αβ). In *in vitro* assays, protofilament polymerization is stimulated by the presence of nucleation sites (preformed microtubule fragments) and the presence of accessory proteins. *In vivo,* α-β tubulin dimers are preferentially added to one end of a microtubule, which "grows" by continued displacement of the preexisting elements by newly added protofilaments. Disassembly also occurs, both *in vivo* and *in vitro,* and seems to depend on the amount of free tubulin and the ratio of GTP to GDP bound to the microtubule (a high proportion of GDP favors disassembly). Unknown factors are responsible for cleaving one of the bound GTP molecules to GDP when the rate of addition of new protofilaments is low. Higher order structures, in which one complete tube and one or two partial tubes are fused together sideways, are found as the basic longitudinal building blocks of cilia, flagella, and centrosomes. Centrosomes serve as nucleation sites for cytoplasmic microtubule reformation after mitosis.

Depending on the cell type and specific structure formed from microtubules, a number of different accessory proteins, the microtubule-associated proteins (MAPs), copolymerize with or attach to the microtubules. The MAPs serve to initiate microtubule formation, to cross link individual microtubules, providing structural stability, and constitute the active components (microtubule-dependent ATPases termed dyneins or kinesins, depending on direction of transport) that not only drive the sliding of microtubules responsible for the movement of cilia and flagella, but also transport macromolecules and cell organelles along the microtubules. The MAPs have been classically divided into groups based on their molecular weight, although it should be emphasized that there is now evidence that each class of MAPs contains early and late members that are differentially expressed during development and presumably are responsible for conferring specific properties on the various structures formed from microtubules.[26] With the advent of higher resolution gel electrophoresis systems and molecular cloning it is now known that the previously designated MAP1 (≈ 350,000 mol wt) and MAP2 (≈ 300,000 mol wt) proteins are actually composed of several different proteins.[27] At least one of these, MAP1c, has been identified as possessing dynein activity.[28,29] Both MAP1 and MAP2 class proteins are thought to promote the polymerization of tubulin subunits into microtubules. The function of the τ proteins (low molecular weight MAPs, ≈ 30 to 50,000 mol wt) includes stimulation *in vitro* of the formation of nucleation sites (short spirals) for microtubule polymerization. Recent studies have shown that at least one τ protein shares antigenic epitopes and displays extensive homology with a larger MAP2 protein.[30] The region of homology between these proteins is rich in basic amino acids and includes three 18 amino acid repeats. Secondary structure predictions suggest that these repeats are appropriately spaced to bind to the highly acidic carboxyl termini of α- and β-tubulin so that several adjoining tubulin molecules might be crosslinked within the microtubules.

2. Degradation of Microtubule Proteins by Calpains

As listed in Table 1, MAP1 and MAP2 proteins are readily degraded *in vitro* and *in vivo* by calpains.[10,31-39] This degradation has been described primarily for MAPs from bovine and rat brain (presumably due to the high concentrations of both microtubules and MAPs) and porcine retina. MAP1, contained in cytoskeleton preparations of SV-3Y1 and HeLa tissue culture cells, is rapidly degraded *in vitro* by calpain, yielding immunoreactive peptides identical to those generated *in vivo,* suggesting that an endogenous calpain also degrades this protein within these cells.[36] At least one high molecular weight microtubular binding protein from mammalian brain is resistant to calpain-mediated proteolysis.[37] This property was exploited during purification to selectively degrade all of the other high molecular weight MAPs. This calpain-resistant microtubular binding protein was shown to possess ATPase activity and may represent a cytoplasmic dynein variant.[37]

Nucleated erythrocytes possess a marginal band of microtubules, thought to be instrumental in maintenance of cell shape. A combination of drug treatments and inhibitor studies[10] showed that calpain degrades chicken erythrocyte MAPs (as well as other proteins that are part of the membrane skeleton, such as α-spectrin) which leads to a dissolution of the marginal band of microtubules and a solubilization of a presumed intermediate filament network. At the cellular level, these changes were correlated with altered erythrocyte morphology, nucleus-plasma membrane fusion, and microvesicle release.

Others have shown that calpains from a variety of tissue sources, such as human platelets,[39] human and pig erythrocytes,[36] or pig kidney[36] efficiently degrade MAPs associated with highly purified brain tubulin in microtubules. As mentioned, brain or other neuronal tissues have been used as a source of microtubules due to their high concentration and ease with which these may be prepared by several cycles of polymerization/depolymerization. Conflicting observations have been published concerning the ability of calpains to digest tubulin itself.[31,32,34-36,38-40] Several of the studies already discussed found no detectable degradation of tubulin by gel analysis or functional (assembly) assay, under conditions where the MAPs were extensively degraded by calpain.[31,32,35,36] Two reports of tubulin degradation are unconvincing on the basis of the data presented[40] or due to the omission of the relevant data from the figures presented.[39] Two recent studies have shown that tubulin, indeed, serves as a substrate for calpain.[34,38] In the more recent study,[38] isolated brain tubulin treated with high concentrations of calpain was degraded to yield a protein with an apparent molecular weight of 50,000. Calpain was largely unable to digest tubulin contained in preformed microtubules, implying that the cleavage site is at least partially masked. The digested tubulin showed no net charge differences relative to native tubulin, suggesting that neutral fragments were removed or that the fragments produced remained associated (as observed with tubulin digested with other proteinases). Significantly, the calpain-digested tubulin was found to spontaneously assemble into polymers with atypical or aberrant morphology.[38]

Calpains have been shown to readily degrade most of the high molecular weight MAPs and can apparently perform limited cleavage of the free tubulin protofilaments themselves. Since microtubule growth is polar and vectorial, there may exist a system for removal of disassembled protofilaments and MAPs from the end of a microtubule in a fully differentiated cell, either by transport or degradation. Calpain might serve a functional role in the irreversible regulation of microtubule structure and function by selectively degrading MAPs and tubulin protofilaments.

B. INTERMEDIATE FILAMENTS

1. Protein Subunits, Occurrence and Structure

Intermediate filaments were identified in electron microscopy as ubiquitous structures in the cytoplasm of higher eukaryotic cells. They were originally thought to represent sample preparation artifacts and to derive from aggregates of actin microfilaments or disassembly products of the thicker microtubules or nonmuscle myosin filaments. Biochemical, immunological, and molecular genetic studies have shown convincingly that the intermediate filaments are composed of unique protein subunits unrelated to the other elements of the cytoskeleton.[2,15] Five classes of cytoplasmic and one class of nuclear intermediate filament subunit proteins have been identified and are listed in Table 1. Different cell types express different cytoplasmic intermediate filament subunit proteins; however, it is not uncommon to find two classes expressed simultaneously at certain stages of development or in tissue culture cells. Indeed, cell typing based on intermediate filament composition is finding widespread application in the differential diagnosis of tumors in human medicine. In contrast, all higher eukaryotic cells investigated express at least one member of the nuclear intermediate filament subunit protein class (nuclear lamins, Table 1). When only one nuclear lamin is expressed, that one is lamin B; lamins A and C seem to be coordinately expressed. Lamins

A and C are translated from two different mRNAs that apparently derive from the same primary gene transcript by differential RNA splicing. Since lamins A and C are not found in all cell types while lamin B is, it is attractive to speculate that lamin B fulfills some essential structural role, while lamins A and C fulfill some regulatory or modulatory role in the structure of the nuclear lamina.

The cytoplasmic intermediate filaments extend from the nucleus throughout the cytoplasm, contacting organelles such as mitochondria and other internal membrane systems, to the plasma membrane in interphase cells. Studies suggest that in interphase cells the ends of the intermediate filaments are anchored to the nucleus, possibly through interactions with the nuclear lamin B, and to the plasma membrane, through interactions with elements of the membrane skeleton or with the membrane itself. During mitosis in some cell types, the cytoplasmic intermediate filaments either disassemble or detach from the nuclear and plasma membranes and reorganize into dense perinuclear tangles or cage-like structures around the spindle. In higher eukaryotic cells, where the nuclear membrane is dissolved during mitosis, the nuclear lamina is disassembled as a result of phosphorylation of its subunit proteins. Lamin B remains associated with nuclear membrane components in the form of vesicles, while lamins A and C are dispersed throughout the cytoplasm. After mitosis, the nucleus and the highly ordered arrays of cytoplasmic intermediate filaments reassemble from the preexisting molecules. This assembly-disassembly process of cytoplasmic intermediate filaments as well as nuclear lamina correlates well with the level of phosphorylation of the individual subunit proteins: a high degree of phosphorylation destabilizes filament structure.

Intermediate filaments have a diameter of 8 to 11 nm and are essentially insoluble in aqueous solution at physiological ionic strength. Within the cell there does not seem to be a pool of free subunit proteins, but rather they seem to be rapidly incorporated into filaments after synthesis. All members of all classes of intermediate filament subunit proteins share a common basic secondary structure, which permits their assembly into higher order structures of similar size and appearance. The typical structure of these proteins consists of three elements:

1. An N-terminal α-helical region of variable length;
2. A central rod domain, consisting of 3-linked regions of α-helical structure of approximately 310 amino acids (or 350 amino acids for the nuclear lamins); and
3. A C-terminal non-α-helical region of variable length. The central rod domain α-helices form coiled-coil intermolecular interactions that are the primary basis for filament assembly.

The central rod domain is highly conserved in both primary and secondary structure among all of the individual subunit proteins. The various proteins are distinguished by the length and composition of their non-α-helical N- and C-termini. These differences in length are responsible for the variations in molecular weight listed in Table 1. It has been shown that the N-terminus modulates or stabilizes filament formation, since proteolytic cleavage,[41] phosphorylation[42,43] or neutralization of the net positive charge[41] inhibits filament formation. Furthermore, the N-terminus possesses an amphiphilic character which is responsible for the association of intermediate filaments with negatively charged lipid bilayers.[44] It has also been shown that the N-terminus of vimentin and desmin can bind to nucleic acids[45] and that vimentin can specifically bind oligonucleotides containing telomere repeats.[46] The precise role or function of the C-terminus is unclear for most of these proteins. The nuclear localization signal for transport of the nuclear lamin proteins into the nucleus has been found to reside in the C-terminus of these proteins. One of the cytokeratins has no additional residues C-terminal to the central rod domain.

Current models propose that the basic building block of intermediate filaments consists of a parallel arrangement of two protein subunits to form dimers. Two dimers associate in an antiparallel fashion to yield protofilaments (tetramers), which in solutions of physiological salt concentration spontaneously assemble into long filaments (Figure 1a) consisting of 8 protofilaments in cross-section, yielding the characteristic 8 to 11 nm diameter. Shadowing isolated filaments with heavy metals reveals a periodicity of 21 nm, the length of the overlap between adjacent α-helical domains in the protofilaments (Figure 1c). Assembly of intermediate filaments does not seem to require energy or additional cofactors.

As for microtubules, a number of predominantly high molecular weight intermediate filament-associated proteins (IFAPs) have been described (reviewed in Reference 47). IFAPs have been found to serve diverse functions such as aggregating and crosslinking filaments, capping the ends of filaments, and anchoring filaments to the membrane skeleton or adhesion plaques. No specific function has been unequivocally demonstrated for intermediate filaments; dogma stipulates that they play a structural role in organizing the cytoplasm or confering shape to the cell, although this may be secondary to an as-yet-undescribed function (this point is raised since the intermediate filament network may be disrupted by microinjection of antibodies or drug treatment and the injected cells retain their shape and seem, by and large, unaffected in their basic functions[2]).

2. Degradation of Intermediate Filament Proteins by Calpains

All of the subunit proteins of intermediate filaments are efficiently degraded by calpains,[24,41,45,48-62] with the exception of nuclear lamin C from Ehrlich ascites tumor cells (which is relatively resistant to digestion by calpain *in vitro*[62]). Digestion of intermediate filament proteins by calpains yields a staircase of degradation products characteristic for each protein that display decreasing molecular weight and an increasingly acidic pI in 2D gel electrophoresis. The kinetics of digestion by calpains of purified vimentin and desmin protofilaments and filaments is shown in Figure 2. These results [60] and protein sequencing experiments[58] have shown that the major (early) cleavage site for calpains is located in the non-α-helical N-terminal region of vimentin and is readily accessible in both protofilaments and filaments. A second major (late) cleavage site is located essentially in the middle of the rod domain and seems to be somewhat protected in intact filaments. Additional minor cleavage sites exist (7 in the N- and 2 in the C-terminus of vimentin) and give rise to the minor bands seen in Figure 2. Brief treatment of vimentin filaments with calpain, such that only a portion of the N-terminus is removed, results in aberrant morphology, with unraveling of the filament and release of material (Figures 1c,d).

The results presented in Figures 1 and 2 were obtained *in vitro* using purified components.[60] Similar studies yielding similar results have been done with all of the intermediate filament subunit proteins,[24] again illustrating the homology that exists between these proteins. Calpains also degrade intermediate filament proteins *in vivo* and may also readily digest these proteins during their isolation. Detailed studies of cellular protein patterns in 2D gel electrophoresis[63,64] show that there is a cell cycle-specific degradation of vimentin (presumably calpain-modulated as the degradation species observed are the same as those observed after *in vitro* treatment of vimentin with calpain) which has been proposed to be an indicator of some important event in the process of cell division (the degradation is highest in G_2 and barely detectable in G_1 and S-phase). Also of potential significance is the relative resistance of vimentin to proteolytic degradation first observed in transformed cells[65] and in a lymphoblastoid cell line.[66] Vimentin has been found to be more highly phosphorylated in fibrosarcoma cells and SV40 virus-transformed human fibroblasts than vimentin of normal human fibroblasts.[67] During preparation of cytoskeletons, the vimentin of fibrosarcoma cells was quantitatively degraded, presumably due not only to higher levels of calpain (relative to normal fibroblasts) but also to the higher degree of phosphorylation of vimentin.[67] α-Spectrin (see Section D 2) was also degraded by calpain in the fibrosarcoma cells.[67] Tissue culture cells from man, mouse, duck, quail, iguana, viper, toad, newt, trout, and carp have

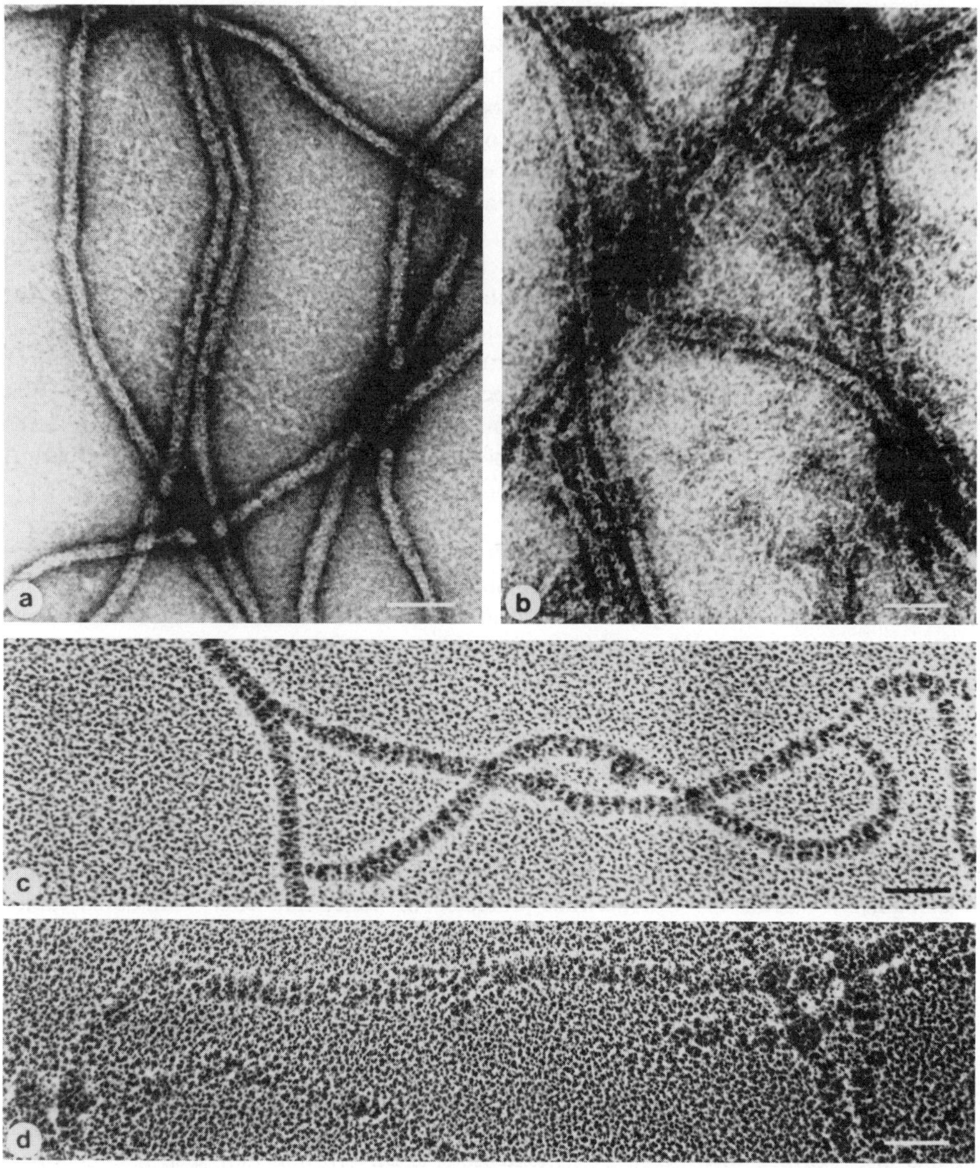

FIGURE 1. High magnification electron micrographs of vimentin-type IFs after a very brief digestion with Ca^{2+}-activated proteinase. (a) Control filaments reconstituted from pure vimentin; (b) vimentin filaments digested with Ca^{2+}-activated proteinase for 0.5 min using only one third of the enzyme concentration normally employed (see Figure 2b). On the basis of an SDS-gel electrophoretic comparison between the filament preparations presented here and those shown in Figure 2b, the sample in (b) would roughly correspond to a degradation pattern normally obtained after a 0.1-min digestion. (c,d) The same samples after spraying with glycerol on freshly cleaved mica and rotary shadowing with tungsten. Bars, 100 nm. = 120,000 ×. (From Perides, G., Kühn, S., Scherbarth, A., and Traub, P., *Eur. J. Cell Biol.*, 43, 450, 1987. With permission.)

all been shown to contain not only vimentin (or vimentin-like) intermediate filaments but also a calpain capable of digesting both the endogenous as well as exogenous intermediate filament proteins.[51]

An article of this length cannot propose to be an exhaustive review of the literature. A few selected examples of degradation of intermediate filaments from diverse systems will suffice to illustrate the wide ranging and ubiquitous nature of calpain-modulated turnover

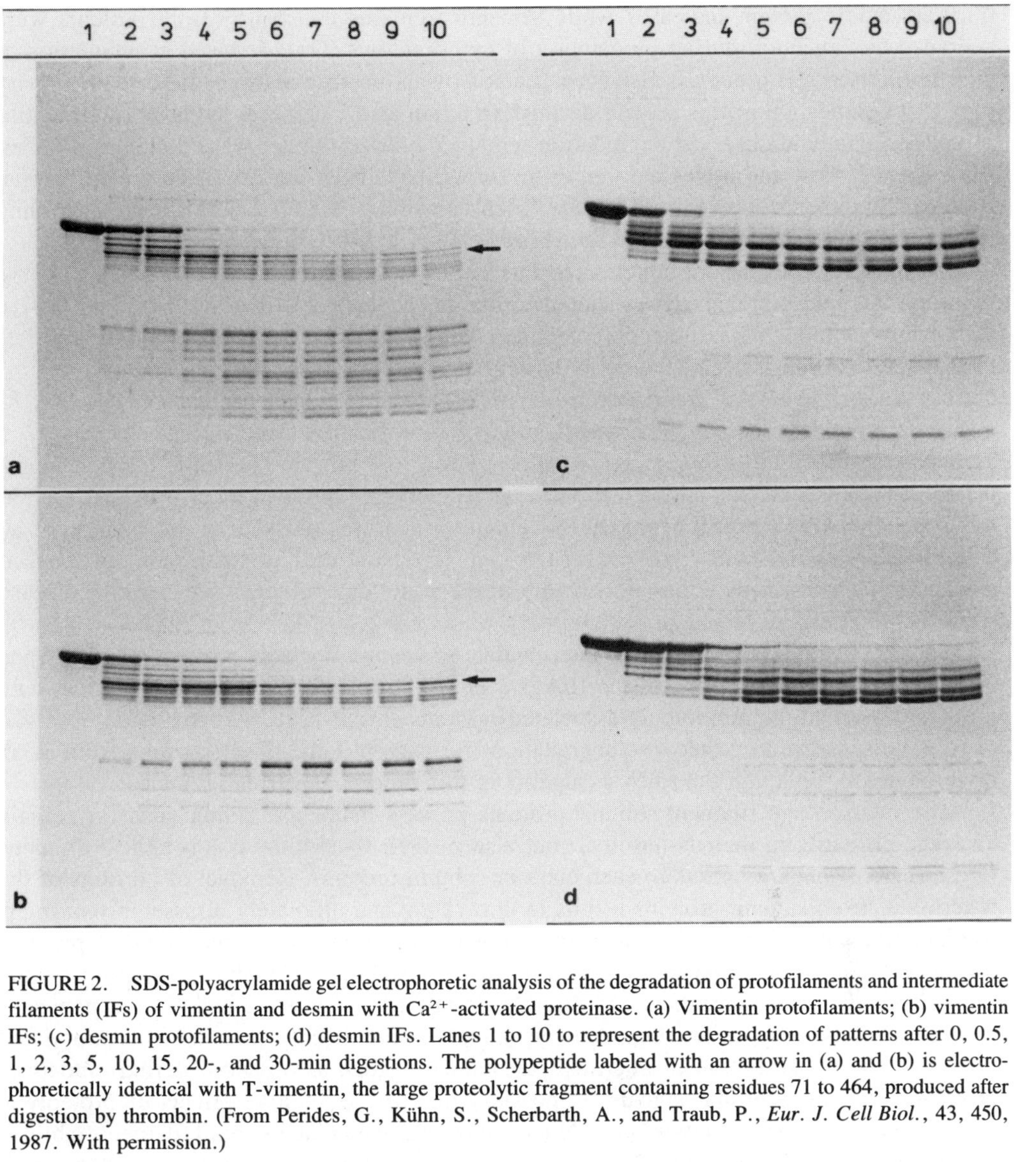

FIGURE 2. SDS-polyacrylamide gel electrophoretic analysis of the degradation of protofilaments and intermediate filaments (IFs) of vimentin and desmin with Ca^{2+}-activated proteinase. (a) Vimentin protofilaments; (b) vimentin IFs; (c) desmin protofilaments; (d) desmin IFs. Lanes 1 to 10 to represent the degradation of patterns after 0, 0.5, 1, 2, 3, 5, 10, 15, 20-, and 30-min digestions. The polypeptide labeled with an arrow in (a) and (b) is electrophoretically identical with T-vimentin, the large proteolytic fragment containing residues 71 to 464, produced after digestion by thrombin. (From Perides, G., Kühn, S., Scherbarth, A., and Traub, P., *Eur. J. Cell Biol.*, 43, 450, 1987. With permission.)

of these proteins. Calpains have been shown to degrade vimentin in an age-dependent fashion in eye lens cells[54,56,57]; this degradation correlates with reduced elasticity of the lens during aging. Several serotypes of human adenovirus travel a transcytoplasmic route to the nucleus during productive viral infection (in contrast to those which reach the nucleus primarily via phagocytotic vesicles). Vimentin and desmin were observed to be cleaved by calpain during infection of tissue culture cells with adenovirus serotypes that traverse the cytoplasm.[59] Other studies with these serotypes showed that the intermediate filament network collapses around the nucleus. Adenovirus serotypes that reach the nucleus via phagocytotic vesicles did not stimulate calpain activity and vimentin breakdown was not observed during infection with these serotypes. Glial fibrillary acidic protein from bovine and rat brain,[45] as well as from mouse spinal cord,[50] was degraded by calpain producing a specific set of increasingly smaller peptides with more acidic pI. Glial fibrillary acidic protein in mouse spinal cord was also degraded *in situ* by calpain.[53] Vimentin and glial fibrillary acidic protein are coexpressed in primary cultures of rat astrocytes; during drug-induced differentiation, the level of glial fibrillary acidic protein increases while vimentin remains constant.[55] Both proteins were degraded by calpain(s) during preparation of cytoskeletons. Drug-induced accumulation of glial fibrillary acidic protein is also accompanied by an increase in a specific calpain activity such that degradation of this protein during extraction also increased; the level of vimentin-specific calpain and observed degradation remained constant under these conditions.[55] Calpains I and II (low and high Ca^{2+}-requiring forms) have been purified from pig epidermis. *In vitro* experiments showed that both purified calpains were capable of rapidly degrading the epidermal keratins into small fragments.[61]

As mentioned in the introduction to this contribution, the action of calpains in nerve tissue on the neurofilament triplet subunit proteins (Table 1) has been well studied *in vivo* and *in vitro*. Although this subject is covered in detail elsewhere in this book, it is appropriate to mention here that calpains cannot only degrade all three neurofilament subunit proteins, but can also remove small fragments as a form of posttranslational modification that may be important in morphogenesis and/or modulation of neurofilament structure and activity.[19,21-23] The membrane skeleton protein spectrin[68] or fodrin[23,69,70] (brain spectrin), through which the intermediate filaments may attach to the plasma membrane, and other proteins, such as tubulin and MAPs, are also degraded by calpain in neuronal tissue. The degradation of all of these proteins has been observed *in vitro, in vivo,* and during experimentally induced damage to neurons; it may also play a role in the pathology of some degenerative disease states.

Spectrin belongs to the group of intermediate filament-associated proteins (IFAPs)[47] and is degraded by calpain. Many of the IFAPs seem to be related to or may be identical with proteins classified as microtubule-associated proteins (MAPs). IFAPs have been observed to be quite sensitive to proteolytic degradation (for example, the IFAP synemin from avian erythrocytes[71]), although the role of calpains in this process is currently unclear.

The intermediate filament subunit proteins possess major and minor sites for calpain cleavage primarily in their N-terminal, but also in their C-terminal non-α-helical domains (those regions that are unique to each class of subunit protein). Removal of portions of the N-terminus from subunit proteins results in unraveling and ultimately disassembly of intermediate filaments (Figure 1). Phosphorylation of intermediate filament subunit proteins probably plays a dual role in influencing filament stability: (1) phosphorylation occurs primarily within the N-terminus and directly destabilizes filaments, and (2) the phosphorylated subunits seem to be degraded by calpains more rapidly than nonphosphorylated subunits. The redistribution of intermediate filaments, such as observed during mitosis or in response to extracellular signals, is possibly due to the additive effects of proteolytic degradation and phosphorylation of the N-terminus, which mediates overall filament stability, interactions with peripheral membrane proteins such as ankyrin and direct interactions with the negatively charged lipid bilayer of membrane systems. Removal of small fragments of neurofilaments as a posttranslational modification has been postulated to play a role in neuron development and differentiation. It is conceivable that such subtle modification of subunit proteins modulates all types of intermediate filaments and may represent another function of calpains, in addition to their role in degradation and turnover of intermediate filaments and their subunit proteins.

C. MICROFILAMENTS

1. Nonmuscle Actin, Accessory Proteins and Structure

Actin is one of the most abundant proteins of eukaryotic cells. Actin, myosin, and various regulatory proteins are highly organized in muscle and provide the motive force for muscle movement. In many organisms, several closely related genes code for actin such that different tissues or cells possess one or more forms of actin. Globular actin molecules bind ATP and polymerize in a polar fashion into long filaments, in which two monomers (or two chains of monomers) are helically arranged into long filaments of approximately 6 to 7 nm diameter. Some time after polymerization, the ATP is cleaved to ADP, a condition which destabilizes the filament structure. Actin may add to either end of a growing filament; however, the rate of addition is lower if the filamentous actin contains bound ADP. Thus, a polarity for filament growth is established by the ATP/ADP content of the filament. A large number of associated proteins bind to both muscle and nonmuscle actin filaments and are responsible for imparting different characteristics to the actin filaments. For example, the regular arrangement of actin filaments, tropomyosin, a complex of three troponins and myosin filaments constitute the basic structure of sarcomeres of muscle tissue responsible for contraction.

Actin filaments are found in such diverse structures as:

1. The central skeletal array that supports the finger-like projections of microvilli
2. Adhesion plaques (sites where cells attach to the substratum)
3. Stress fibers (bundles of actin filaments thought to play a role in attachment to the substratum and in modulating cell shape)
4. The interdigitated microfilament network at the leading edge of lamellipodia of moving cells
5. A peripheral ring subjacent to the plasma membrane attached to the membrane skeleton in erythrocytes, platelets and probably other cell types (thought to stabilize or modulate cell shape)
6. A diffuse cytoplasmic system of polymerized/unpolymerized actin molecules (modulates cytoplasmic viscosity and is involved in the gel/sol transitions seen in ameboid movement), and
7. A contractile ring that is responsible for cytokinesis.

2. Degradation of Microfilaments and Accessory Proteins by Calpains

Microfilaments and actin subunits are generally quite resistant to degradation by calpains. One study[57] has shown that purified calpain II (high Ca^{2+}-requiring form) from bovine lens is capable of degrading bovine lens actin, through specific intermediates of molecular weight 42,000 and 40,000, to small peptides. This finding suggests that calpains may be responsible for the loss of actin from lens fiber cells as they age. Muscle cell α-actinin binds to the ends of actin filaments and anchors them to the cell membrane, probably through another protein, vinculin (a membrane-associated protein). Calpain activity disrupts these structures and has been reported to degrade vinculin[72] and to release undegraded α-actinin.[73-76] Another study[77] has found that purified bovine calpains can degrade purified α-actinin *in vitro* and α-actinin has been observed to be degraded by a Ca^{2+}-dependent mechanism during autolysis of isolated myofibrils.[78] These latter results suggest that calpain may not have access to α-actinin *in situ* unless myofiber degradation is quite advanced. One study[76] has found an additional form of actin in muscle cells that is cleaved by calpain and whose cleavage is associated with the release of α-actinin and breakdown in cell architecture. Thus, nonmyofibrillar actin in muscle cells may play an important role in stabilizing the myofibrillar structures. α-Actinin also anchors actin filaments to the membrane in other cells, such as platelets. In contrast to the situation generally observed in muscle cells, caplains readily degrade platelet α-actinin,[79] thereby permitting the dynamic rearrangements in microfilament organization that accompany platelet activation.

Filamin, fodrin and spectrin are similar proteins that crosslink actin filaments in a variety of tissues; all of these crosslinking proteins have been observed to be degraded by calpains[68-70,80-83] (Table 1). Filamin, also termed actin-binding protein, has recently been shown to be protected from degradation by calpain, but not by trypsin, papain or thermolysin, when phosphorylated *in vitro* by a cyclic AMP-dependent kinase.[84] Thus, phosphorylation of filamin may regulate filamin-actin interactions by preventing calpain proteolysis of filamin.

Talin (Table 1) is a cytoplasmic protein found in adhesion plaques that extends through the plasma membrane and binds to both fibronectin on the external surface as well as to actin on the internal surface of the plasma membrane and thus provides a link between actin filaments bound to vinculin and the external environment. Talin has been shown to be very similar or identical to the platelet protein P235.[85] Several studies have demonstrated that talin and P235, as well as filamin, are excellent substrates for calpains;[85-89] cleavage of platelet P235 by calpain is also associated with the reorganization of microfilaments that accompanies platelet activation. Calpain was found to degrade membrane plaques (as well as intermediate filaments) in smooth muscle *in situ,* thereby releasing microfilaments from these sites of anchorage to the plasma membrane.[82] Calpain and talin have been colocalized in adhesion plaques of several established tissue culture cell lines by immunofluorescence microscopy.[90] Purified calpain was found to preferentially degrade purified talin *in vitro,* whereas other adhesion plaque proteins such as vinculin and α-actinin were unaffected.[90] Calpain has also been shown to be responsible for the formation of plasma membrane blebs in hepatocytes,[91] probably by disrupting the attachment of the cytoskeleton to the plasma membrane.

As with microtubules, microfilaments have a wide variety of associated proteins that confer unique properties to the complexes, giving rise to a variety of skeletal and dynamic structures. Calpains regulate the disassembly of these structures primarily by degrading the accessory proteins (Table 1), leaving the actin subunits intact (usually) for recycling into new structures.

D. ADDITIONAL CYTOSKELETON COMPONENTS

1. Nonmuscle Myosin

Myosin is a filamentous protein that possesses actin-dependent ATPase activity. A myosin molecule is composed of two identical heavy subunits (≈ 200,000 mol wt) and two different pairs of light subunits (each with ≈ 20,000 mol wt). The heavy subunits are wound around each other for most of their length in a coiled-coil conformation and terminate in two globular headpieces; the two pairs of light subunits are bound to these globular headpieces, forming a lollipop-like structure. In muscle, bundles of hundreds of myosin molecules are arranged in a regular array with actin filaments to form the contractile apparatus. In nonmuscle cells, many individual myosin molecules bind to individual actin microfilaments. Although the precise mechanism is not known, it is presumed that the interaction of myosin and microfilaments results in a net movement of the microfilaments, in a manner similar to that seen for muscle actin-myosin complexes.

Stimulation of human neutrophils by phorbol esters results in the calpain-catalyzed activation of protein kinase C (a subject covered elsewhere in this book). This kinase has been shown to be responsible for the phosphorylation of two cytoskeletal proteins of 20,000 and 48,000 mol wt.[92] The 20,000-mol wt species has been tentatively identified as a myosin light chain molecule. Phosphorylation of the myosin light chain serves as a signal for degradation of this molecule by calpain in the stimulated neutrophil, resulting in the disappearance of the phosphorylated myosin light chain from the cytoskeletal fraction.[93] This cytoskeletal reorganization seems to be associated with the degranulation response of stimulated neutrophils and is affected by calpain at two levels: (1) calpain activates protein kinase C in response to external stimuli, and (2) calpain degrades the phosphorylated form of the (presumed) myosin light chain components of the cytoskeleton.

2. Membrane Skeleton Proteins

The membrane skeleton of mammalian erythrocytes represents the best-described and understood membrane skeleton system. This skeleton forms a uniform shell in the erythrocyte and is attached at many places to the interior of the plasma membrane by specific anchoring proteins. Two glycoproteins, glycophorin and band 3 protein, are membrane proteins with exposed regions on both surfaces of the plasma membrane. (Erythrocyte membrane skeleton proteins have classically been numbered in order of decreasing molecular weight). α- and β-spectrin are the major proteins in the membrane skeleton and form αβ-dimers; two dimers associate head to head to form tetramers. The ends of spectrin tetramers are linked to the sides of short actin fibers; band 4.1 protein apparently not only participates in the interaction of spectrin and actin but also connects them to the membrane-bound glycophorin and band 3 protein. The meshwork that results from these interactions is bound to the membrane by the protein ankyrin, which links the spectrin tetramers to the integral membrane protein band 3.

The membrane skeleton of erythrocytes is unique in its regular structure and multiple attachment sites to the plasma membrane. However, it seems that many, maybe most, cells possess similar membrane skeletons, albeit with fewer attachment sites to the plasma membrane. For example, fodrin seems to be the brain cell equivalent of erythrocyte spectrin. Fodrin and spectrin also interact with and anchor the cytoplasmic intermediate filaments to the membrane. Calpains have been observed to be responsible for the degradation of glycophorin,[8] ankyrin,[6,9,12] band 3 protein,[6-9,11,13] band 4 protein[7,11-13] and spectrin[9,10,13,68] in erythrocytes (Table 1), and fodrin in brain [69,70] (Table 1). Thus, evidence begins to accumulate that calpains are probably also involved in cleavage of connections between the cytoplasmic cytoskeleton and the membrane skeleton.

3. Nebulin, Titin, and Dystrophin in Muscle Cells

Four high molecular weight proteins, titin (a pair of proteins of ≈ 1,000,000 mol wt), nebulin (≈ 700,000 mol wt), and dystrophin (427,000 mol wt) are thought to be components of the muscle cell cytoskeleton.[94-96] Immunoelectron microscopy[95] suggests that titin and myosin and nebulin and actin may form composite filaments and thus the contractile apparatus would consist of four filaments, with the titin, nebulin, and actin filaments directly attached to the plasma membrane at the Z-line. The postulated role for these additional cytoskeletal proteins, particularly nebulin, is in the assembly, stability, and function of the composite filaments.[95] These proteins have been difficult to study due to their large size, unique solubility properties, and, germane to this review, their sensitivity to degradation by calpains.[94] Failure to observe or isolate these proteins has been correlated to high endogenous levels of calpain.[94,95]

A rapidly advancing field of muscle research deals with the molecular basis of muscular dystrophy. Muscular dystrophy in man is an X-linked degenerative disease of muscle. A genetic locus has been identified that, when its gene product is missing, results in the generally fatal Duchenne muscular dystrophy or, when it codes for an altered product, results in the less severe Becker muscular dystrophy. The gene product has been termed dystrophin and the complete primary structure of this protein has been deduced from the complementary DNA sequence.[96] Secondary structure analysis predicts several domains:

1. A 240-amino-acid-long N-terminus highly homologous to the actin binding domain of α-actinin
2. 25 repeats of a triple helical segment similar to those of spectrin
3. a cysteine-rich segment similar to the C-terminus of α-actinin and
4. a unique 420 amino acid C-terminus.[96]

Dystrophin is predicted to be a large, rod-shaped structural protein and possibly forms a component of the muscle cell skeleton.[96] In the *mdx* mouse system (an animal model of Duchenne muscular dystrophy), it has been observed that a defect exists in the regulation of the intracellular Ca^{2+} level so that *mdx* muscle cells contain higher than normal levels of Ca^{2+} when the extracellular Ca^{2+} concentration is at physiological levels.[97] This increased Ca^{2+} concentration in *mdx* cells correlates well with total protein degradation in *mdx* cells (measured by release of tyrosine).[97] Calpains normally degrade their substrates into large peptides rather than releasing individual amino acids. Thus activation of calpain is not the only cause of muscle protein degradation in muscular dystrophy. It is, however, attractive to speculate that the activation of calpain in muscular dystrophy by increased intracellular Ca^{2+} initiates the process of protein degradation by not only hydrolyzing the proteins listed in Table 1, but possibly also by degrading dystrophin. This speculation is based on the homology of dystrophin to α-actinin and spectrin,[96] both of which are calpain substrates. (It seems clear that the defect responsible for muscle degeneration in muscular dystrophy results in altered regulation of intracellular Ca^{2+} concentrations and is not primarily due to any causal role of calpains). This hypothesis must of course be carefully tested, but is at least in accordance with many experimental observations. This also illustrates that there may be many roles for calpains in degenerative diseases or pathological conditions.

III. FUNCTIONAL SIGNIFICANCE AND CONSEQUENCES OF DEGRADATION OF THE CYTOSKELETON PROTEINS BY CALPAINS

From the foregoing discussion it is quite clear that calpains possess the ability to degrade one or more components crucial to the structure and function of the three classes of filaments found in the cytoskeleton, as well as additional cytoskeletal elements and crosslinking proteins. As a general rule, two forms of calpain coexist within cells and are classified as type I (low Ca^{2+}-requiring, μ*M* range) and type II (high Ca^{2+}-requiring, m*M* range). In the presence of high Ca^{2+}-concentrations, calpain II can undergo autolysis to a lower Ca^{2+}-requiring form. Although not dealt with at all in this contribution, many cytological studies have shown that calpains are commonly localized to membrane systems and, furthermore, often colocalize with their protein substrates within specialized structures such as the Z-disks of muscle cells or in adhesion plaques. Despite this colocalization of proteinase and substrate, the cytoskeletal proteins are normally quite stable. This is presumably due to the low endogenous free Ca^{2+}-concentration (approximately 10^{-7} *M*), so that even the low Ca^{2+}-requiring calpain is inactive, and to the activity of endogenous inhibitors of calpains, calpastatins (covered in detail elsewhere in this book). Local increases in Ca^{2+}-concentration are required for activation of calpains *in situ* and degradation of cytoskeletal elements. It is therefore not surprising that most of the degradation catalyzed by calpains *in vivo* is observed to be in close proximity to Ca^{2+}-sequestering membrane systems, such as the sarcolemma in muscle cells, the plasma membrane of erythrocytes and platelets, mitochondria, and the endoplasmic reticulum. Thus, the most obvious consequence of calpain action on the cytoskeleton *in vivo* would be to sever the connections of the cytoskeleton to membranes and organelles in response to transient increases of Ca^{2+}-concentration. Since this effect would be limited by the local nature of the calpain activation, the bulk of cytoskeletal proteins would not be degraded and new interconnections of the cytoskeleton might form after the Ca^{2+}-concentration returns to normal levels. Naturally, long duration increases in Ca^{2+}-concentration due to specific, irreversible stimulations (such as thrombin or ionophore activation of platelets), to aging or to pathological conditions (such as seen in muscle degenerative diseases) may be responsible for more complete degradation of cytoskeletal elements permitting specific, dramatic changes in cell shape (for platelets) or pathological degradation of erythrocyte or muscle cytoskeleton.

A more subtle and potentially more significant role of the ability of calpains to digest cytoskeletal elements is suggested by the observations that calpains specifically remove short fragments from the N-terminus of intermediate filament proteins, producing a characteristic staircase of lower molecular weight, more acidic proteins. This processing by calpains has been proposed to represent a specific posttranslational modification of neurofilaments and may modulate neurofilament structure and/or function. A somewhat more specific and more radical model of the role of calpain-mediated release of the N-terminus of intermediate filament subunit proteins has recently been put forth by our research group.[98,99] Intermediate filament subunit proteins have been observed to bind nucleic acids not only with base composition specificity but also with sequence specificity *in vitro* (vimentin binds oligonucleotides containing telomere repeat sequences with high affinity and specificity). The nucleic acid binding site is located in the N-terminus of these proteins and peptides derived from the N-terminus retain the ability to bind nucleic acids. Cytoplasmic intermediate filaments traverse the entire cytoplasm, from plasma membrane to nuclear membrane, and are possibly directly attached to the nuclear lamina and/or insert into the nucleus. Thus, it is conceivable that intermediate filaments may participate in or influence gene expression, DNA replication, recombination and repair by binding to specific nucleic acid sequences (1) directly in the form of filaments that traverse the nuclear membrane or (2) in the form of free N-terminal fragments, released from filaments in the vicinity of membranes as a result of local increases in Ca^{2+}-concentration (and therefore increased calpain activity) brought about by transmembrane signal transduction. The liberated, N-terminal fragments might be envisioned as intracellular second messengers that, due to their release from filaments, would be free to diffuse to the nucleus and interact with specific nucleic acid sequences and other proteins to modulate nuclear events. In this model, calpain plays a pivotal role in releasing the putative ''messenger'' fragments from the intermediate filaments in response to extracellular signals that give rise to increased intracellular Ca^{2+}-concentrations.

IV. SUMMARY

Calpains are ubiquitous proteinases capable of specific degradation of key elements of the eukaryotic cytoskeleton. Calpains possess the substrate specificity to degrade most of the components of the cytoskeleton, yet the cytoskeleton *in vivo* is notably stable. Thus, an understanding of the regulation of calpain activity, through modulation of Ca^{2+}-concentration, endogenous calpastatin activity and intracellular localization, is essential to a complete understanding of the dynamics of cytoskeletal structure and function. Our understanding of many fundamental cell biology processes, as mentioned in the introduction to this contribution, will be greatly enhanced if as many groups devote as much effort to elucidating the *role* of calpain degradation on the cytoskeleton as has been devoted to the isolation of the enzymes and description of their substrates.

REFERENCES

1. **Zimmerman, U.-J. P. and Schlaepfer, W. W.,** Calcium-activated neutral protease (CANP) in brain and other tissues, *Progr. Neurobiol.*, 23, 63, 1984.
2. **Traub, P.,** *Intermediate Filaments. A Review,* Springer-Verlag, Berlin, Heidelberg, New York, Tokyo, 1985.
3. **Bershadsky, A. D. and Vasiliev, J. M.,** *Cytoskeleton,* Plenum Press, New York, 1988.
4. **Schliwa, M.,** *The Cytoskeleton,* Springer-Verlag, Wien, 1986.
5. **Rousset, B. A. F., Ed.,** *Structure and Functions of the Cytoskeleton — Biological and Physiopathological Aspects,* John Libbey, London, 1988.
6. **Ahkong, Q. F., Botham, G. M., Woodward, A. W., and Lucy, J. A.,** Calcium-activated thiol-proteinase activity in the fusion of rat erythrocytes induced by benzyl alcohol, *Biochem. J.*, 192, 829, 1980.
7. **Murakami, T., Hatanaka, M., and Murachi, T.,** The cytosol of human erythrocytes contains a highly Ca^{2+}-sensitive thiol protease (Calpain I) and its specific inhibitor protein (Calpastatin), *J. Biochem.*, 90, 1809, 1981.

8. **Lorand, L., Bjerrum, O. J., Hawkins, M., Lowe-Krentz, L., and Siefring, G. E., Jr.,** Degradation of transmembrane proteins in Ca^{2+}-enriched human erythrocytes, *J. Biol. Chem.*, 258, 5300, 1983.
9. **Pant, H. C., Virmani, M., and Gallant, P. E.,** Calcium-induced proteolysis of spectrin and band 3 protein in rat erythrocyte membranes, *Biochem. Biophys. Res. Commun.*, 117, 372, 1983.
10. **Thomas, P., Limbrick, A. R., and Allan, D.,** Limited breakdown of cytoskeletal proteins by an endogenous protease controls Ca^{2+}-induced membrane fusion events in chicken erythrocytes, *Biochim. Biophys. Acta*, 730, 351, 1983.
11. **Hatanaka, M., Yoshimura, N., Murakami, T., Kannagi, R., and Murachi, T.,** Evidence for membrane-associated calpain I in human erythrocytes. Detection by an immunoelectrophoretic blotting method using monospecific antibody, *Biochemistry*, 23, 3272, 1984.
12. **Pontremoli, S., Melloni, E., Sparatore, B., Salamino, F., Michetti, M., Sacco, O., and Horecker, B. L.,** Binding to erythrocyte membrane is the physiological mechanism for activation of Ca^{2+}-dependent neutral proteinase, *Biochem. Biophys. Res. Commun.*, 128, 331, 1985.
13. **Croall, D. E., Morrow, J. S., and DeMartino, G. N.,** Limited proteolysis of the erythrocyte membrane skeleton by calcium-dependent proteinases, *Biochim. Biophys. Acta*, 882, 287, 1986.
14. **Gilbert, D. S., Newby, B. J., and Anderton, B. H.,** Neurofilament disguise, destruction and discipline, *Nature*, 256, 586, 1975.
15. **Lazarides, E.,** Intermediate filaments as mechanical integrators of cellular space, *Nature*, 283, 249, 1980.
16. **Malik, M. N., Meyers, L. A., Iqbal, K., Sheikh, A. M., Scotto, L., and Wisniewski, H. W.,** Calcium-activated proteolysis of fibrous proteins in central nervous system, *Life Sci.*, 29, 795, 1981.
17. **Schlaepfer, W. W. and Zimmerman, U.-J. P.,** Calcium-mediated breakdown of glial filaments and neurofilaments in rat optic nerve and spinal cord, *Neurochem. Res.*, 6, 243, 1981.
18. **Schlaepfer, W. W., Zimmerman, U.-J. P., and Micko, S.,** Neurofilament proteolysis in rat peripheral nerve. Homologies with calcium-activated proteolysis of other tissues, *Cell Calcium*, 2, 235, 1981.
19. **Nixon, R. A., Brown, B. A., and Marotta, C. A.,** Posttranslational modification of a neurofilament protein during axoplasmic transport: implications for regional specialization of CNS axons, *J. Cell Biol.*, 94, 150, 1982.
20. **Zimmerman, U.-J. P., and Schlaepfer, W. W.,** Characterization of a brain calcium-activated protease that degrades neurofilament proteins, *Biochemistry*, 21, 3977, 1982.
21. **Nixon, R. A., Brown, B. A., and Marotta, C. A.,** Limited proteolytic modification of a neurofilament protein involves a proteinase activated by endogenous levels of calcium, *Brain Res.*, 275, 384, 1983.
22. **Schlaepfer, W. W., Lee, C., Lee, V. M.-Y., and Zimmerman, U.-J. P.,** An immunoblot study of neurofilament degradation *in situ* and during calcium-activated proteolysis, *J. Neurochem.*, 44, 502, 1985.
23. **Nixon, R. A., Quackenbush, R., and Vitto, A.,** Multiple calcium-activated neutral proteinases (CANP) in mouse retinal ganglion cell neurons: specificities for endogneous neuronal substrates and comparison to purified brain CANP, *J. Neurosci.*, 6, 1252, 1986.
24. **Vorgias, C. E. and Traub, P.,** Efficient degradation *in vitro* of all intermediate filament subunit proteins by the Ca^{2+}-activated neutral thiol proteinase from Ehrlich ascites tumor cells and porcine kidney, *Biosci. Rep.*, 6, 57, 1986.
25. **Banik, N. L., Happel, R. D., Sostek, M. B., Chiu, F. C., and Hogan, E. L.,** Ca^{2+}-mediated degradation of central nervous sytem (CNS) proteins: topographic and species variation, *Metab. Brain Dis.*, 2, 117, 1987.
26. **Tucker, R. P. and Matus, A. I.,** Microtubule-associated proteins characteristic of embryonic brain are found in the adult mammalian retina, *Develop. Biol.*, 130, 423, 1988.
27. **Bloom, G. S., Schoenfeld, T. A., and Vallee, R. B.,** Widespread distribution of the major polypeptide component of MAP1 (micro-tubule-associated protein 1) in the nervous system, *J. Cell Biol.*, 98, 320, 1984.
28. **Paschal, B. M. and Vallee, R. B.,** Retrograde transport by the microtubule-associated protein MAP 1C, *Nature*, 330, 181, 1987.
29. **Vallee, R. B., Wall, J. S., Paschal, B. M., and Shpetner, H. S.,** Microtubule-associated protein 1C from brain is a two-headed cytosolic dynein, *Nature*, 332, 561, 1988.
30. **Lewis, S. A., Wang, D., and Cowan, N. J.,** Microtubule-associated protein MAP2 shares a microtubule binding motif with Tau protein, *Science*, 242, 936, 1988.
31. **Sandoval, I. V. and Weber, K.,** Calcium-induced inactivation of microtubule formation in brain extracts, *Eur. J. Biochem.*, 92, 463, 1978.
32. **Klein, I., Lehotay, D., and Gondek, M.,** Characterization of a calcium-activated protease that hydrolyzes a microtubule-associated protein, *Arch. Biochem. Biophys.*, 208, 520, 1981.
33. **Burgoyne, R. D. and Cumming, R.,** Evidence for the presence of high-M_r microtubule-associated proteins and their Ca^{2+}-dependent proteolysis in synaptosomal cytosol, *FEBS Lett.*, 146, 273, 1982.
34. **Malik, M. N., Fenko, M. D., Iqbal, K., and Wisniewski, H. M.,** Purification and characterization of two forms of Ca^{2+}-activated neutral protease from calf brain, *J. Biol. Chem.*, 258, 8955, 1983.

35. **Yoshimura, N., Tsukahara, I., and Murachi, T.,** Calpain and calpastatin in porcine retina, *Biochem. J.*, 223, 47, 1984.
36. **Sato, C., Nishizawa, K., Nakayama, T., Nakamura, H., Yoshimura, N., Takano, E., and Murachi, T.,** Rapid proteolysis of brain MAP-1 related cytoskeleton-associated 350 kd protein by purified calpain, *Cell Struct. Funct.*, 11, 253, 1986.
37. **Koszka, Ch., Foisner, R., Seyfert, H. M., and Wiche, G.,** Isolation of a Ca^{2+}-protease resistant high M_r microtubule binding protein from mammalian brain: characterization of properties partially expected for a dynein-like molecule, *Protoplasma*, 138, 54, 1987.
38. **Billger, M., Wallin, M., and Karlsson, J.-O.,** Proteolysis of tubulin and microtubule-associated proteins 1 and 2 by calpain I and II. Difference in sensitivity of assembled and disassembled microtubules, *Cell Calcium*, 9, 33, 1988.
39. **Malik, M. N., Ramaswamy, S., Tuzio, H., Shiekh, A. M., Fenko, M. D., Wisniewski, H. M., and Howard, R. G.,** Micromolar Ca^{2+}-requiring protease from human platelets: purification, partial characterization and effect on the cytoskeletal proteins, *Life Sci.*, 40, 593, 1987.
40. **Banik, N. L., Hogan, E. L., Jenkins, M. G., McDonald, J. K., McAlhaney, W. W., and Sostek, M. B.,** Purification of a calcium-activated neutral proteinase from bovine brain, *Neurochem. Res.*, 8, 1389, 1983.
41. **Traub, P. and Vorgias, C. E.,** Involvement of the N-terminal polypeptide of vimentin in the formation of intermediate filaments, *J. Cell Sci.*, 63, 43, 1983.
42. **Geisler, N. and Weber, K.,** Phosphorylation of desmin *in vitro* inhibits formation of intermediate filaments; identification of three kinase A sites in the aminoterminal head domain, *EMBO J.*, 7, 15, 1988.
43. **Evans, R. M.,** Phosphorylation of vimentin in mitotically selected cells. *In vitro* cyclic AMP-independent kinase and calcium-stimulated phosphatase activities, *J. Cell Biol.*, 108, 67, 1989.
44. **Perides, G., Harter, C., and Traub, P.,** Electrostatic and hydrophobic interactions of the intermediate filament protein vimentin and its amino terminus with lipid bilayers, *J. Biol. Chem.*, 262, 13742, 1987.
45. **Nelson, W. J. and Traub, P.,** Proteolysis of vimentin and desmin by the Ca^{2+}-activated proteinase specific for these intermediate filament proteins, *Mol. Cell. Biol.*, 3, 1146, 1983.
46. **Shoeman, R. L., Wadle, S., Scherbarth, A., and Traub, P.,** The binding *in vitro* of the intermediate filament protein vimentin to synthetic oligonucleotides containing telomere sequences, *J. Biol. Chem.*, 263, 18744, 1988.
47. **Steinert, P. M. and Roop, D. R.,** Molecular and cellular biology of intermediate filaments, *Annu. Rev. Biochem.*, 57, 593, 1988.
48. **Dahl, D.,** Glial fibrillary acidic protein from bovine and rat brain degradation in tissues and homogenates, *Biochim. Biophys. Acta*, 420, 142, 1976.
49. **Traub, P. and Nelson, W. J.,** Occurrence in various mammalian cells and tissues of the Ca^{2+}-activated protease specific for the intermediate-sized filament proteins vimentin and desmin, *Eur. J. Cell Biol.*, 26, 61, 1981.
50. **Nelson, W. J. and Traub, P.,** Purification and further characterization of the Ca^{2+}-activated proteinase specific for the intermediate filament proteins vimentin and desmin, *J. Biol. Chem.*, 257, 5544, 1982.
51. **Nelson, W. J. and Traub, P.,** Intermediate (10 nm) filament proteins and the Ca^{2+}-activated proteinase specific for vimentin and desmin in the cells from fish to man: an example of evolutionary conservation, *J. Cell Sci.*, 57, 25, 1982.
52. **Virtanen, I., Vartio, T., and Lehto, V.-P.,** Low-ionic strength induces degradation of vimentin in cultured human fibroblasts, *Biochem. Biophys. Res. Commun.*, 105, 730, 1982.
53. **DeArmond, S. J., Fajardo, M., Naughton, S. A., and Eng, L. F.,** Degradation of glial fibrillary acidic protein by a calcium dependent proteinase: an electroblot study, *Brain Res.*, 262, 275, 1983.
54. **Roy, D., Chiesa, R., and Spector, A.,** Lens calcium-activated proteinase: degradation of vimentin, *Biochem. Biophys. Res. Commun.*, 116, 204, 1983.
55. **Ciesielski-Treska, J., Goetschy, J.-F., and Aunis, D.,** Proteolytic degradation of vimentin and glial fibrillary acidic protein in rat astrocytes in primary culture, *Eur. J. Biochem.*, 138, 465, 1984.
56. **Ireland, M. and Maisel, H.,** Evidence for a calcium activated protease specific for lens intermediate filaments, *Current Eye Res.*, 3, 423, 1984.
57. **Yoshida, H., Murachi, T., and Tsukahara, I.,** Degradation of actin and vimentin by calpain II, a Ca^{2+}-dependent cysteine proteinase, in bovine lens, *FEBS Lett.*, 170, 259, 1984.
58. **Fischer, S., Vandekerckhove, J., Ampe, C., Traub, P., and Weber, K.,** Protein-chemical identification of the major cleavage sites of the Ca^{2+} proteinase on murine vimentin, the mesenchymal intermediate filament protein, *Biol. Chem. Hoppe-Seyler*, 367, 1147, 1986.
59. **Belin, M.-T. and Boulanger, P.,** Processing of vimentin occurs during the early stages of adenovirus infection, *J. Virol.*, 61, 2559, 1987.
60. **Perides, G., Kühn, S., Scherbarth, A., and Traub, P.,** Probing of the structural stability of vimentin and desmin-type intermediate filaments with Ca^{2+}-activated proteinase, thrombin and lysine-specific endoproteinase Lys-C, *Eur. J. Cell Biol.*, 43, 450, 1987.

61. **Ando, Y., Miyachi, Y., Imamura, S., Kannagi, R., and Murachi, T.,** Purification and characterization of calpains from pig epidermis and their action on epidermal keratin, *J. Invest. Dermatol.*, 90, 26, 1988.
62. **Traub, P., Scherbarth, A., Willingale-Theune, J., Paulin-Levasseur, M., and Shoeman, R.,** Differential sensitivity of vimentin and nuclear lamins from Ehrlich ascites tumor cells toward Ca^{2+}-activated neutral thiol proteinase, *Eur. J. Cell Biol.*, 46, 478, 1988.
63. **Bravo, R. and Celis, J. E.,** A search for differential polypeptide synthesis throughout the cell cycle of HeLa cells, *J. Cell. Biol.*, 84, 795, 1980.
64. **Bravo, R., Small, J. V., Fey, S., Larsen, P. M., and Celis, J. E.,** Architecture and polypeptide composition of HeLa cytoskeletons. Modification and cytoarchitectural polypeptides during mitosis. *J. Mol. Biol.*, 154, 121, 1982.
65. **Ochs, D. C., McConkey, E. H., and Guard, N. L.,** Vimentin-derived proteins. Differences between normal human fibroblasts and transformed human cells, *Exp. Cell Res.*, 135, 355, 1981.
66. **Giometti, C. S., Willard, K. E., and Anderson, N. L.,** Cytoskeletal proteins from human skin fibroblasts, peripheral blood leukocytes, and a lymphoblastoid cell line compared by two-dimensional gel electrophoresis, *Clin. Chem.*, 28, 955, 1982.
67. **Virtanen, I., Närvänen, O., and Lehto, V.-P.,** Differential immunoreactivity and Ca^{2+}-dependent degradation of vimentin in human fibroblasts and fibrosarcoma cells, *Int. J. Cancer*, 42, 256, 1988.
68. **Fox, J. E. B., Reynolds, C. C., Morrow, J. S., and Phillips, D. R.,** Spectrin is associated with membrane-bound actin filaments in platelets and is hydrolyzed by the Ca^{2+}-dependent protease during platelet activation, *Blood*, 69, 537, 1987.
69. **Siman, R., Baudry, M., Lynch, G.,** Brain fodrin: substrate for calpain I, an endogenous calcium-activated protease, *Proc. Natl. Acad. Sci. U.S.A.*, 81, 3572, 1984.
70. **Nixon, R. A.,** Fodrin degradation by calcium-activated neutral proteinase (CANP) in retinal ganglion cell neurons and optic glia: preferential localization of CANP activities in neurons, *J. Neurosci.*, 6, 1264, 1986.
71. **Granger, B. L., Repasky, E. A., and Lazarides, E.,** Synemin and vimentin are components of intermediate filaments in avian erythrocytes, *J. Cell Biol.*, 92, 299, 1982.
72. **Evans, R. R., Robson, R. M., and Stromer, M. H.,** Properties of smooth muscle vinculin, *J. Biol. Chem.*, 259, 3916, 1984.
73. **Dayton, W. R., Goll, D. E., Stromer, M. H., Reville, W. J., Zeece, M. G., and Robson, R. M.,** Some properties of a Ca^{2+}-activated protease that may be involved in myofibrillar protein turnover, in *Proteases and Biological Control*, Vol. 2, Reich, E., Rifkin, D. B. and Shaw, E., Eds., Cold Spring Harbor Conf. Cell Proliferation, Cold Spring Harbor Laboratory, Cold Spring Harbor, New York, 1975, 551.
74. **Reddy, M. K., Etlinger, J. D., Rabinowitz, M., Fischman, D. A., and Zak, R.,** Removal of Z-lines and α-actinin from isolated myofibrils by a calcium-activated neutral protease, *J. Biol. Chem.*, 250, 4278, 1975.
75. **Dayton, W. R., Reville, W. J., Goll, D. E., and Stromer, M. H.,** A Ca^{2+}-activated protease possibly involved in myofibrillar protein turnover. Partial characterization of the purified enzyme, *Biochemistry*, 15, 2159, 1976.
76. **Nagainis, P. and Wolfe, F. H.,** Calcium-activated neutral protease hydrolyzes Z-disc actin, *J. Food Sci.*, 47, 1358, 1982.
77. **Croall, D. E. and DeMartino, G. N.,** Comparison of two calcium-dependent proteinases from bovine heart, *Biochim. Biophys. Acta*, 788, 348, 1984.
78. **Kulesza-Lipka, D., and Jakubiec-Puka, A.,** Degradation of α-actinin during Ca^{2+}-sensitive proteolysis of myofibrils, *FEBS Lett.*, 187, 354, 1985.
79. **Gache, Y., Landon, F., Touitou, H., and Olomucki, A.,** Susceptibility of platelet α-actinin to a Ca^{2+}-activated neutral protease, *Biochem. Biophys. Res. Commun.*, 124, 877, 1984.
80. **Davies, P. J. A., Wallach, D., Willingham, M. C., Pastan, I., Yamaguchi, M., and Robson, R. M.,** Filamin-actin interaction, *J. Biol. Chem.*, 253, 4036, 1978.
81. **Truglia, J. A. and Stracher, A.,** Purification and characterization of a calcium-dependent sulfhydryl protease from human platelets, *Biochem. Biophys. Res. Commun.*, 100, 814, 1981.
82. **Haeberle, J. R., Collican, S. A., Evan, A., and Hathaway, D. R.,** The effects of a calcium-dependent protease on the ultrastructure and contractile mechanics of skinned uterine smooth muscle, *J. Muscle Res. Cell Motil.*, 6, 347, 1985.
83. **Onji, T., Takagi, M., and Shibata, N.,** Calpain abolishes the effect of filamin on the actomyosin system in platelets, *Biochim. Biophys. Acta*, 912, 283, 1987.
84. **Zhang, Z., Lawrence, J., and Stracher, A.,** Phosphorylation of platelet actin-binding protein protects against proteolysis by calcium-dependent sulfhydryl protease, *Biochem. Biophys. Res. Commun.*, 151, 355, 1988.
85. **Beckerle, M. C., O'Halloran, T., and Burridge, K.,** Demonstration of a relationship between talin and P235, a major substrate of the calcium-dependent protease in platelets, *J. Cell. Biochem.*, 30, 259, 1986.

86. **Nachmias, V. T., Sullender, J. S., and Fallon, J. R.,** Effects of local anesthetics on human platelets: filopodial suppression and endogenous proteolysis, *Blood,* 53, 63, 1979.
87. **Collier, N. C. and Wang, K.,** Purification and properites of human platelet P235, *J. Biol. Chem.,* 257, 6937, 1982.
88. **Tsujinaka, T., Sakon, M., Kambayashi, J., and Kosaki, G.,** Cleavage of cytoskeletal proteins by two forms of Ca^{2+} activated neutral proteases in human platelets, *Thrombosis Res.,* 28, 149, 1982.
89. **Fox, J. E. B., Goll, D. E., Reynolds, C. C., and Phillips, D. R.,** Identification of two proteins (actin-binding protein and P235) that are hydrolyzed by endogenous Ca^{2+}-dependent protease during platelet aggregation, *J. Biol. Chem.,* 260, 1060, 1985.
90. **Beckerle, M. C., Burridge, K., DeMartino, G. N., and Croall, D. E.,** Colocalization of calcium-dependent protease II and one of its substrates at sites of cell adhesion, *Cell,* 51, 569, 1987.
91. **Nicotera, P., Hartzell, P., Davis, G., and Orrenius, S.,** The formation of plasma membrane blebs in hepatocytes exposed to agents that increase cytosolic Ca^{2+} is mediated by the activation of a nonlysosomal proteolytic system, *FEB Lett.,* 209, 139, 1986.
92. **Pontremoli, S., Melloni, E., Michetti, M., Sparatore, B., Salamino, F., Sacco, O., and Horecker, B. L.,** Phosphorylation by protein kinase C of a 20-kDa cytoskeletal polypeptide enhances its susceptibility to digestion by calpain, *Proc. Natl. Acad. Sci. U.S.A.,* 84, 398, 1987.
93. **Pontremoli, S., Melloni, E., Michetti, M., Sparatore, B., Salamino, F., Sacco, O., and Horecker, B. L.,** Phosphorylation and proteolytic modification of specific cytoskeletal proteins in human neutrophils stimulated by phorbol 12-myristate 13-acetate, *Proc. Natl, Acad. Sci. U.S.A.,* 84, 3604, 1987.
94. **Wang, K.,** Purification of titin and nebulin, *Methods Enzymol.* 85, 264, 1982.
95. **Wang, K. and Wright, J.,** Architecture of the sarcomere matrix of skeletal muscle: Immunoelectron microscopic evidence that suggests a set of parallel inextensible nebulin filaments anchored at the 7 line, *J. Cell Biol.,* 107, 2199, 1988
96. **Koenig, M., Monaco, A. P., and Kunkel, L. M.,** The complete sequence of dystrophin predicts a rod-shaped cytoskeletal protein, *Cell,* 53, 219, 1988.
97. **Turner, P. R., Westwood, T., Regen, C. M., and Steinhardt, R. A.,** Increased protein degradation results from elevated free calcium levels found in muscle from mdx mice, *Nature,* 335, 735, 1988.
98. **Traub, P.,** Are intermediate filament proteins involved in gene expression? *Ann. N.Y. Acad. Sci.,* 455, 68, 1985.
99. **Traub, P., Plagens, U., Kühn, S., and Perides, G.,** Function of intermediate filaments. A novel hypothesis, *Fortschritte der Zoologie,* 34, 275, 1987.

Chapter 15

CALPAINS AND HEMOSTASIS

Robert W. Colman and Ian Hoffman

TABLE OF CONTENTS

The process of hemostasis involves the interaction of plasma proteins, blood cells (including the platelet, monocyte, neutrophil, and under special circumstances, erythrocytes), and vascular cells such as the endothelial cells. Although all blood cells contain calpain(s),[1] most studies have concentrated on platelets. Thus, this review is limited to platelet calpains and their interaction with plasma proteins. Like most vertebrate cells, human platelets contain two distinct calcium-dependent neutral proteases (calpains): type I, active at micromolar Ca^{2+}, and type II, requiring millimolar Ca^{2+} for its activity.[2] In addition, at least two inhibitors of calpain(s) have been described in platelets — calpastatin[3] and kininogen.[4] While the roles of calpains are not fully understood, this chapter will review the current knowledge and specific functions of these ubiquitous enzymes in hemostasis. The hypothesis will be presented that each of these enzymes plays a critical role in hemostasis. Calpain I appears to be a candidate for participation in intracellular proteolysis of both cytoskeletal proteins and regulatory enzymes. Calpain II may function on the external membrane surface to process coagulation proteins and to cleave critical glycoproteins mediating platelet activation by physiological agonists.

I. PLATELETS AND HEMOSTASIS

To understand the roles of the calpains, it is important to review the modes of participation of platelets in the hemostatic (or when pathologic) the thrombotic process. The protean reactions of platelets include adhesion of platelets to the exposed subendothelial surface following blood vessel injury, shape change from a disc to a spiculated sphere, secretion of intracellular constituents (many with biological activity), and processing of secreted α-granule proteins. These early changes are followed by aggregation to form a mass of platelets to occlude the vessel, and exposure of procoagulant phospholipids with concomitant acceleration of coagulation, resulting in the formation of a fibrin network to stabilize the platelet plug. Each of these processes involves critical protein components, and in each case some have been identified as substrates of the calpains.

Platelets do not adhere to normal endothelial cells, but will adhere to altered cells or to subendothelial collagen in areas of endothelial disruption. At the high shear rates which occur in arterioles and the microvasculature, a protein, von Willebrand factor (vWF), which binds specifically to glycoprotein Ib (GPIb), a surface membrane component of platelets[5] is required for adhesion. *In vitro*, vWF, in the presence of ristocetin (a basic polypeptide antibiotic), causes platelet agglutination, providing a useful model for vWF interactions with GPIb. The adherent platelets then undergo a series of changes in response to specific physiological agonists which, in addition to collagen, include ADP, epinephrine, prostaglandin cyclic endoperoxides (or thromboxane A_2,) and proteolytic enzymes including thrombin and plasmin. The first morphological change is shape change from disc to spiny sphere, which is mediated by a moderate increase in intracellular calcium and accompanied by phosphorylation of platelet myosin light chain by a calmodulin stimulated protein kinase.[6] These altered platelets then secrete substances into the extracellular milieu from at least three types of granules which serve as "storage pools"[7] for biologically potent agonists. The lysosomal granules contain acid hydrolases, the dense granules ADP, ATP, pyrophosphate, and serotonin, and the α-granules, platelet-specific proteins and coagulation factors.

The platelets then undergo aggregation or platelet-platelet interaction. This process is mediated by the unmasking of latent receptors for fibrinogen by all of the physiological agonists mentioned above. The fibrinogen receptor is the heterodimer of platelet glycoproteins IIb and IIIa associated with ionic calcium to which fibrinogen can bind. The dimeric structure of fibrinogen facilitates the formation of molecular bonds between platelets. In the course of activation, extensive cytoskeletal remodeling occurs, with such consequences as disappearance of the marginal bond of microtubules, centralization of storage granules, and formation of pseudopodia.[8] In addition, activated platelets develop binding sites for coagulation proteins. Acidic phospholipids, normally present on the inside of the platelet membrane, become translocated to the outside and may serve as binding sites for activated coagulation cofactors such as factors VIIIa and Va. These cofactors then bind serine proteases such as factors IXa and Xa, as well as the substrates, factor X and prothrombin. The conversion of factor X to Xa and prothrombin to thrombin then takes place on the platelet surface.[9]

Each physiological agonist triggers some or all of these changes by several independent but potentially interacting mechanisms. Agonists may trigger phospholipase A_2 (a calcium-requiring enzyme) which liberates free arachidonic acid, predominantly from phosphatidyl choline. This highly unsaturated fatty acid is then converted by cyclooxygenase to prostaglandin cyclic endoperoxides and then by thromboane synthetase to thromboxane A_2.[10] Both of these lipids serve as potent positive feedback mechanisms to further activate platelets. Agonists may also trigger phospholipase C, which hydrolyzes phosphatidyl inositol diphosphate to two intracellular mediators,[11] diacylglycerol (DAG) and inositol triphosphate (IP3). DAG is the physiological stimulator of protein kinase C, which phosphorylates a 47-kDa protein closely linked to the secretory response. IP3 is the physiologic ionophore which mobilizes calcium from intracellular stores to the cytoplasm. All of these processes for the activation of platelets are inhibited by a rise in platelet intracellular cAMP, which is formed by stimulated adenylate cyclase and degraded by cAMP phosphodiesterases.

II. PROPERTIES OF CALPAINS AND THEIR INHIBITORS

The presence of calcium-activated neutral proteases (calpains) in human platelets was first suggested by Phillips and Jakabova.[12] Calpain II was purified from human platelets by Truglia and Stracher.[13] More recently, homogeneous preparations of calpain II[14] and calpain I[15] have been achieved. Remarkably, platelet calpains may be involved in most, if not all, of the processes of activation of platelets and coagulation proteins. Since these enzymes absolutely require Ca^{2+}, it is important first to consider when they might be active. Calpain I requires 1 to 10 μM ionic calcium for half maximum activity. In platelets, the resting intracellular calcium concentration is about 100 nM. Agonists such as thrombin, plasmin, and collagen (in high concentration), as well as the calcium ionophore A23187, in the presence of extracellular Ca^{2+} can stimulate a rise of intracellular Ca^{2+} to a concentration of 1 to 5 μM and thus activate calpain I. Calpain II, which in its native state requires 150 μM for half maximal activity,[16] could not undergo intracellular activation. However, although cleavage of calpain II (by calpain I) leads to an increase in calcium affinity and increased activity, the activation of calpain II cannot lower the calcium requirement low enough to permit activation by intracellular ionized calcium concentrations. In the presence of phosphatidyl inositol vesicles,[17] however, significant further lowering of the Ca^{2+} requirement takes place. Moreover, autoproteolysis converts calpain II to a more hydrophobic calpain I-like form which further allows activity at even lower Ca^{2+} concentrations. Finally, the possibility that calpain II may be exposed on external platelet membranes (see below) permits its full activity to be expressed at extracellular Ca^{2+} concentrations.

Another control on calpain activity is the presence of calpastatin, the specific calpain inhibitor protein in platelets.[3] However, 30 to 60% of the calpain activity persists, despite the presence of calpastatin, by virtue of its association with membranes.[18] Calpain II, the protease isozyme active at extracellular Ca^{2+} concentrations, is controlled by plasma inhibitors including α_2-macroglobulin and kininogens.[16] The amino acid sequence of two of the cysteine protease inhibitors in plasma was shown to be identical to the two kininogens.[19] In human plasma, low molecular weight kininogen (LK) and high molecular weight kininogens (HK) are known to exist and are products of a single gene which by alternate splicing of the primary transcript produces two different mRNAs. The heavy chain of both molecules which contains cysteine protease inhibitory sites, is identical in both HK and LK. In human plasma, HK is the major inhibitor of platelet calpain II.[16] Kinetic analysis indicates that it is a mixed noncompetitive tight-binding inhibitor with a K_i of 0.5 to 2 nM.[16] Experiments using normal and deficient plasma indicate HK is the predominant inhibitor in plasma and is more important than LK or α_2-macroglobulin. Furthermore, HK is present in the α-granules of platelets and is secreted following exposure to thrombin.[4] HK not only binds to platelets but is expressed on the surface of the activated platelets. Calpain can increase HK coagulant activity but extensively hydrolyzes HK at enzyme to substrate ratios greater than 2,[20] while when HK is in excess it acts as a potent inhibitor. A monoclonal antibody to the heavy chain of HK blocked the ability of HK to inhibit platelet calpain.[21] Thus, kininogen in plasma and platelets can be expected to regulate calpain expression, especially on the platelet surface.

III. PLATELET PROTEIN SUBSTRATES FOR CALPAINS

Any action of the platelet calpains should be understood in the context of hydrolysis of its substrates. However, the calpains are not fastidious enzymes, and have been described to hydrolyze a wide variety of platelet proteins. These substrates fall into four major classes: cytoskeletal proteins, α-granule proteins, intracellular proteins involved with platelet activation, and membrane receptors. Historically, the first substrates described were cytoskeletal proteins. When platelets are extracted with Triton X-100 and the insoluble residue subjected to SDS gel electrophoresis, the proteins visualized are defined as the platelet cytoskeleton which includes actin, myosin, actin-binding protein, talin, and other muscle proteins or proteins binding to cytoskeletal components. Tsujunaka et al.[22] were the first investigators to describe the cleavage of cytoskeletal proteins of human platelets by partially purified platelet calpains I and II. They documented that actin-binding protein (M_r = 260 kDa), also known as filamin, and a second protein of 235 kDa were completely degraded. The observation that platelet calpain I can cleave an endogenous substrate led the authors to suggest that this enzyme is involved in the stimulus-linked platelet reactions that are associated with a rise in intracellular Ca^{2+} to the micromolar range. Fox et al[23] confirmed this suggestion by demonstrating that calcium-dependent proteolysis occurred during platelet aggregation with the formation of new polypeptides of M_r = 200, 100, and 91 kDa. These polypeptides were subsequently identified as deriving from actin-binding protein,[24] and these investigators also demonstrated hydrolysis of P235 to M_r = 200 and 46 kDa during platelet aggregation by thrombin. P235 has recently been demonstrated to be identical to talin.[25] A third platelet cytoskeletal protein, α-actinin, has been identified as a substrate for calpain II derived from smooth muscle.[26] Purified platelet α-actinin from fresh platelets consist of two dimer isoforms aa and cc. On proteolysis with muscle calpain II, aa is converted to bb, but cc is not a substrate. The conversion of the aa form of α-actinin to bb also occurs on storage of platelets, but whether this conversion occurs during platelet activation is unknown, and therefore its physiological relevance is unclear. Finally, Fox and co-workers[27] have demonstrated that platelets contain spectrin (M_r 240 and 235 kDa) associated with actin in the Triton-insoluble cytoskeleton. Platelet spectrin is hydrolyzed to 160 kDa during exposure of platelets to thrombin or the calcium ionophore A23187, a process blocked by calpain inhibitors such as leupeptin, EGTA, or *N*-ethylmalamide (NEM), and is a putative substrate of calpain I.

The second class of endogenous platelet proteins hydrolyzed by calpain are those present in α-granules. Thus far, the recognized substrates of calpain which can be secreted into the extracellular vascular space are all hemostatic proteins also present in plasma. The first recognized substrate was fibrininogen, (M_r = 330 kDa) which is composed of three polypeptide chains: Aα (71 kDa), Bβ (58 kDa) and γ (54 kDa). Kunicki et al.[28] found that platelet fibrinogen was degraded by a soluble protease in platelet sonicates or in platelets solubilized using sonicates or nonionic detergents. The enzyme(s) were inhibited by leupeptin, EGTA, and NEM, but not by serine protease inhibitors. The core derivative was composed of partially degraded Aα (M_r 47, 46, 34 kDa), Bβ (56 kDa), and intact γ (54 kDa) with an unreduced M_r of 290 kDa. This cleavage pattern is unique for calpain, and an identical derivative results from calpain hydrolysis of plasma fibrinogen. The second hemostatic protein recognized as a substrate for platelet calpain was vWF, a multimeric protein composed of 230-kDa subunits, important for platelet adhesion to vessel walls. Kunicki et al.[29] demonstrated that both platelet and plasma vWF were degraded with a decrease in size of the reduced monomer to 205 kDa in platelet lysates prepared by sonication or addition of nonionic detergent. Since EGTA, leupeptin, and NEM all inhibited the proteolysis, the authors concluded that calpain was responsible for this proteolytic cleavage. Gralnick et al.[30] also found that platelet lysates could degrade unreduced radiolabeled vWF in normal plasma with reduction in the larger multimers.

As noted earlier, HK, an α-granule protein, can be cleaved and activated by calpain when the enzyme is in excess.[20] A fourth protein known to be a substrate for calpain is coagulation factor V. Factor V is a single chain plasma protein of M_r 330 kDa, critical for the conversion of prothrombin to thrombin. Factor V has been localized to the α granules of platelets, and is secreted in response to both collagen and ADP.[31] Our laboratory demonstrated that, following secretion, platelet factor V was cleaved to a more positively charged form,[32] that that change could be reproduced by a platelet lysate that activated factor V by two-fold, and that the reaction was blocked by protease inhibitors. Tracy et al.[33] found that at least one protease that cleaved factor V was inhibited by leupeptin and EDTA but not by DFP. We have recently demonstrated[34] that factor V is activated and cleaved by purified platelet calpain. The cleavage pattern differs markedly from thrombin since it does not result in the usual calcium-linked light chain (M_r = 74 kDa) and heavy chain (M_r = 105 kDa) which makes up factor Va. Instead, one of the activation peptides (M_r = 150 kDa) is extensively proteolyzed. Moreover, while thrombin hydrolysis of factor V results in a 20-fold activation, calpain only achieves a 2.5-fold activation, and the resultant form remains susceptible to subsequent activation by thrombin.

The third group of calpain substrates in platelets are calcium-binding intracellular proteins thought to function in platelet activation. Kosaki et al.[35] first demonstrated that two of the platelet proteins which were capable of binding calmodulin (as demonstrated by a gel overlay technique) were cleaved by a platelet lysate under conditions (1 to 10 μM Ca^{2+}) known to activate calpain I. The 90-dKa and 60-dKa calmodulin-binding polypeptides were degraded to components of M_r = 76, 70 and 54 kDa. The cleavage was blocked by leupeptin but not pepstatin or DFP, suggesting that calpain was responsible. The identity of these two proteins is uncertain. However, Tallant et al.[36] have recently demonstrated that a 60-kDa calmodulin-dependent phosphatase (calcineurin), present in both brain and platelets, is degraded by calpain I to a calmodulin-binding 58-dKa fragment and a 45-kDa peptide which does not bind calmodulin. Although not demonstrable by the gel overlay technique, two other calmodulin-binding proteins are known to be present in platelet. Kosaki et al.[35] found that purified caldesmon (M_r = 150 and 147 kDa) is degraded by calpain I in platelet lysates to multiple low M_r components. Similarly, purified myosin light chain kinase (M_r = 135 kDa) was extensively degraded to products of 108, 98, 87, 68, and 27 kDa. Follow-up studies by Kajiwara et al.[37] demonstrated that specific calpain inhibitors not only inhibited degradation of actin-binding protein and talin, but also inhibited the phosphorylation of myosin light chain by about 50% in both platelet lysates and electropermeabilized platelets. In order to demonstrate this in intact platelets, Tsujunaka et al.[38] employed a cell penetrating thiol protease inhibitor. The compound inhibited myosin light chain phosphorylation in platelets stimulated by a calcium ionophore, thrombin, or collagen, but did not inhibit phosphorylation induced by phorbol ester, which does not increase Ca^{2+}. Not all Ca^{2+}-binding proteins involved in platelet activation are calmodulin dependent. As previously indicated, a major regulatory step in stimulus-response coupling is the hydrolysis of phosphoinositides by a specific phospholipase C which requires Ca^{2+} and is sensitive to the lipid microenvironment. Low et al.[39] have demonstrated different molecular size forms of the enzyme on gel filtration of the supernatant fraction from freshly sonicated human platelets. When phospholipase C activity was measured after sonication in the presence of leupeptin or EDTA, the predominant molecular weight form was 290 kDa. When both these inhibitors were omitted, the predomoninant form was 100 kDa. Though this observation provided preliminary evidence for the cleavage of platelet phospholipase C, it does not define the physiological occurrence of the hydrolysis. A second calcium-dependent enzyme which is independent of calmodulin is platelet factor XIII. This protein is a transaminidase which catalyzes the formation of glutamyl-lysyl isopeptide cross-links. Factor XIII is converted to a cleaved form by thrombin and then activated by calcium. Recently, Ando et al.[40] demonstrated that platelet factor XIII can also be activated by calpain, thus providing a thrombin-independent mechanism for cross-linking of platelet proteins.

The final group of substrates for calpains are membrane receptor proteins. Glycoprotein Ib is a major glycoprotein present at about 25,000 copies per platelet, comprised of two disulfide-linked subunits Ibα (M_r = 143 kDa) and Ibβ (M_r = 22 kDa). If platelets are solubilized in the presence of Ca^{2+} by nonionic detergents, then glycocalicin, a single chain glycopeptide (M_r = 130 kDa), is liberated, which immunologically and functionally has been shown to derive from GPIbα.[41,42] Since the formation of glycocalicin from GPIb is inhibited by leupeptin and EDTA, platelet calpain is thought to be responsible.[44] Yamamoto et al.[43] have shown that purified GPIb can be cleaved to glycocalicin and a 100-kDa fragment by purified platelet calpain I and II. Activation of endogeneous calpain by stimulation of platelets with calcium ionophore produced only glycocalicin, which could then be hydrolyzed to the 100-kDa frament by calpain, suggesting that calpain can hydrolyze GPIbα at at least two sites. Glycoprotein V is cleaved by both calpain[45] and thrombin,[46] but the physiological significance of this reaction is not clear. Recently, we have identified another membrane receptor, aggregin, as a substrate for calpain II.[46a] Aggregin[47] is a 100-kDa platelet membrane protein that is labeled with radiolabeled fluorosulfonylbenzoyl adenosine (^{3}H-FSBA) after exposure of intact platelets. ^{3}H-FSBA is an affinity reagent which labels adenosine nucleotide (ADP, ATP) binding sites on a wide variety of proteins. ADP or ATP prevents the labeling of this protein on intact peptides. Purified platelet calpain II digests ^{3}H-aggregin on the intact platelet surface or platelet membranes to small peptide fragments. The cleavage is inhibited by the calpain inhibitors, leupeptin, EDTA, or HK.[48]

IV. ROLES OF CALPAIN IN PLATELET ACTIVATION

Having reviewed the mechanisms of platelet activation and identified the substrates of platelet calpain, it is possible to suggest some roles for calpain in platelet activation. The presence of as many as 20 different endogenous substrates for the two platelet calpains indicates both the importance and the complexity of trying to define the physiological actions of these enzymes. Thus, any roles assigned to these enzymes must be considered tenative, and are more accurately termed working hypotheses than firmly established mechanisms. We will therefore attempt to indicate where calpain could influence each of the steps in platelet activation.

The adhesion of platelets to the subendothelial basement membrane exposed by loss of endothelial cells requries vWF and glycoprotein I_b, both substrates of calpain(s) (Figure 1). Since the absence of large multimers of vWF in type IIa von Willebrand (vW) disease may lead to a hemorrhagic state, it is inferred that the large multimers secreted by perturbed endothelial

ADHESION

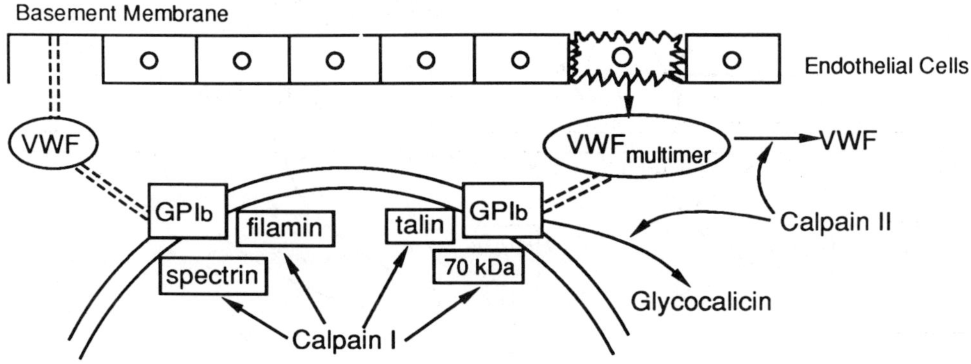

FIGURE 1. Role of calpain in adhesion. The mechanisms of adhesion of platelets to endothelial cells are schematized. Double dashed line represents binding. Solid arrows represent enzyme actions or conversions. The jagged cell represents a stimulated endothelial cell. Boxed proteins are calpain substrates. vWF = Von Willebrand factor; GPI_b = platelet glycoprotein I_b, a transmembrane receptor. The description of these interactions is in the text.

cells are critical for the participation of vWF in normal hemostasis. Gralnick et al.[30] postulated that patients with type IIa von Willebrand disease might synthesize an abnormal vWF more susceptible to proteolysis by calpain than the normal gene product. Thus, by degrading vWF multimers, calpain might be responsible for defective hemostasis or might reverse the adhesive reaction when it is no longer required in normal individuals. However, Berkowitz et al.[49] after digesting human vWF with purified calpain, performed epitope mapping with monoclonal antibodies, and determined the amino-terminal sequences of the calpain-cleaved vWF fragments. The results indicate that the proteolyzed fragments present in normal and type IIa vW disease are not the same as those produced by digestion by calpain. Thus, calpain is probably not responsible for the loss of large multimers seen in type IIa vW disease.

However, the effects of calpain on GPIb, the receptor for vW factor, are perhaps of greater importance. Okita et al.[50] have demonstrated the GPIb is associated with actin-binding protein, provided that GPIb is purified in the presence of calpain inhibitors from nonactivated platelets. The authors suggest that noncovalent linkage betwen GPIb and actin-binding protein exists in intact platelets and may limit the mobility of GPIb, and thus its receptor function. This postulate is consistent with the finding by Coller[51] that dibucaine, which activates intracellular calpain allowing cleavage of actin-binding protein, initially enhances the rate of vWF-dependent platelet agglutination. In contrast, diamide, which inhibits calpain and cross-links GPIb to actin-binding protein, inhibits vWF-induced platelet agglutination.[52] These findings also explain the reversal by dibucaine of the effects of diamide. The agglutination produced by vWF is inhibited after prolonged incubation with dibucaine.[51] This observation probably reflects the eventual cleavage of GPIb by calpain[43] to release glycocalicin, a reaction which occurs less rapidly than cleavage of actin-binding protein. A further consequence of cleavage of actin-binding protein (filamin) could be to enhance the concentration of the activated platelets. Normally, filamin cross-links F actin and inhibits actomyosin ATPase activity. When degraded by calpain, filamin loses both its cross-linking ability and its ability to inhibit actomyosin function,[53] thus enhancing G actin polymerization important for shape change and secretion. Recently, Meyer[54] has described a protein of M_r = 70 kDa hydrolyzed rapidly by calpain, which also binds to GPIb. However,

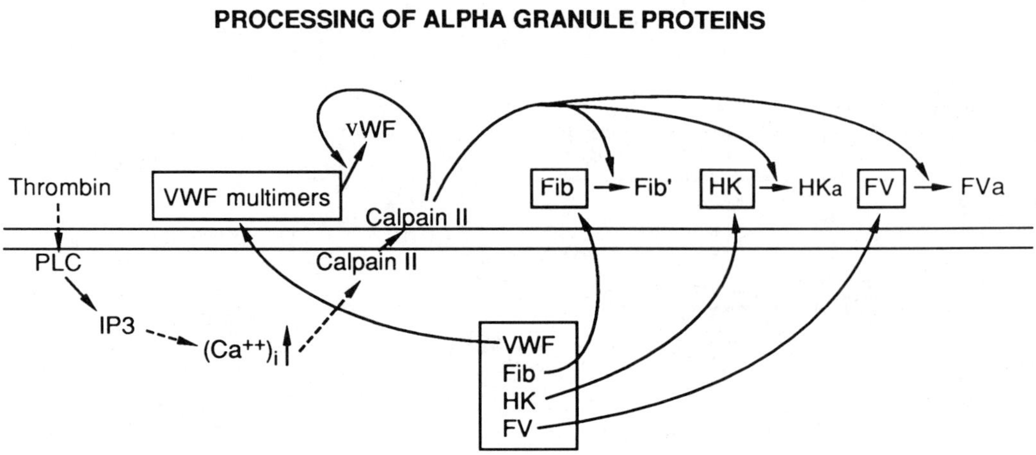

FIGURE 2. Calpain processing of α-granule proteins. The double line represents the platelet membrane with the cytosol in the lower section. Dashed lines represent agonists or stimuli influencing transport or activity. Fib = fibrinogen; Fib′ = fibrinogen degradation products; HK = high molecular weight kininogen; HK_a = activated HK; FV = factor V; FVa = activated FV; PLC = phospholipase C; IP3 = inositol triphosphate; (Ca^{++}) = cytosolic ionized calcium. Other symbols as in Figure 1.

the role of this component is not yet clear. It may function to reorganize actin filaments to filipodin involved in clot retraction.

The functional significance of hydrolysis of talin by calpain is also somewhat obscure. In a variety of cells, calpain II is localized with talin (P235) at adhesion plaques or focal contacts of the cell with the extracellular matrix.[55] Moreover, talin is the most sensitive of the cytoskeletal proteins present at this location to proteolysis by calpain II. However, in platelets, adhesion occurs early, while proteolysis of talin accompanies clot retraction[23] and may facilitate that process. Spectrin, a major component of erythrocytes but a trace component of platelets,[27] may serve the same purpose.

What is the role of calpain in the hydrolysis of the α-granule proteins (Figure 2)? We have already discussed the effects of calpain on vWF. Fibrinogen plays a critical role in platelet aggregation, and plasmin-degradation of fibrinogen diminishes or abolishes its ability to bind to platelets. Since, in the resting platelet, fibrinogen is protected from calpain by its α-granule localization, this degradation could only be important if calpain were active in the extracellular milieu, as will be discussed below. Were calpain exposed on the platelet membrane, it could represent a potential focus for cellular-associated fibrinolysis. Although, classically, HK is an inhibitor of calpain, our studies clearly show that in enzyme excess, calpain can cleave HK, increasing its procoagulant activity[4] and liberating kinins.[56] Since calpain is present at 2 to 4 μg/10^8 platelets and the total HK is 60 ng/10^8, it is clear that this condition may apply on the platelet surface.[4] In contrast, HK in plasma exists at a concentration of 80 μg/ml and would be expected to be a potent inhibitor of platelet calpain.[16] Thus, the balance between HK and calpain is critically regulated by the relative concentrations.

For factor V, calpain may play a critical role. As it exists in plasma, factor V is a procofactor which increases in activity at least 20-fold when proteolysed by thrombin.[57] However, essentially no thrombin can form in the absence of activated factor V. Thus, the effect of thrombin on factor V is one of extremely powerful positive feedback, but the question of how the first molecules of factor V become activated remains unanswered. It is well accepted that the locus of factor V action is the platelet surface. The plasma procofactor binds poorly compared to factor Va. It is thus likely that factor V is secreted onto the platelet surface, where it is processed to a partially activated form by platelet calpain.[34] This form

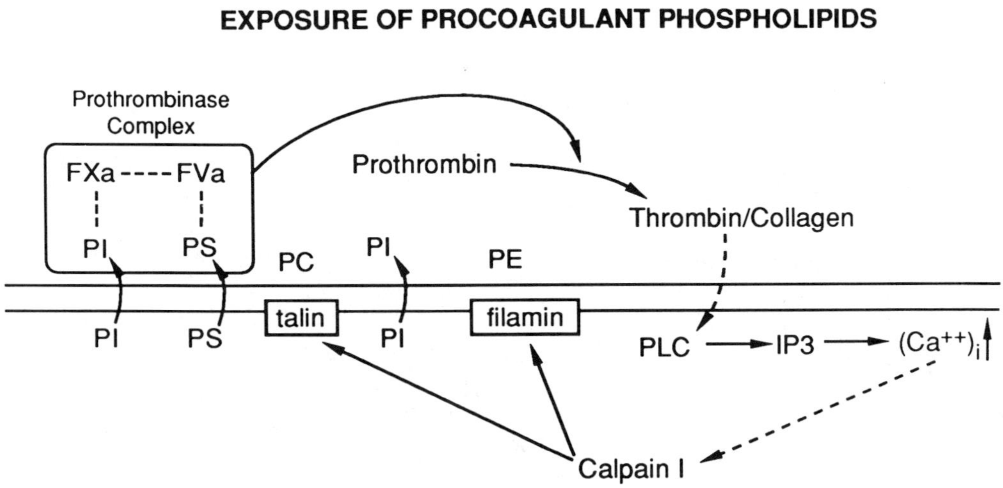

FIGURE 3. Exposure of procoagulent phospholipids by calpain. FXa = factor Xa; PI = phosphatidyl inositol; PS = phosphatidyl serine; PE = phosphatidyl ethanolamine. Other symbols as in the preceding figures.

of factor Va can then catalyze the formation of a small amount of thrombin to operate its potent feedback reaction.

Calpain may play an additional critical role in the development of coagulant activity on the surface of the platelet. For binding of factor Va and to enhance conversion of prothrombin to thrombin by factor Xa, exposure of the phospholipids (such as PS) bearing net negative charges on the platelet surface is crucial (Figure 3). This transbilayer movement of phospholipids may be mediated by the action of calpain on its cytoskeletal substrates. The evidence for this concept is four-fold. First, the order of potency of platelet activators such as thrombin and collagen is similar for degradation of cytoskeletal proteins, formation of procoagulant activity and the exposure of acidic phospholipids.[58] Second, proteolysis of filamin and talin temporally parallels the increase of prothrombinase.[59] Third, dibucaine and tetracaine stimulate calpain activity and prothrombinase activity with a good correlation over time.[59,4] Fourth, fluoride-dependent calcium-induced platelet procoagulant activity closely parallels calpain activity, while leupeptin blocks both calpain and the coagulant activity.[60]

Platelet shape change, the transition from a discoid appearance to a spiculated sphere, occurs quickly following adhesion of platelets to an injured vessel (Figure 4). The biochemcial correlate of platelet shape change is phosphorylation of the light chain of myosin. The activity of the protein kinase responsible is dependent on the presence of calmodulin and calcium. Inhibitors of calpain markedly decrease the cleavage of myosin light chain kinase and prevent the phosphorylation of the light chain of myosin without lowering the intracellular calcium. Thus, calpain may be enhancing platelet shape change by increasing the activity of the myosin light chain kinase.[37,38] Caldesmon, which plays a less clear role in platelet contraction, is also cleaved by calpain.[35] Maintenance of the phosphorylated state may be mediated by degradation of the phosphatase calcineurin.[36] The function of other calmodulin-binding sites of M_r 90 and 60 kDa is unknown.

Secretion by platelets is closely related to changes in calcium (Figure 4). Since phospholipase C is cleaved by calpain, this appeared to be an alternative feedback regulation mechanism. However, the cleavage did not alter the sensitivity of phospholipase C to lipid or calcium. Since activation of phospholipase C must precede the elevation of calcium, thought to be induced by the formation of IP_3, this finding is not so surprising. Thus, calcium elevation is an early change, but calpain cleavage is delayed and is more likely to regulate later events, such as clot retraction and aggregation.

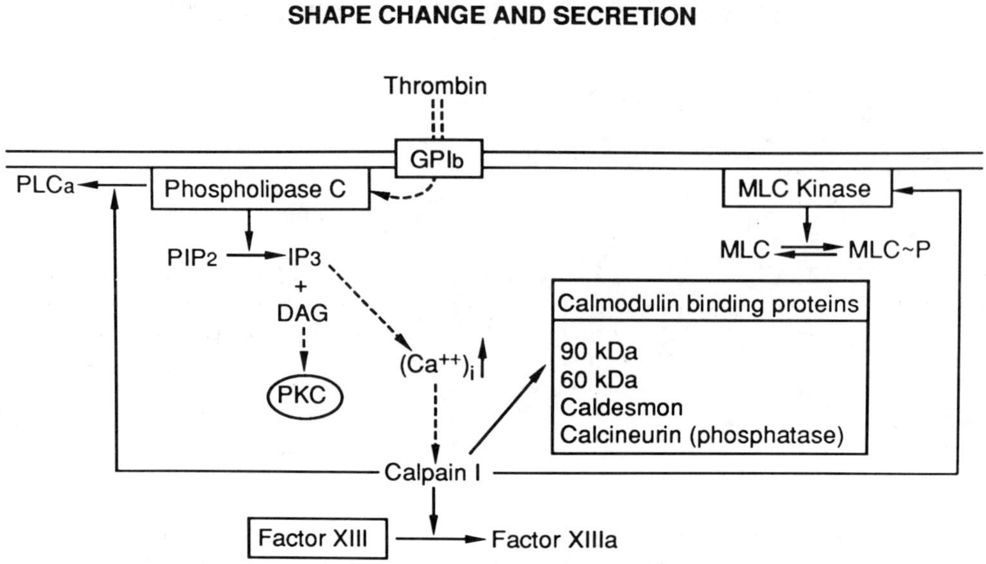

FIGURE 4. Role of calpain in platelet shape change and secretion. MLC = myosin light chain; MLC ~ P = phosphorylated MLC; PIP2 = phosphatidyl inositol diphosphate; DAG = diacylglycerol; PLC_a = activated phospholipase C; PKC = protein kinase C; FXIII = factor XIII; $FXIII_a$ = activated FXIII.

Involvement of calpain with platelet aggregation was first suggested by Fox et al.[23] (Figure 5). They found that following exposure of 0.1 μ/ml of thrombin to aggregated washed platelets, several new polypeptides of M_r 200, 100, and 93 kDa were visualized on SDS, which were shown to originate from filamin and talin. Chelation with EGTA, incubation without stirring, leupeptin (100 μg/ml) and NEM all prevented both aggregation and the cleavage of filamin and talin. The authors conclude that calcium-dependent protease (calpain) is activated during platelet aggregation by thrombin. Ruggiero and Lapetina[61] found that leupeptin (50 to 200 μg/ml), in addition to inhibiting platelet aggregation, inhibited serotinin release, activation of phospholipase C, protein kinase C, and myosin light chain kinase. However, leupeptin did not block response to other agonists, such as collagen, arachidonic acid, and calcium ionophore. They conclude that calpain may be critical in thrombin-induced platelet activation. Baldassere et al.[62] also confirmed that leupeptin (and antipain) inhibited thrombin-induced binding of fibrinogen to platelets, as did the calpain inhibitors iodoacetamide and NEM. However, leupeptin did not inhibit thrombin binding or thrombin hydrolysis of fibrinogen.

Recent studies from our laboratory have been directed at the mechanism by which fibrinogen receptors are exposed by thrombin, thus facilitating platelet aggregation[63] (Figure 5). We demonstrated that at thrombin concentrations of greater than 0.2 units/ml, platelet aggregation was independent of ADP, since it occurred in FSBA-labeled platelets which could not respond to ADP. Since aggregin was labeled by FSBA, we tested the hypothesis that cleavage of aggregin, after exposure to thrombin, might be responsible for exposure of fibrinogen receptors and platelet aggregation. ^{3}H-FSBA-labeled aggregin was cleaved after exposure of intact platelets to thrombin. But since aggregin was not cleaved in platelet membranes or in ATP-depleted platelets, we concluded that the effect of thrombin requires metabolic energy and proceeds by an indirect mechanism. We previously found that calpain could cleave aggregin in the intact platelet, but unlike thrombin, calpain directly hydrolyzed aggregin in the platelet membrane. These results suggested that thrombin acts by increasing intracellular Ca^{2+}, thereby activating calpain, which then cleaved aggregin. Calpain II also exposed latent fibrinogen-binding sites, thus facilitating aggregation when fibrinogen was

AGGREGATION

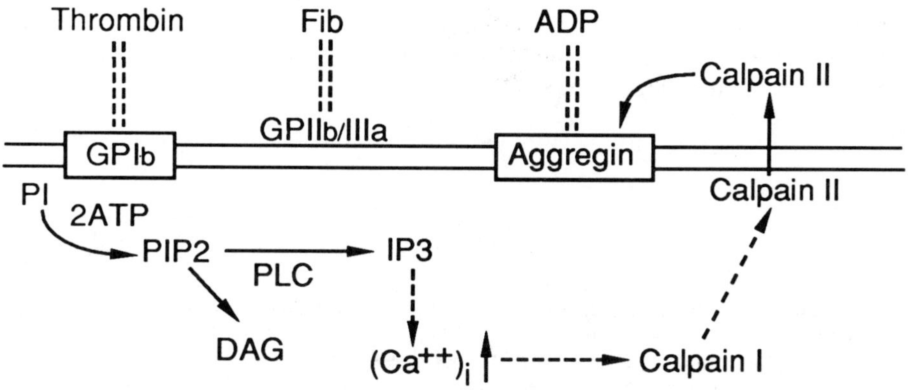

FIGURE 5. Action of calpain in platelet aggregation. $GPII_b/III_a$ = platelet glycoprotein II_b/III_a; Aggregin = ADP receptor responsible for the latency of fibrinogen binding.

added to platelets. Thus, it appeared that the effects of thrombin on platelets to produce aggregation, expose fibrinogen binding-sites, and cleave aggregin, were all mediated by the proteolytic activity of calpain. We confirmed that leupeptin and antipain (10 to 20 μM), which did not directly inhibit thrombin amidolytic activity at these concentrations, blocked platelet aggregation, and we demonstrated that these calpain inhibitors also prevented the cleavage of aggregin. The effect was specific, since leupeptin and antipain failed to inhibit ADP, epinephrine, collagen, or phorbol ester-induced platelet aggregation.

To further verify this hypothesis, we used a naturally occurring inhibitor of calpain, HK, which has no inhibitory action on thrombin.[47] HK, at physiological concentration (80 μ/ml), completely inhibited thrombin-induced aggregation and cleaving of aggregin, but did not inhibit aggregation induced by ADP, collagen, U46619 (a thromboxane A_2 analog), or phorbol ester. Moreover, in a plasma environment, the major inhibitor of γ-thrombin induced platelet aggregation was HK. This conclusion was based on the observation that the ten-fold increase in α-thrombin concentration needed for aggregation in normal plasma was abolished in plasma deficient in kininogen.

Since it is known that HK binds to platelets but it is not internalized, we hypothesized that calpain was being expressed on the external membrane of platelets following exposure to thrombin.[64] We directly verified this hypothesis by demonstrating the following:

1. About half of platelet calpain is located on the internal platelet membrane in unstimulated platelet;
2. On exposure to thrombin, rim immunofluorescence is detected on paraformaldehyde-fixed platelets with a monospecific polyclonal antibody to platelet calpain;
3. About 15 to 20% of the calpain is expressed on the membrane as calculated by the binding of a radiolabeled anticalpain Fab[1]′ after exposure to thrombin; and
4. Both dibucaine and PAF stimulate platelets to bind increased Fab′ of an anticalpain antibody.

These combined studies indicate that a substantial portion of calpain is membrane associated and is expressed on the external platelet surface after exposure to thrombin. Calpain then cleaves aggregin, thus exposing fibrinogen binding-sites and facilitating platelet aggregation. HK regulates platelet aggregation by thrombin in plasma.

This review has indicated the multiple potential actions of calpain in platelet activation, including the intracellular hydrolysis of cytoskeletal proteins and enzymes, and the extracellular processing of coagulation proteins and membrane receptors. With the increasing realization that calcium is one of the primary regulators of cellular function, calpains are uniquely positioned to irreversibly modify key regulatory proteins, and thus allow the platelet to participate in the formation of the hemostatic plug or thrombus.

REFERENCES

1. **Murachi, T.,** Distribution of calpains and calpastatin in human blood cells *Biochem. Int.*, 16, 391, 1988.
2. **Murachi, T.,** Calpain and calpastatin, *Trends Biochem. Sci.*, 8, 167, 1983.
3. **Shiba, E., Tsujinaka, T., Kambayashi, J., and Kosaki, G.,** Purification and characterization of Ca^{2+}-activated neutral protease inhibitor from human platelets, *Thromb. Res.*, 32, 207, 1983.
4. **Schmaier, A. H., Zuckerberg, A., Silverman, C., Kuchibhotla, J., Tuszynski, G., and Colman, R. W.,** High molecular weight kininogen: a secreted platelet protein, *J. Clin. Invest.*, 71, 1477,1983.
5. **Weiss, J. H., Turitto, V. T., and Baumgartner, H. R.,** Effect of shear rate on platelet interaction with subendothelium in activated and native blood. I. Shear-rate dependent decrease in von Willebrand's disease and the Bernard Soulier syndrome, *J. Lab. Clin. Med.*, 92, 750, 1978.
6. **Adelstein, R. S. and Conti, M. A.,** Phosphorylation of platelet myosin increases actin-activated myosin ATPase activity. *Nature,* 256, 597,1975.
7. **Holmsen, H. and Weiss, J. H.** Secretable storage pools in platelets, *Annu. Rev. Med.*, 30, 119, 1979
8. **White, J. G.,** Current concepts of platelet structure, *Am. J. Clin. Path.*, 71, 363, 1979.
9. **Mann, K. C.,** Membrane-bound enzyme complexes in blood coagulation, *Prog. Hemost. Thromb.*, 7, 1, 1984.
10. **Hamberg, M., Svensson, J., and Samuelsson, B.,** Thromboxanes: a new group of biologically active compounds derived from prostaglandin endoperoxides, *Proc. Natl. Acad. Sci. U.S.A.*, 72, 2994, 1975.
11. **Berridge, M. J.** Inositol trisphosphate and diacylglycerol as second messengers, *Biochem. J.*, 220, 345, 1984.
12. **Phillips, D. R. and Jakabova, M.,** Ca^{++}-dependent protease in human platelets, *J. Biol. Chem.*, 252, 1358, 1977.
13. **Truglia, J. A. and Stracher, A.,** Purification and characterization of a calcium-dependent sulfhydryl protease for human platelets, *Biochem. Biophys. Res. Commun.*, 100, 814, 1981.
14. **Yoshida, N., Weksler, B., and Nachman, R.,** Purification of human platelet calcium-activated protease. Effect on platelet and endothelial function *J. Biol. Chem.*, 258, 7168, 1983.
15. **Malik, M. N., Ramaswamy, S., Tuzio,H., Shiekh, A. M., Fenko, M. D., Wisniewski, H. M., and Howard, R. G.,** Micromolar Ca^{2+} requiring protease from human platelets: purification, partial characterization and effect on the cytoskeletal proteins, *Life Sci.*, 40, 593, 1987.
16. **Schmaier, A. H., Bradford, H., Silver, L. D., Farber, A., Scott, C. F., Schutsky, D., and Colman, R. W.,** High molecular weight kininogen is an inhibitor of platelet calpain, *J. Clin. Inves.*, 77, 1565, 1986.
17. **Suzuki, K.,** Calcium-activated neutral protease: domain structure and activity regulation, *Trends Biochem. Sci.*, 12, 103, 1987.
18. **Gopalakrishna, R. and Barsky, S. H.,** Hydrophobic association of calpains and subcellular organelles compartmentalization of calpains and the endogenous inhibitor calpastatin in tissues, *J. Biol. Chem.*, 261, 13936, 1986.
19. **Okubo, I., Kurachi, T., Takasawa, T., Shiokawa, H., and Sasaki, M.,** Isolation of human cDNA for alpha-2-thiol protease inhibitor and its identity with low molecular weight kininogen, *Biochemistry,* 23, 3891, 1984.
20. **Schmaier, A. H., Smith, P. M., Purdon, A. D., White, J. G., and Colman, R. W.,** High molecular weight kininogen: localization in the unstimulated and activated platelet and activation by a platelet calpain(s), *Blood,* 67, 119, 1986.
21. **Schmaier, A. H., Schutsky, D., Farber, A., Silver, L. D., Bradford, H. N., and Colman, R. W.,** Determination of the bifunctional properties of high molecular weight kininogen by studies with monoclonal antibodies directed to each of its chains, *J. Biol. Chem.*, 262, 1405, 1987.
22. **Tsujinaka, T., Sakon, M., Kambayashi, J., and Kosaki, G.,** Cleavage of cytoskeletal proteins by two forms of Ca^{2+}-activated neutral proteases in human platelets, *Thromb. Res.*, 28, 149, 1982.

23. **Fox, J. E., Reynolds, C. C., and Phillips, D. R.,** Calcium-dependent proteolysis occurs during platelet aggregation, *J. Biol. Chem.*, 258, 9973, 1983.
24. **Fox, J. E., Goll, D. E., Reynolds, C. C., and Phillips, D. R.,** Identification of two proteins (actin-binding protein and P235) that are hydrolyzed by endogenous Ca^{2+}-dependent protease during platelet aggregation, *J. Biol. Chem.*, 260, 1060, 1985.
25. **O'Halloran, T., Beckerle, M. C., and Burridge, K.,** Identification of talin as a major cytoplasmic protein implicated in platelet activation, *Nature*, 317, 449, 1985.
26. **Gache, Y., Landon, F., Touitou, H., and Olomucki, A.,** Susceptibility of platelet alpha-actinin to a Ca^{2+}-activated neutral protease, *Biochem. Biophys. Res. Commun.*, 124, 877, 1984.
27. **Fox, J. E., Reynolds, C. C., Morrow, J. S., and Phillips, D. R.,** Spectrin is associated with membrane-bound actin filaments in platelets and is hydrolyzed by the Ca^{2+}-dependent protease during platelet activation, *Blood*, 69, 537, 1987.
28. **Kunicki, T. J., Mosesson, M. W., and Pidard, D.,** Cleavage of fibrinogen by human platelet calcium-activated protease, *Thromb. Res.*, 35, 169, 1984.
29. **Kunicki, T. J., Montgomery, R. R., and Schullek, J.,** Cleavage of human von Willebrand factor by platelet calcium-activated protease, *Blood*, 65, 352, 1985.
30. **Gralnick, H. R., Williams, S. B., McKeown, L. P., Maisonneuve, P., Jenneau, C., Sultan, Y., and Rick, M. E.,** *In vitro* correction of the abnormal multimeric structure of von Willebrand factor in type IIa von Willebrand's disease, *Proc. Natl. Acad. Sci. U.S.A.*, 82, 6435, 1985.
31. **Chesney, C. M., Pifer, D., and Colman, R. W.,** Subcellular localization and secretion of factor V from human platelets, *Proc. Natl. Acad. Sci. U.S.A.*, 78, 5180, 1981.
32. **Itteryah, T. R., Rawala, R., and Colman, R. W.,** Factor V binding to human platelets: the role of divalent cations and of platelet preparatory methods, *Thromb. Res.*, 26, 59, 1982.
33. **Tracy, P. B., Nesheim, M. E., and Mann, K. G.,** Proteolytic alterations of factor Va bound to platelets, *J. Biol. Chem.*, 258, 662, 1983.
34. **Bradford, H. N., Annamalai, A., Doshi, K., and Colman, R. W.,** Factor V is activated and cleaved by platelet calpain: comparison with thrombin proteolysis, *Blood*, 71, 388, 1988.
35. **Kosaki, G., Tsujinaka, T., Kamayashi, J., Morimoto, K., Yamamoto, K., Yamagami, K., Sobue, K., and Kakiuchi, S.,** Specific cleavage of calmodulin-binding proteins by low Ca^{2+}-requiring form of Ca^{2+}-activated neutral protease in human platelets, *Biochem. Int.*, 6, 767, 1983.
36. **Tallant, E. A., Brumley, L. M., and Wallace, R. W.,** Activation of a calmodulin-dependent phosphatase by a Ca^{2+}-dependent protease, *Biochemistry*, 27, 2205, 1988.
37. **Kajiwara, Y., Tsujinaka, T., Sakon, M., Kambayashi, J., Ohshiro, T., Murachi, T., and Mori, T.,** Elucidation of calpain dependent phosphorylation of myosin light chain in human platelets, *Biochem. Int.*, 15, 935, 1987.
38. **Tsujinaka, T., Kajiwara, Y., Kambayashi, J., Sakon, M., and Mori, T.,** Studies on light chain phosphorylation in intact platelets utilizing a cell-penetrating thiol protease inhibition, *Thromb. Res.*, 51, 365, 1988.
39. **Low, M. G., Carroll, R. C., and Weglicki, W. B.,** Multiple forms of phosphoinositide-specific phospholipase C of different relative molecular masses in animal tissues. Evidence for modification of the platelet enzyme by Ca^{2+}-dependent proteinase, *Biochem. J.*, 221, 813, 1984.
40. **Ando, Y., Imamura, S., Yamagata, Y., Kitahara, A., Saji, H., Murachi, T., and Kannagi, R.,** Platelet factor XIII is activated by calpain, *Biochem. Biophys. Res. Comm.*, 144, 484, 1987.
41. **Solum, N. O., Hagen, I., Filion-Myklebust, C., and Stabaek, T.,** Platelet glycocatian: its membrane association and solubilization in aqueous media. *Biochim. Biophys. Acta*, 597, 235, 1980.
42. **Okumura, T., Hasitz, M., and Jamieson, G. A.,** Platelet glycocatian. Interaction with thrombin and role as thrombin receptor of the platelet surface, *J. Biol. Chem.*, 253, 3435, 1978.
43. **Yamamoto, K., Kosaki, G., Suzuki, K., Tanoue, K., and Yamazaki, H.,** Cleavage site of calcium-dependent protease in human platelet membrane glycoprotein Ib, *Thromb. Res.*, 43, 41, 1986.
44. **Clemetson, K. J., Naim, H. J., and Lüscher, E. F.,** Relationship between glycocatian and glycoprotein Ib of human platelet, *Proc. Natl. Acad. Sci. U.S.A.*, 78, 2712, 1981.
45. **McGowan, E. B., Yeo, K. T., and Detwiler, T. C.,** The action of calcium-dependent protease on platelet surface glycoproteins, *Arch. Biochem. Biophys.*, 227, 284, 1983.
46. **McGowan, E. B., Ding, A., and Detwiler, T. C.,** Correlation of thrombin-induced glycoprotein V hydrolysis and platelet activation, *J. Biol. Chem.*, 258, 11243, 1983.
47. **Colman, R. W., Figures, W. R., Wu, Q.-X., Chung, S.-Y., Morinelli, T. A., Tuszynski, G. P., and Niewiarowski, S.,** Distinction between glycoprotein IIIa and aggregin. The 100 kDa membrane protein mediating ADP-induced platelet activation, *Arch. Biochem. Biophys.*, 262, 298, 1988.
48. **Puri, R. N., Gustafson, E. J., Zhou, F.-X., Bradford, H., Colman, R. F., and Colman, R. W.,** Inhibition of thrombin-induced platelet aggregation by high molecular weight kininogen, *Trans. Assoc. Am. Physicians*, 100, 232, 1987.

49. **Berkowitz, S. D., Nozaki, H. Titani, K., Murachi, T., Plow, E. F., and Zimmerman, T. S.,** Evidence that calpain and elastase do not produce the von Willebrand factor fragments in normal plasma and IIa von Willebrand disease, *Blood,* 72, 721, 1988.
50. **Okita, J. R., Pidard, D., Newman, P. J., Montgomery, R. R., and Kunicki, T. J.,** On the association of glycoprotein Ib and actin-binding protein in human platelets, *J. Cell Biol.,* 100, 317, 1985.
51. **Coller, B. S.** Effects of tertiary amine local anesthetics on von Willebrand factor-dependent platelet function: alteration of membrane reactivity and degradation of GP Ib by a calcium-dependent protein(s), *Blood,* 60, 731, 1982.
52. **Solum, N. O. and Olsen, T. M.,** Effects of diamide and dibucaine on platelet glycoprotein Ib, actin-binding protein and cytoskeleton, *Biochim. Biophys. Acta,* 817, 249, 1985.
53. **Onji, T., Takagi, M., and Shibata, N.,** Calpain abolishes the effect of filamin on the actomyosin system in platelets, *Biochim. Biophys. Acta,* 912, 298, 1987.
54. **Meyer, M.,** Endogenous calcium-activated neutral protease in human platelets. Identification of a M_r 70,000 substrate, *Biomed. Biochem. Acta,* 47, 125, 1988.
55. **Beckerle, M. C., Barridge, K., DeMartino, G. N., and Croall, D. E.,** Colocalization of calcium-dependent protease II and one of its substrates at sites of cell adhesion, *Cell,* 51, 569, 1987.
56. **Higashiyama, S., Ishiguro, H., Ohkubo, I., Fujimoto, S., Matsuda, T., and Sasaki, M.,** Kinin release for kininogens by calpains, *Life Sci.,* 39, 1639, 1986.
57. **Chiu, H. C., Whitaker, E., and Colman, R. W.,** Heterogeneity of human factor V deficiency, *J. Clin. Invest.,* 72, 493, 1983.
58. **Comfurius, P., Bevers, E. M., and Zwaal, R. F. A.,** The involvement of cytoskeleton in the regulation of the transbilayer movement of phospholipids in human blood platelets, *Biochim. Biophys. Acta,* 815, 143, 1985.
59. **Verhallen, P. F. Bevers, E. M., Comfurius, P., and Zwaal, R. F.,** Correlation between calpain-mediated cytoskeletal degradation and expression of platelet procoagulant activity. A role for the platelet membrane-skeleton in the regulation of membrane lipid asymmetry?, *Biochim. Biophys. Acta,* 903, 206, 1987.
60. **Verhallen, P. F., Bevers, E. M. Comfurius, P., and Zwaal, R. F.,** Fluoride-dependent calcium-induced platelet procoagulant activity shows that calpain is involved in increased phospholipid transbilayer movement, *Biochim. Biophys. Acta,* 942, 150, 1988.
61. **Ruggiero, M. and Lapetina, E. G.,** Leupeptin selectivity inhibits human platelet responses induced by thrombin and trypsin: a role for proteolytic activities of phospholipase C, *Biochem. Biophys. Res. Comm.,* 131, 1198, 1985.
62. **Baldassare, J. J., Bakshian, S., Knipp, M. A., and Fisher, G. J.,** Inhibition of fibrinogen receptor expression and serotonin release by leupeptin and antipain, *J. Biol. Chem.,* 260, 10531, 1985.
63. **Gustafson, E. J. Schutsky, D., Knight, L. C., and Schmaier, A. H.,** High molecular weight kininogen binds to unstimulated platelets, *J. Clin. Invest.,* 78, 310, 1986.
64. **Schmaier, A. H., Bradford, H. N., Lundburg, D., Farber, A., and Colman, R. W.,** Membrane expression of platelet calpain, *Blood,* 75, 1273, 1990.

Chapter 16

ERYTHROCYTE CALPAIN IN HEALTH AND DISEASE

S. Pontremoli and E. Melloni

TABLE OF CONTENTS

I. MOLECULAR AND CATALYTIC PROPERTIES OF ERYTHROCYTE CALPAINS

In erythrocytes of various species so far examined calpain has been found to be present as a single enzyme type composed by two unequal subunits of approximately 80 and 30 kDa.[1-3] Following the current definition, erythrocyte calpains appear to belong to the μM or type I class, which indicates that they are all, although with some differences, active at micromolar concentrations of Ca^{2+}. The molecular and catalytic properties of the two erythrocyte calpains that will be considered in the present article are listed in Table 1.

An important point in understanding the physiological function of calpain is represented by the mechanism of activation of the proteinase. Although erythrocyte calpains belong to the μM type, still their Ca^{2+} requirement (see Table 1) exceeds the physiological intracellular concentration of the metal ion and therefore calpain is presumably inactive in red cell. Two mechanisms of activation have been described, both of which promote increase in the sensitivity of calpain for Ca^{2+}, so that the proteinase acquires the property to express activity within more physiological concentrations of the metal activator.[4-9]

According to our results,[1,10] in human erythrocytes the basic mechanism for activation involves autoproteolysis of the 80 kDa catalytic subunit, resulting in the removal of a fragment of approximate 5 kDa and in the formation of a 75-kDa catalytic subunit active within 1 micromolar Ca^{2+} (or less) concentration. Thus, activation of calpain can be regarded as the conversion of an inactive enzyme form (procalpain) to an active form. The autoproteolytic formation of an active catalytic subunit can occur also in the cytosol, although various experimental evidence suggests that this activation process probably follows translocation and association of procalpain to the plasma membrane[11-13] (Table 2).

On the basis of the results obtained in our laboratory we propose the two models for activation of calpain. Each model considers that dissociation of the dimer and formation of the free catalytic subunit precedes autoproteolysis of the 80-kDa subunit, representing either an obligatory or a facilitating step.[1,14]

According to Model 1, a marked increase in Ca^{2+} concentration or the presence of a digestible substrate in association with a much less pronounced elevation in the concentration of Ca^{2+} can initiate, in the cytosol, the process of activation of calpain which then proceeds in both circumstances through common steps. Alternatively, as represented in Model 2A, as a result of fluctuations (within μM concentrations) of intracellular Ca^{2+}, calpain can be translocated and associated to the red cell plasma membrane, where favorable conditions (localized increased in Ca^{2+} concentration, interaction with phospholipids, or others) promote its autoproteolytic activation.[13]

The activated calpain, located at the cell membranes, can digest membrane-cytoskeleton anchoring proteins;[14-18] or following its release from the membranes[13] can diffuse in other cell compartments. In the mechanisms represented in Model 1 and 2A, activation of human erythrocyte calpain is an irreversible process.

On the contrary, rat erythrocyte calpain (Model 2B) does not undergo autoproteolytic activation. Recent experimental results have indicated that in the presence of μM concentrations of Ca^{2+}, calpain associates to the plasma membrane fraction. In this bound form the proteinase expresses catalytic activity within 1 μM Ca^{2+} concentrations.[2] When calpain is released from the membranes, it reacquires the original high Ca^{2+} requirement[2] suggesting that activation of rat erythrocyte calpain is a site directed reversible process that presumably results in proteolytic modifications of membrane-cytoskeleton proteins.

II. REGULATION OF ERYTHROCYTE CALPAIN ACTIVITY

It is generally accepted that regulation of calpain activity is accomplished by a natural endogenous inhibitor,[1,19,20] calpastatin, a protein containing repetitive domains,[21-23] each one

TABLE 1
Molecular and Catalytic Properties of Human and Rat Erythrocyte Calpains

Source	Type	Molecular mass (kDa)	Subunit Composition (kDa)	K_{50}
Human	I	110	~ 80 + 30	35
Rat	I	110	~ 80 + 30	30

Note: ~ 80 kDa, catalytic subunit; 30 kDa, regulatory subunit. Type I, calpain with low (μ*M*) Ca^{2+} requirement. K_{50} refers to the concentration of Ca^{2+} required to express 50% of maximum activity.

For experimental details, see Melloni et al.[16] and Pontremoli et al.[2]

TABLE 2
Activation of Erythrocyte Calpain in Various Systems

Source	Activation System	Calpain activity[a] assayed in μ*M* Ca^{2+} (% of V_{max})	K_{50} (μ*M*)	Molecular mass of catalytic subunit (kDa)
Human	None	0	35	80
	1 m*M* Ca^{2+}, 1 min	100	<1	75
	1 μ*M* Ca^{2+} + substrate, 30 min	100	<1	75
	1 μ*M* Ca^{2+} + phospholipid vesicles, 10 min	60	2	80—75
	1 μ*M* Ca^{2+} + isolated plasma membranes, 10 min	90	<1	75
Rat	None	0	30	80
	1 m*M* Ca^{2+}, 1 min	0	30	80
	1 μ*M* Ca^{2+} + substrate, 30 min	0	30	80
	1 m*M* Ca^{2+} + substrate, 1 min	0	30	80
	1 μ*M* Ca^{2+} + phospholipid vesicles, 10 min	80	1	80
	1 μ*M* Ca^{2+} + isolated plasma membranes, 10 min	95	<1	80

Note: K_{50} refers to the concentration of Ca^{2+} required to express 50% of maximum activity.

[a] The activity expressed by the enzyme in the presence of 1 m*M* Ca^{2+} was taken as 100%.

For experimental details, see Pontremoli et al.,[2] Pontremoli and Melloni,[7] and Melloni et al.[16]

potentially capable of interaction with one molecule of the proteinase.[24,25] On the basis of our results, in both models we have indicated the requirements for calpastatin activity, the steps at which it interacts with the proteinase and the resulting effects on the activation and on the catalytic activity of calpain. Using purified enzyme and inhibitor preparations from human erythrocytes, we have in fact observed that calpastatin can inhibit both autoproteolytic conversion of procalpain to calpain, as well as the catalytic activity of active calpain[1] (see Model 1). Furthermore, once the calpain-calpastatin complex has been formed, the proteinase can no longer be associated to the membrane,[13] and thus its autoproteolytic activation at this specific cellular site is inhibited (see Model 2A). In spite of this potential highly efficient mechanism for the regulation of calpain activity in the cytosol and at the membrane level, the exact modulatory effect of calpastatin has still to be clarified. A first limitation derives from the observation[16] according to which the formation of the calpain-calpastatin complex

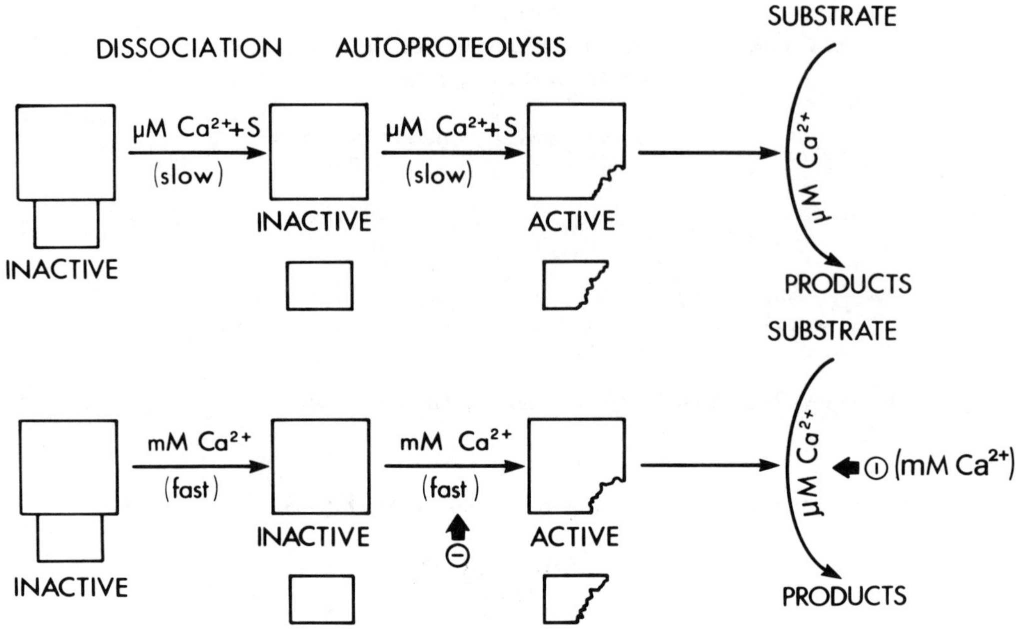

MODEL 1. Proposed model for the activation of human erythrocyte calpain. The arrows indicate the steps susceptible to inhibition by calpastatin. The large and the small quadrangle indicate the catalytic and the regulatory subunit of calpain, respectively.

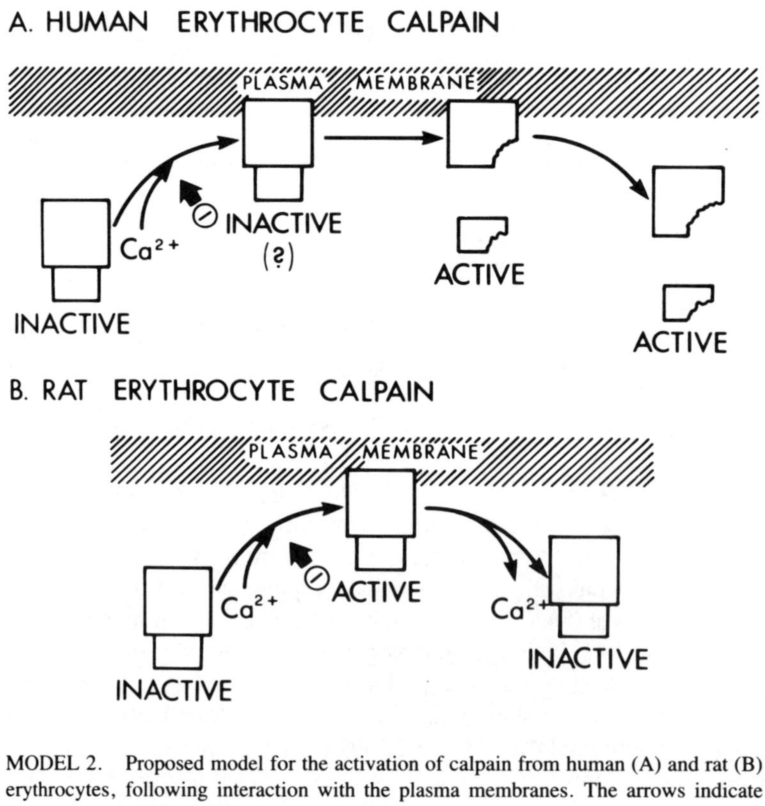

MODEL 2. Proposed model for the activation of calpain from human (A) and rat (B) erythrocytes, following interaction with the plasma membranes. The arrows indicate the steps susceptible to inibition by calpastatin.

requires high (m*M*) concentrations of Ca^{2+}, these being presumably necessary to promote dissociation of the oligomeric structures of both molecules in order to create single monomeric highly reactive species. However, calpain has been shown to undergo activation and to express activity at μM Ca^{2+} concentration, a condition in which calpastatin seems to have a limited inhibitory effect.[1,16] For this and other reasons that will be discussed later, we think that the mechanism of action of calpastatin requires further investigation, unless we consider that the only regulatory effect of calpastatin is to prevent a massive activation of calpain in conditions in which the intracellular free concentration of Ca^{2+} rapidly increases reaching high values. As an attractive hypothesis it might also be considered that less Ca^{2+} is needed for calpastatin inhibition, when the inhibitor is associated to biologic cell membranes.[8,19,26]

With respect to rat erythrocyte calpain, inhibition by calpastatin appears to be somehow simplified since this type of proteinase does not undergo autoproteolytic activation and interacts with the inhibitor[2] at μM concentrations of Ca^{2+}.

Also rat erythrocyte calpain following interaction with calpastatin can no longer associate to the plasma membrane and thereby activation can be efficiently modulated (see Model 2B).

III. THE ROLE OF CALPAIN IN ERYTHROCYTE FUNCTIONS

In spite of significant progress achieved in the elucidation of the structure-function relationship of calpain and of calpastatin, little is known concerning the biological role of this intracellular proteolytic system in physiological and pathological conditions. At present, evidence seems to indicate that calpain activity is preferentially directed to proteolytic modifications of cytoskeletal-membrane proteins and, in more general terms, of proteins located at the inner surface of the plasma membrane.[16,17]

Thus a controlled activation of calpain which satisfies the requirement of specific localization, timing and intensity can mediate important physiological functions which also include defense mechanisms. Among these are membrane fusion;[27-29] initiation of the sequential degradation of excess globin chains;[17] modification of the catalytic and regulatory properties of ion pumps, such as the Ca^{2+}-ATPase;[18,30-32] and the defense mechanism against oxidant-induced Heinz bodies formation.[33,34]

However, if activation of calpain occurs in an uncontrolled spatial and temporal manner, then calpain activity could severely damage the structural and functional organization of the cell membrane and, more generally, the cell architecture.

These considerations have suggested the hypothesis that structural and functional cell alterations characteristic of the hypertensive state, could result from an uncontrolled calpain activity. It is in fact well known that essential hypertension in humans or the genetically determined hypertension in rats of Milan strain (MHS) is characterized at cell level by structural and functional abnormalities of membrane-cytoskeleton, well documented in kidney cells[35] and erythrocytes.[36] In human and rat red cells, such abnormalities involve changes in cell shape and volume;[37] modifications in cation transport;[38-40] and deficiency of the Ca^{2+}-ATPase activity.[41] The reduced activity of the erythrocyte Ca^{2+}-ATPase may sustain a general defect in cellular Ca^{2+} handling and, in association with the altered Na^+ transport, can cause an increase[41-44] in the level of intracellular free Ca^{2+}, which might represent an important event in the development of hypertension.[42-45] High concentrations of intracellular Ca^{2+} have been observed in lymphocytes[46] and in platelets[47] of hypertensive subjects. Thus, the type of molecular and functional lesions observed in both cells of animals with spontaneous hypertension, and of patients with essential hypertension have suggested the involvement of calpain in the genesis of hypertension or in the development of the molecular alterations which characterize the hypertensive state. This hypothesis has stimulated our studies aimed to verify the existence of such a correlation.

TABLE 3
Level of Calpain and of Calpastatin Activity in Erythrocytes from Normotensive and Hypertensive Rats

	Calpain Activity (A)		Calpastatin activity	
Rat strain	U/ml RBC	U/mg	U/ml RBC	U/mg
MNS	41.2 ± 3	4500	79.4 ± 6	7560
MHS	40.2 ± 2	4350	7.5 ± 1	7440

Note: MNS, Milan Normotensive Strain; MHS, Milan Hypertensive Strain. For the experimental details, see Pontremoli, S., et al.[48,49]

IV. THE CALPAIN-CALPASTATIN SYSTEM IN ERYTHROCYTES OF HYPERTENSIVE RATS OF THE MILAN STRAIN (MHS)

In mature red cells from hypertensive (MHS) and normotensive (MNS) rats, calpain is present in equal amounts and with identical catalytic properties. On the contrary, the amount of the natural endogenous inhibitor (calpastatin) is ten times less in red cells of MHS rats with the result that the inhibitor/calpain ratio (unit/unit) is 0.2 in these cells as compared to a value of 2.0 in red cells from MNS rats. The decreased amount of calpastatin has been established on the basis of: (1) decrease in total activity of the inhibitor;[48] (2) decrease amount of the inhibitor protein;[49] and (3) identical specific activity of calpastatins isolated and purified from erythrocytes of MNS and MHS rats.[49]

All the data concerning the properties of the calpain/calpastatin system in red cells of MHS rats are summarized in Table 3. The defect in calpastatin, whatever its origin, is certainly responsible for an uncontrolled or unmasked calpain activity. This conclusion is supported by data indicating that, although the amount of calpain is identical in MNS and MHS rats, in cell lysates from MNS rats calpain activity is almost undetectable (~ 2.5 units/ml of packed cells), whereas in lysates from MHS rats the amount that can be detected (~ 40 units/ml of packed cells) corresponds to the total activity present (see Table 3). Full calpain activity can be evaluated in MNS red cells only after complete chromatographic resolution of calpain from calpastatin.[48]

In order to attribute a physiological significance to our findings, it might be useful to summarize the previously discussed general properties of the calpain-calpastatin system of rat erythrocytes:

1. Calpain does not undergo autoproteolytic conversion to a low Ca^{2+}-requiring enzyme;
2. In its free form the proteinase requires approximately 0.1m*M* Ca^{2+} to express maximum proteolytic activity;[2]
3. Its activation involves interaction with the plasma membrane phospholipids,[2] in which form calpain expresses full activity at 1 μ*M* Ca^{2+};
4. This activation process is prevented by the presence of calpastatin;[2,49]
5. With respect to the proteinase, calpastatin is normally present in 2 to 1 ratio, and in different oligomeric forms; but only the monomeric one interacts with calpain at ~ 1 μ*M* Ca^{2+} concentration.[48,49]

On the basis of the properties of this proteolytic system and due to the profound decrease in the amount of calpastatin, it can be concluded that in MHS red cells calpain is exposed to an uncontrolled activation being thereby responsible for proteolytic modifications of integral membrane proteins and membrane-bound cytoskeletal proteins.

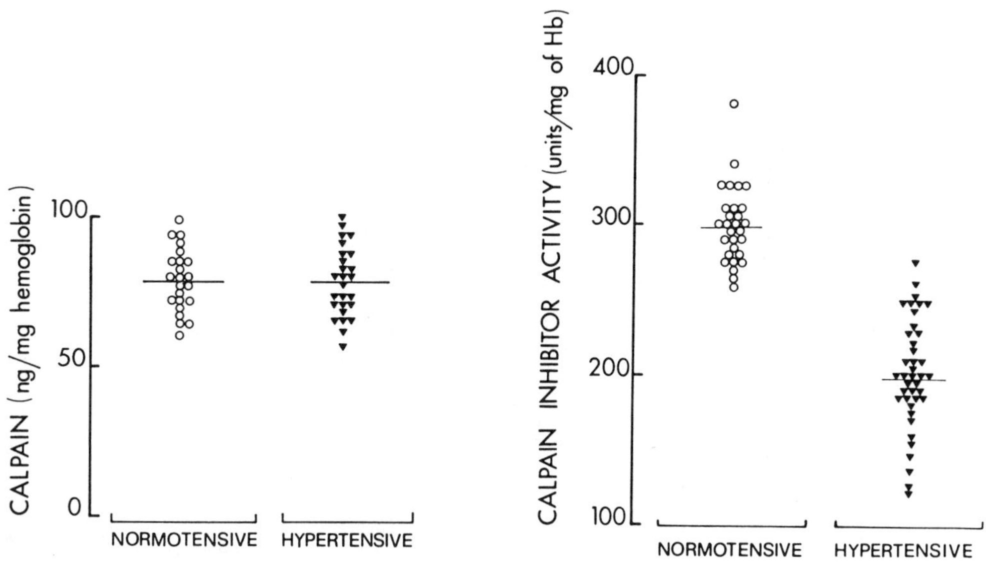

FIGURE 1. Level of calpain and of calpastatin in erythrocytes from normotensive and hypertensive subjects. For details on the selection of the two groups of subjects and on the assay of calpain and calpastatin, see Pontremoli et al.[51]

In further support of these conclusions, a similar alteration in the calpain-calpastatin system has been observed in renal tissues of MHS rats.[50]

V. THE CALPAIN-CALPASTATIN SYSTEM IN ERYTHROCYTES OF PATIENTS WITH ESSENTIAL HYPERTENSION

This proteolytic system has been investigated also in erythrocytes of patients with essential hypertension. None of the patients were on drug treatment and their selection was performed using criteria for excluding secondary hypertension.[51]

As shown in Figure 1 the level of calpain is comparable in red blood cells from hypertensive and normotensive subjects. On the contrary, the levels of calpastatin activity in erythrocytes of hypertensives is lower than that of normotensive subjects. The mean values for inhibitory activity is 301 $\pm$ 26 unit/mg Hb in normal subjects and 220 $\pm$ 14 units/mg Hb in hypertensive patients with a modest individual overlapping among the two groups. The decrease in calpastatin has been established on the basis of: (1) a decrease in the total activity of the inhibitor;[51] (2) a decreased amount of calpastatin protein;[52] and (3) an identical specific activity of the two calpastatins isolated and purified from red cells of the two groups of subjects.[52] Thus, as in red cells of genetically hypertensive MHS rats, also in red cells of patients with essential hypertension an unbalanced intracellular calpain-calpastatin system is present. The decrease in the amount of calpastatin is, however, significantly different, corresponding to approximately 20% to 50% in erythrocytes from hypertensive subjects as compared to 90% in cells from MHS rats. On the basis of these observations it is tempting to suggest that in human essential hypertension an uncontrolled activation of calpain, due to a decreased amount of calpastatin, may be responsible for those molecular lesions characteristic of the hypertensive state and which involve primarily integral membrane proteins and membrane-bound cytoskeletal proteins.

It must, however, also be considered that in human erythrocytes calpastatin is present in large excess with respect to calpain.[53] Each calpastatin molecule contains three inhibitory domains which correspond to the potential number of molecules of calpain that can be

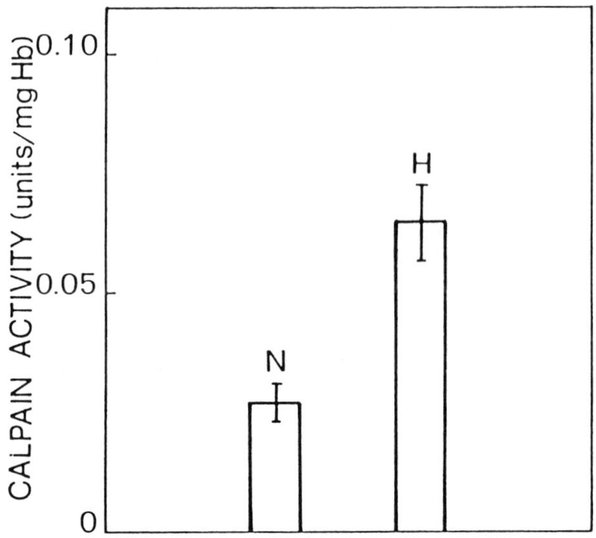

FIGURE 2. Calpain activity detectable in hemolysates from normotensive and hypertensive subjects. The values reported are the means of three independent experiments, carried out on three different subjects. N refers to normotensive subjects; H refers to hypertensive patients. For experimental details, see Pontremoli et al.[54]

inhibited.[9] Accordingly, the observed approximate decrease of 20% to 50% in calpastatin activity should not reduce the efficiency of the regulatory system in controlling calpain activation and activity.

In spite of these considerations, several experimental observations seem to indicate an impairment in the control of calpain activity in red cells of hypertensives and suggest the implication of a defective calpastatin activity in this phenomenon. Consistently, we have observed that in crude lysates of these cells calpain activity is expressed (or can be measured) at a rate at least twofold higher with respect to cell lysates from normotensive subjects[54] (Figure 2). Furthermore, red cells from hypertensive patients exposed to A23187 ionophore and increasing concentrations of Ca^{2+} show a significantly higher degree of membrane fragility[54] (Figure 3).

A significant increase (1.5-fold) in the extent of phosphorylation of band 3 region has been observed in inside out vesicles from hypertensive patients as compared to normotensive subjects (Figure 4). An identical rate and extent of phosphorylation of band 3 has been induced in vesicles from normotensive cells exposed to a controlled digestion with small amounts of calpain (Figure 5) or in vesicles[55] prepared from normotensive red cells preloaded with 5 μM Ca^{2+} and A23187 ionophore (Table 4). In these conditions no increase in ^{32}P incorporation was observed if erythrocytes were previously loaded with the thiol proteinase inhibitor leupeptin, or with an anticalpain monoclonal antibody, directed to the catalytic site of the enzyme (Table 4).

The increased rate of phosphorylation of band 3 represents a constitutive property of red cells from hypertensive patients. It does not occur during cell lysis; nor is it the result of a proteolytic activation of the membrane-bound protein kinase.[55]

The hypothesis that a normal control of the intracellular calpain-calpastatin system might be one of the relevant conditions that prevent development of the hypertensive state is supported by other observations. Among these, of particular interest is the finding of a significant inverse correlation in humans ($r = -0.43$, $p < 0.001$) between mean arterial pressure and the level of erythrocyte calpastatin.[51] Furthermore, although previous reports

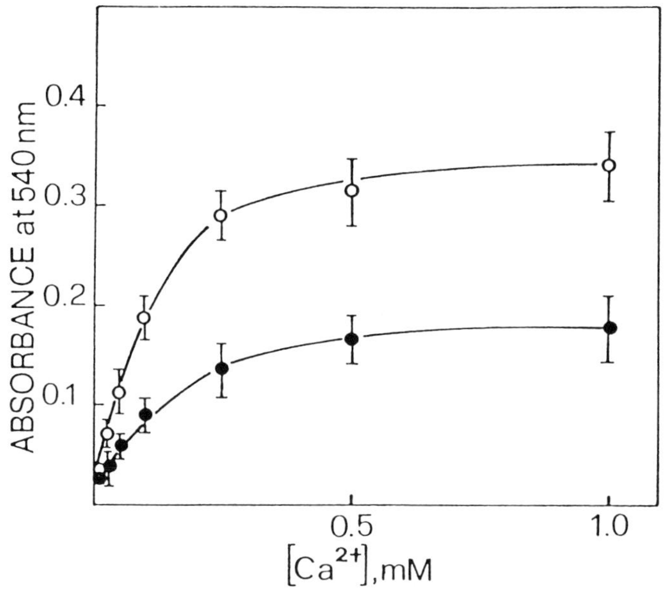

FIGURE 3. Effect of Ca^{2+} loading on the lysis of erythrocytes from normotensive (closed symbols) and hypertensive (open symbols) subjects. Erythrocytes from three normotensive and three hypertensive subjects were incubated with 5 μ*M* A 23187, in the presence of the indicated amounts of Ca^{2+}. The release of hemoglobin from the cells was measured, after 1 h incubation at 37°C. For more experimental details, see pontremoli et al.[54]

have indicated the absence of calpastatin in isolated erythrocyte membrane preparations,[26] we have detected a small but constant amount of calpastatin[52] in plasma membranes of erythrocytes from normotensive subjects and significantly lower amounts or complete absence of the inhibitor in membrane preparations from hypertensive patients (Figure 6).

With respect to the molecular mechanisms responsible for the decreased level of calpastatin, it has been shown that the inhibitor does not undergo an accelerated decay during the life span of the erythrocyte.[52]

Finally it must be carefully considered that in the spontaneous hypertension of MHS rats, which show a high degree of similarity with human essential hypertension, both erythrocytes and renal tissue cells[48-50] show a profound decrease in the amount of calpastatin and that a similar situation occurs in red cells and in other blood cells of hypertensive subjects (unpublished observations).[57]

VI. GENERAL REMARKS

Although we are well aware that further work is needed to clarify and to prove in more detail the pathophysiological role of the calpain-calpastatin system in the development of essential hypertension, we have selected this topic because we consider it at present one of the most appropriate models to discuss the function of "calpain" in health and disease.

Accordingly, it is the purpose of the present article to suggest the hypothesis that an uncontrolled activation of calpain, promoting proteolytic alterations; in the membrane-cytoskelton, might be responsible for those structural and functional abnormalities which have been observed in red blood cells and in other cell types and which are characteristic of the hypertensive state. Consistent with this hypothesis are the reported experimental evidences indicating the existence of an impairment in the control of calpain activity in red cells of

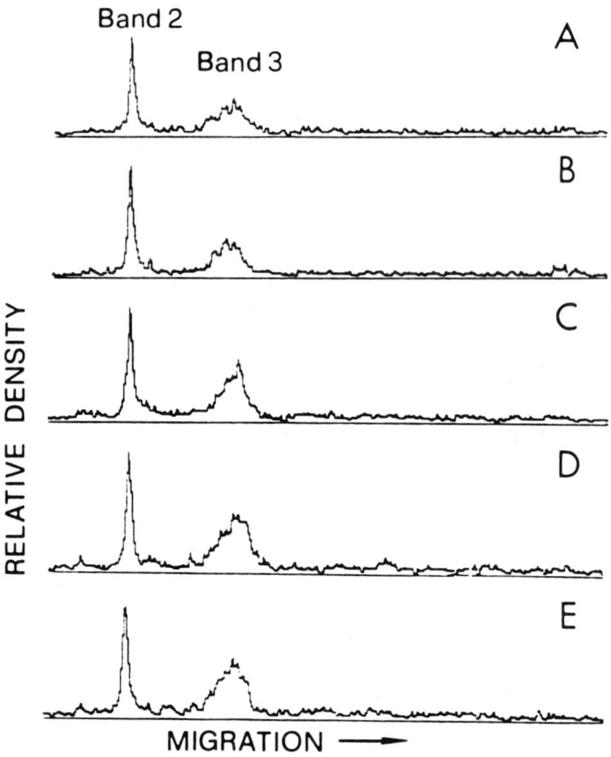

FIGURE 4. Protein phosphorylation in inside out vesicles from erythrocytes of normotensive and hypertensive subjects. The inside-out vesicles prepared from two normotensive subjects (A and B) and three hypertensive patients (C, D and E) were incubated with γ-^{32}P-ATP and the labeled proteins were separated on PAGE in the presence of SDS. The radiograms and the densitometric analysis of the dried gels were performed with an Ambis beta-scanning system. For more experimental details, see Pontremoli et al.[54]

hypertensive patients, resulting in an enhanced calpain activity. These conclusions are supported by the observation that membrane alterations comparable to those identified in hypertensive cells can be reproduced by controlled digestion by calpain and prevented by calpain inhibition. As far as the mechanism for this enhancement in calpain activity is concerned, we have suggested that this is probably the result of the decrease in the amount of calpastatin which thereby appears to be the relevant defect in hypertensive cells. It is not possible at present to explain how and to what extent this decrease in the amount of calpastatin can affect the activity of calpain which, however, is significantly altered in cells of hypertensive subjects.

From a more general point of view we suggest that in cells of hypertensives due to transient intracellular localized increase in Ca^{2+}, calpain is moved and activated at the plasma membrane.

In normal conditions the overall activation process should be modulated by the inhibitor eventually present in two different compartments.[26] In the cytosol, calpastatin could prevent translocation and activation of calpain, whereas in its membrane-associated form (if present), calpastatin could prevent activation of calpain at this specific site. These modulatory effects are presumably altered to a different extent in hypertensive cells as a result of a decrease in the inhibitor concentration, or of the absence of the inhibitor at the membrane level, or eventually, of possible alterations in the molecular structure of the inhibitor not directly

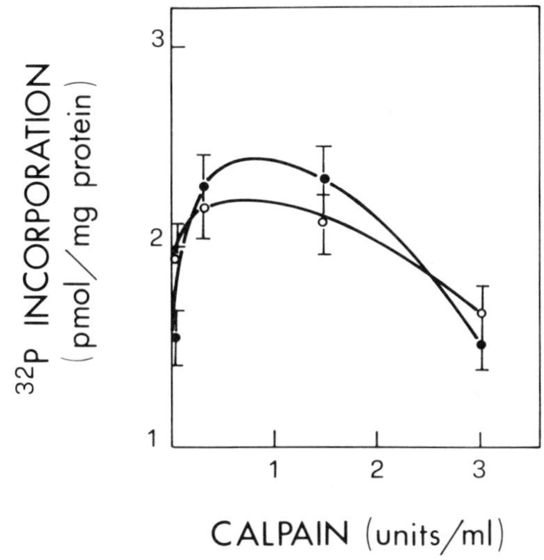

FIGURE 5. Effect of treatment with calpain on the ^{32}P-incorporation of band 3 in inside out vesicles from normotensive (closed symbols) and hypertensive (open symbols) subjects. The inside-out vesicles were incubated at 20°C with the indicated amounts of purified erythrocyte calpain, in the presence of 50 μ*M* Ca^{2+}. After 10 min, the digestion was stopped by the addition of 0.1 m*M* EDTA (final concentration) and the vesicles were submitted to a second incubation in the presence of γ-^{32}P-ATP. The ^{32}P incorporation in band 3 was determined as described in Figure 4, using an Ambis beta-scanning system. For more experimental details, see Pontremoli et al.[54]

TABLE 4
Effect of Inhibition of Intracellular Proteinases on the Phosphorylation of Band 3 Protein

Additions		Normotensive Subjects	Hypertensive Subjects	
Proteinase Inhibitor	**Ca^{2+} (μ*M*)**	**Band 3 (A) (cpm)**	**Band 3 (B) (cpm)**	**Ratio (B/A)**
None	0	4600	7250	1.58
None	5	7480	8800	1.18
Leupeptin	5	4390	7100	1.62
Pepstatin	5	7100	9080	1.28
Leupeptin + Pepstatin	5	4520	7800	1.73
mAb C 56.3	5	4750	7250	1.53

Note: Erythrocytes were previously loaded with proteinase inhibitors and resealed as previously described.[55] The concentration of leupeptin and pepstatin was 0.2 m*M* and that of mAb C 56.3 0.1 mg/ml. Aliquots (0.2 ml) of the resealed cells were incubated with 5 μ*M* A 23187 and 5 μ*M* Ca^{2+}. Inside-out vesicles were prepared and incubated with γ-^{32}P-ATP. ^{32}P incorporation in band 3 was evaluated as described in Figure 4. For more experimental details, see Pontremoli et al.[55]

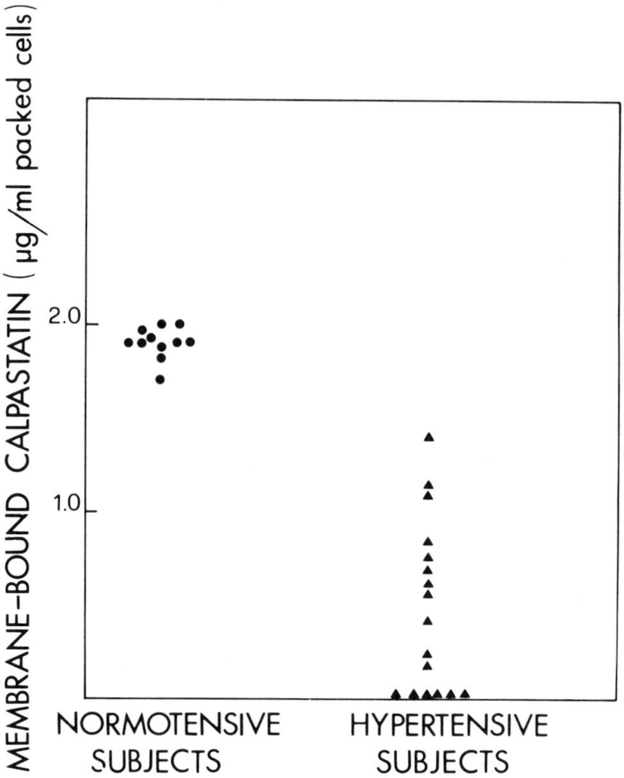

FIGURE 6. Level of membrane-associated calpastatin in erythrocytes from normotensive and hypertensive subjects. The calpastatin level was determined by a Western blot analysis, using a specific monoclonal antibody raised against human erythrocyte calpastatin. For experimental details, see Pontremoli, et al.[52]

related with the interaction with calpain but with other functions including its association to the plasma membrane.

Whatever might be the origin of the defect in the amount of calpastatin and its relevance in the development as well as in the complications of essential hypertension, our observations suggest a new area of investigations aimed at defining the molecular basis of this disease.

ACKNOWLEDGMENTS

This work was supported in part by grants from the Ministero della Pubblica Istruzione, from Consiglio Nazionale delle Ricerche, Progetto Finalizzato Oncologia, Sottoprogetto Biologia Molecolare, and from the Associazione Italiana per la Ricerca sul Cancro.

REFERENCES

1. **Pontremoli, S. and Melloni, E.,** Regulation of Ca^{2+}-dependent proteinase of human erythrocytes, in *Calcium and Cell Function,* Vol.6, Academic Press, New York, 1986, 159.
2. **Pontremoli, S, Melloni, E., Salamino, F., Sparatore, B., Michetti, M., Sacco, O., and Bianchi, G.,** Characterization of the defective calpain-endogenous calpain inhibitor system in erythrocytes from Milan hypertensive rats, *Biochem. Biophys. Res. Commun.,* 139, 341, 1986.
3. **Kitahara, A., Sasaki, T., Kikichi, T., Yumoto, N., Yoshimura, N., Hatanaka, M., and Murachi, T.,** Large-scale purification or porcine calpain I and calpain II and comparison of proteolytic fragments of their subunits, *J. Biochem.,* 95, 1759, 1984.
4. **Coolican, S. A. and Hathaway, D. R.** Effect of L-α-phosphatidylinositol on a vascular smooth muscle Ca^{2+}-dependent protease, *J. Biol. Chem.,* 259,11627, 1984.
5. **Murachi, T.,** *Calcium and Cell Function,* Cheung, W. Y. Ed., Academic Press, New York, 1983, 377.
6. **Kay, J.,** *Proteases: Potential Role in Health and Disease,* Horl, W. H. and Heidland, A., Eds., Plenum, New York, 1984, 519.
7. **Pontremoli, S. and Melloni, E.,** Extralysosomal protein degradation, *Annu. Rev. Biochem.,* 55, 455, 1986.
8. **Mellgren, R. L.,** Calcium-dependent proteases: an enzyme system active at cellular membranes?, *FASEB, J.,* 1, 110, 1987.
9. **Suzuki, K., Imajoh, S., Emori, Y., Kawasaki, H., Minami, Y., and Ohno, S.,** Calcium-activated neutral protease and its endogenous inhibitor, *FEBS Lett.,* 220, 271, 1987.
10. **Pontremoli, S., Sparatore, B., Melloni, E., Michetti, M., and Horecker, B. L.,** Activation by hemoglobin of the Ca^{2+}-requiring neutral proteinase of human erythrocytes: structural requirements, *Biochem. Biophys. Res. Commun.,* 123, 331, 1984.
11. **Pontremoli, S., Melloni, E., Sparatore, B., Salamino, F., Michetti, M., Sacco, O., and Horecker, B. L.,** Binding to erythrocyte membrane is the physiological mechanism for activation of Ca^{2+}-dependent neutral proteinase, *Biochem. Biophys. Res. Commun.,* 128, 331, 1985.
12. **Pontremoli, S., Melloni, E. Sparatore, B., Salamino, F., Michetti, M., Sacco, O., and Horecker, B. L.,** *Biochem. Biophys. Res. Commun.,* 129, 389, 1985.
13. **Pontremoli, S., Salamino, F., Sparatore, B., Michetti, M., Sacco, O., and Melloni, E.,** Following association to the membrane, human erythrocyte procalpain is converted and released as fully active calpain, *Biochim. Biophys. Acta,* 831, 335, 1985.
14. **Lorand, L., Bjerrum, O. J., Hawkins, M., Lowe-Krentz, L., and Siegfring, G. E.,** Degradation of transmembrane proteins in Ca^{2+}-enriched human erythrocytes, *J. Biol. Chem.,* 258, 5300, 1983.
15. **Pant, H. P. Virmani, M., and Gallant, P. E.,** Calcium-induced proteolysis of spectrin and band 3 protein in rat erythrocyte membranes, *Biochem. Biophys. Res. Commun.,* 117, 372, 1983.
16. **Melloni, E., Salamino, F., Sparatore, B., Michetti, M., and Pontremoli, S.,** Ca^{2+}-dependent neutral proteinase from human erythrocytes: activation by Ca^{2+} ions and substrate and regulation by the endogenous inhibitor, *Biochem. Int.,* 8, 477, 1984.
17. **Pontremoli, S., Melloni, E., Sparatore, B., Michetti, M., and Horecker, B. L.,** A dual role for the Ca^{2+}-requiring proteinase in the degradation of hemoglobin by erythrocyte membrane proteinases, *Proc. Natl. Acad. Sci. U.S.A.,* 81, 6714, 1984.
18. **Wang, K. K. W., Roufogalis, B. D., and Villalobo, A.,** Further characterization of calpain-mediated proteolysis of the human erythrocyte plasma membrane Ca^{2+}-ATPase, *Arch. Biochem. Biophys.,* 267, 317, 1988.
19. **Mellgren, R. L., Lane, R. D. and Kakar, S. S.,** Isolated bovine myocardial sarcolemma and sarcoplasmic reticulum vesicles contain tightly bound calcium-dependent protease inhibitor, *Biochem. Biophys. Res. Commun.,* 142, 1025, 1987.
20. **Takano, E., Kitahara, A., Sasaki, T., Kannagi, R., and Murachi, T.,** Two different molecular species of pig calpastatin, *Biochem. J.,* 235, 97, 1986.
21. **Takano,E., Maki, M., Hatanaka, M., Mori, H., Zenita, K., Sakihama, T., Kannagi, R., Marti, T., Titani, K., and Murachi, T.,** Evidence for the repetitive domain structure of pig calpastatin as demonstrated by cloning of complementary DNA, *FEBS Lett.,* 208, 199, 1986.
22. **Emori, Y., Kawasaki, H., Imajoh, S., Imahori, K., and Suzuki, K.,** Endogenous inhibitor for calcium-dependent cysteine protease contains four internal repeats that could be responsible for its multiple reactive sites, *Proc. Natl. Acad. Sci. U.S.A.,* 84, 3590, 1987.
23. **Maki, M., Takano, E., Mori, H., Kannagi, R., Murachi, T., and Hatanaka, M.,** Repetitive region of calpastatin is a functional unit of the proteinase inhibitor, *Biochem. Biophys. Res. Commun.,* 143, 300, 1987.
24. **Imajoh, S., Kawasaki, H., Emori, Y., and Suzuki, K.** Calcium-activated neutral protease inhibitor from rabbit erythrocytes lacks the N-terminal region of the liver inhibitor but retains three inhibitory units, *Biochem. Biophys. Res. Commun.,* 146, 630, 1987.

25. **Imajoh, S., Kawasaki, H., Emori, Y., Ishiura, S., Minami, Y., Sugita, H., Imahori, K., and Suzuki, K.,** A fragment of an endogenous inhibitor produced in *Escherichia coli* for calcium-activated neutral protease (CANP) retains an inhibitory activity. *FEBS Lett.*, 215, 274, 1987.
26. **Mellgren, R. L.,** On the mechanism of binding of calpastatin the protein inhibitor of calpains, to biologic membranes, *Biochem. Biophys. Res. Commun.*, 150, 170, 1988.
27. **Kosower, N. S., Glaser, T., and Kosower, E. M.,** Membrane-mobility agent-promoted fusion of erythrocytes: fusibility is correlated with attack by calcium-activated cytoplasmic proteases on membrane proteins, *Proc. Natl. Acad. Sci. U.S.A.*, 80, 7542, 1983.
28. **Glaser, T. and Kosower, N. S.,** Fusion of rat erythrocytes by membrane-mobility agent A_2C depends on membrane proteolysis a cytoplasmic calpain, *Eur. J. Biochem.*, 159, 387, 1986.
29. **Glaser , T. and Kosower, N. S.,** Calpain-calpastatin and fusion. Fusibility of erythrocytes is determined by a protease-protease inhibitor (calpain-calpastatin) balance, *FEBS Lett.*, 206, 155, 1986.
30. **Au, K. S.,** Activation of erythrocyte membrane Ca^{2+}-ATPase by calpain, *Biochim. Biophys. Acta,* 905, 273, 1987.
31. **Wang, K. K. W., Villalobo, A., and Roufogalis, B. D.,** Activation of the Ca^{2+}-ATPase of human erythrocyte membrane by an endogenous Ca^{2+}-dependent neutral protease, *Arch. Biochem. Biophys.*, 260, 696, 1988.
32. **James, P. Vorherr, T., Krebs, J., Morelli, A., Castello, G., Damonte, G., McCormick, D. J., Penniston, J. T., De Flora, A., and Carafoli, E.,** Structural and functional modifications of erythrocyte Ca^{2+}-ATPase by limited digestion with calpain, in *Advances in Biotechnology of Membrane Ion Transport,* Jorgensen, P. L. and Verna, R., Eds., Raven Press, New York, 1988, 51.
33. **Morelli, A., Grasso, M., and De Flora, A.,** Oxidative inactivation of the calcium-stimulated neutral proteinase from human red blood cells by divicine and intracellular protection by reduced glutathione, *Arch. Biochem. Biophys.*, 251, 1, 1986.
34. **Morelli, A., Grasso, M., Meloni, T, Forteleoni, G., Zocchi, E., and De Flora, A.,** Favism: impairment of proteolytic systems in red blood cells, *Blood,* 69, 1753, 1987.
35. **Hanozet, G. M. Parenti, P., and Salvati, P.,** Presence of a potential-sensitive Na+ transport across renal brush-border membrane vesicles from rats of the Milan hypertensive strain, *Biochim, Biophys. Acta,* 819, 179, 1985.
36. **Bianchi, G., Ferrari, P., Trizio, D., Ferrandi, M., Torielli, L., Barber, B. R., and Polli, E.,** Red blood cell abnormalities and spontaneous hypertension in the rat a genetically determined link, *Hypertension,* 7, 319, 1985.
37. **Ferrari, P., Ferrandi, M., Torielli, L., Canessa, M., and Bianchi, G.,** Relationship between erythrocyte volume and sodium transport in the Milan hypertensive rate and age-dependent changes, *J. Hypertens.*, 6, 199, 1987.
38. **Garay, R. and Meyer, P.,** A new test showing abnormal net Na and K fluxes in erythrocytes of essential hypertensive patients, *Lancet,* 1, 349, 1979.
39. **Swales, J. D.,** Abnormal ion transport by cell membranes in hypertension, in *Handbook of Hypertension, Clinical Aspects of Essential Hypertension,* Robertson, J. I. S., Ed., Elsevier, Amsterdam, 1983, 239.
40. **Cusi, D., Barlassina, C., Ferrandi, M., Lupi, P., Ferrari, P., and Bianchi, G.,** Familial aggregation of cation transport abnormalities and essential hypertension, *Clin. Exp. Hypertens.*, 3, 871, 1981.
41. **Vezzoli, G., Elli, A. A., Tripodi, G., Bianchi, G., and Carafoli, E.,** Calcium ATPase in erythrocytes of spontaneously hypertensive rats of the Milan strain, *J. Hypertens.*, 3, 645, 1985.
42. **Blaustein, M.,** Sodium ions, calcium ions, blood pressure regulation and hypertension: a reassessment and a hypothesis, *Am. J. Physiol.*, 232, c165, 1977.
43. **Kesteloot, H. and Geboers, J.,** Calcium and blood pressure, *Lancet,* 1, 813, 1982.
44. **Bukoski, R. D., Xue, H., and McGarron, D. A.,** Effect of 1,25 $(OH)_2$ vitamin D_3 and ionized Ca^{2+} on ^{45}Ca uptake by primary cultures of aortic myocytes of spontaneously hypertensive and wistar kyoto normotensive rats, *Biochem. Biophys. Res. Commun.*, 146, 1330, 1987.
45. **Blaustein, M.,** The role of Na-Ca exchange in the regulation of tone in vascular smooth muscle, in *Excitation Contraction Coupling in Smooth Muscle,* Casteels, R., Godfraind, T., and Ruegg, J. C., Eds., Elsevier/North Holland, Biomedical, Amsterdam, 1977, 101.
46. **Zidek, W., Knorr, M., and Losse, H.,** *J. Hypertens.*, 1 (Suppl. 2), 3, 1983.
47. **Erne, P., Bolli, P., and Burgisser, E.,** Correlation of platelet calcium with blood pressure. Effect of antihypertensive therapy, *N. Engl. J. Med.*, 310, 1084, 1984.
48. **Pontremoli, S., Melloni, E., Salamino, F., Sparatore, B., Viotti, P., Michetti, M., Duzzi, L., and Bianchi, G.,** Decreased level of calpain inhibitor activity in red blood cells from Milan hypertensive rats, *Biochem. Biophys. Res. Commun.*, 138, 1370, 1986.
49. **Pontremoli, S., Melloni, E., Salamino, F., Sparatore, B., Michetti, M., Sacco, O., and Bianchi, G.,** Characterization of the defective calpain-endogenous calpain inhibitor system in erythrocytes from Milan hypertensive rats, *Biochem. Biophys. Res. Commun.*, 139, 341, 1986.

50. **Pontremoli, S., Melloni, E., Salamino, F., Sparatore, B., Viotti, P., Michetti, M., Duzzi, L., and Bianchi, G.,** Decreased level of calpain inhibitor activity in kidney from Milan hypertensive rats, *Biochem. Biophys. Res. Commun.,* 145, 1287, 1987.
51. **Pontremoli, S., Melloni, E., Sparatore, B., Salamino, F., Pontremoli, R., Tizianello, A., Barlassina, C., Cusi, D., Colombo, R., and Bianchi, G.,** Erythrocyte deficiency in calpain inhibitor activity: a new biochemical abnormality in essential hypertension, *Hypertension,* 12, 474, 1988.
52. **Pontremoli, S., Salamino, F., Sparatore, B., De Tullio, R., Pontremoli, R., and Melloni, E.,** Characterization of the calpastatin defect in erythrocytes from patients with essential hypertension, *Biochem. Biophys. Res. Commun.,* 157, 867, 1988.
53. **Murachi, T.,** Calpain and calpastatin, *Trends Biochem. Sci.,* 8, 167, 1983.
54. **Pontremoli, S., Melloni, E., Sparatore, B., Salamino, F., Pontremoli, R., and Tizianello, A.,** Increased phosphorylation in red cell membranes of subjects affected by essential hypertension, *Biochem. Biophys. Res. Commun.,* 145, 1329, 1987.
55. **Pontremoli, S., Sparatore, B., Salamino, F., De Tullio, R., Pontremoli, R., and Melloni, E.,** The role of calpain in the selective increased phosphorylation of the anion-transport protein in red cell of hypertensive subjects, *Biochem. Biophys. Res. Commun.,* 151, 590, 1988.
56. **Kaplan, N. M.,** Primary (essential) hypertension: pathogenesis, in *Clinical Hypertension,* 4th ed., Williams & Wilkins, Baltimore, 1986, chapt. 3.
57. **Pontremoli, S. and Melloni, E.,** unpublished observations.

Chapter 17

THE DEGRADATION OF NEUROFILAMENTS BY CALPAINS

William W. Schlaepfer and U.-J. P. Zimmerman

TABLE OF CONTENTS

I. INTRODUCTION

The roles of calpains in neuronal metabolism are undoubtedly complex and may be modified in different subsets of neurons, in different species as well as at different stages of neuronal growth and development. Indeed, differences in calpain metabolism may be an important factor in determining the nature of the neuronal phenotypes. Different functional roles of calpain may also be derived from the same or similar signals, such as the sudden influxes of calcium that are associated with the generation and conduction of nerve impulses or with the transmission of impulses to target cells across the synaptic gap. Although our overall understanding of calpain metabolism in neurons is quite limited, there is a growing body of information to indicate that calpains play an important role in degrading the neuronal cytoskeleton and in interfacing this breakdown with influxes of calcium, primarily at nerve endings. By altering the shape of the axon, calpain metabolism may play a major role in determining the individual components of the neuronal circuitries that underlie neuronal function.

While the cytoskeleton provides structural support and integrates space within many cell types,[1] the neuronal cytoskeleton is unique due to the presence of cell processes of extraordinary size. In most cells, for example, cytoskeletal proteins are confined to a limited cytoplasmic domain, occupying the space between the outer nuclear membrane and the inner surface of the plasma membrane. In neurons, the cytoplasmic domain has been markedly expanded into several cell processes or neurites. Large neurons elaborate dendritic trees that branch extensively and extend over an area that is 20-fold larger than the diameter of the cell body itself. The axons of large neurons are even much more asymmetric, extending for distances that may be 20,000-fold greater than the diameter of the parent cell body. Hence, the cytoplasmic extensions into axons and dendrites are the major cytoplasmic compartments of the neuron, comprising well over 99% of the neuronal cell volume.

Both dendrites and axons are filled with cytoskeletal proteins that are synthesized in the cell body and transported into the respective cell processes, thereby providing a continuous and abundant supply of cytoskeletal proteins for the establishment and maintenance of the axonal and dendritic processes. While microtubules are the predominant cytoskeletal component of most dendrites, neurofilaments are the major structural support for large axons.[2] Indeed, neurofilaments have evolved as a unique form of intermediate filament that are specialized for support of the axonal cytoskeleton. Within the axon, the density of neurofilaments remains constant[3-6] so that the size and shape of the axon is determined by neurofilament metabolism.[7-9] Hence, the breakdown of neurofilaments at nerve endings, or elsewhere along the neuraxis, is highly instrumental in regulating the overall configuration of the axon.

This review will attempt to highlight the studies which indicate (1) that neurofilament proteins are a natural substrate of calpain, and (2) that the breakdown of neurofilaments by calpain occurs *in vitro* and *in vivo*, in mammalian tissues and in lower forms, and (3) that degradation of the neurofilaments by calpain is an important and necessary determinant in regulating the precise size and shape of the axon, the fundamental component of neuronal circuitries.

II. BREAKDOWN OF MAMMALIAN NEUROFILAMENTS BY CALPAINS

The first indications that mammalian neurofilaments are substrates for calpain degradation came from morphological studies on neurofilaments in rat peripheral nerve following exposure to calcium prior to fixation and processing of the tissues for electron microscopy. Incubations in calcium caused a granular disintegration of neurofilaments that did not occur in the absence of calcium and could be prevented by the addition of calcium-chelating agents.[10] The same calcium-dependent granular disintegration of neurofilaments occurred more quickly in nerve tissues that were manipulated in order to promote the penetration of calcium into the tissues, such as by incubating with a calcium ionophore (A23187),[11] by freeze-thawing[12] or even by incubating under energy deprivational conditions.[13] Identical patterns of calcium-induced granular disintegration of neurofilaments occurred in different regions of the rat nervous system, such as in rat peripheral nerve,[14] in rat optic nerve,[15] in rat spinal cord,[15] and in rat neurons that were grown *in vitro*.[16] The calcium-induced granular disintegration of neurofilaments occurred in all nerve fibers examined, more rapidly in unmyelinated and small myelinated fibers, did not progess in the absence of calcium and could be inhibited by thiol reagents.[14] Also, the calcium-dependent disintegration of neurofilaments appeared to be due to factors that could be inactivated or lost during prolonged preincubations in a calcium-free state.[14,17]

Comparative biochemical and morphological analyses showed that the calcium-induced granular disintegration of neurofilaments in rat peripheral nerve,[18] optic nerve,[15] and spinal cord,[15] were accompanied by a progressive loss of the three neurofilament subunits, including the light (NF-L), mid-sized (NF-M) and heavy (NF-H) neurofilament proteins, with the appearance of breakdown products. These findings indicated that the calcium-dependent disintegration of neurofilaments was due to the breakdown rather than simple disassembly of the neurofilament subunits. In all instances, the same pattern of neurofilament protein fragmentation was seen, namely, a preferential loss of NF-M and an intermediate loss of NF-L, both preceding the loss of NF-H, in spite of the fact that the large carboxyl domain of NF-H is believed to protrude from the surface of neurofilament.[19]

Neurofilaments isolated from rat peripheral nerve or spinal cord also undergo a calcium-dependent degradation that could be monitored by morphological and biochemical methods.[20] The breakdown of isolated neurofilaments was shown to be mediated by soluble factor(s) that could be separate from neurofilaments by centrifugation. Addition of these factor(s) restored the susceptibility of isolated neurofilaments to calcium-dependent degradation. The tissue factors were inactivated by heat, inhibited by thiol reagents and optimally active at pH 6.8 to 8.0. Under certain conditions, some of these factors may sediment with neurofilaments.[21,22] Nevertheless, both the soluble and cytoskeletal-associated factors seem to produce identical patterns of neurofilament fragmentation.

Soluble tissue factors from neural tissues that degrade neurofilament proteins have the biochemical and physicochemical attributes of calpain I and calpain II,[23-29] as present in most mammalian tissues.[30] Calpain I has also been isolated directly from rat peripheral nerve and can be identified in immunoblots of nerve tissues.[31-32] Some calpain also copurifies with the lipid-rich myelin fractions[33-35] of neural tissues, possibly reflecting the hydrophobic affinities of the enzyme,[36] although calpains have also been implicated directly in the breakdown of myelin proteins.[37-39] Histochemical studies have noted that calpain I is enriched in neuronal perikarya and dendrites,[40,41] while antibodies to calpain II preferentially stained glial elements, endothelial cells and myelinated axons.[41]

Degradation of neurofilaments by calpain generates a series of neurofilament fragments that retain their immunoreactivity to neurofilament proteins and can be identified by immunoblotting the digests with monospecific antibodies to defined epitopes on each of the neurofilament proteins.[42] Digestion of isolated rat neurofilaments with purified calpain generates the same pattern of peptide fragmentation as that produced *in situ* by influxing calcium into rat peripheral nerve, nerve roots, or spinal cord.[42] A similar pattern in the loss of neurofilament proteins is seen when neurofilaments are digested with different calpain preparations.[21,26,31] A different pattern of neurofilament protein fragmentation is generated when neurofilaments are digested by other enzymes, such as cathepsin B,[31], cathepsin D,[43-45] trypsin,[46] or chymotrypsin.[46]

III. BREAKDOWN OF INVERTEBRATE NEUROFILAMENTS BY CALPAINS

The susceptibility of invertebrate neurofilaments to a calpain-like enzyme was first observed by Gilbert et al.[47] while working with neurofilaments from the giant axons of the seaworm, *Myxicola infundibulum*. Crude preparations of neurofilaments were rapidly degraded by calcium. The calcium-dependent changes did not occur with washed neurofilaments and could be prevented by several enzyme inhibitors, including thiol reagents. Subsequently, the incubation of *Myxicola* neurofilaments with calcium was shown to cause the disappearance of radiolabeled neurofilament proteins and the generation of smaller peptide fragments.[48]

Neurofilaments from squid giant axons are also susceptible to a calcium-dependent degradation.[49] The calpain-like enzyme of squid axoplasm is a soluble enzyme that selectively degraded neurofilament proteins.[50] The enzyme is activated by m*M* levels of calcium, has an optimal pH range between 7.0 and 8.0, and is inhibited by thiol reagents, TLCK, and leupeptin. Calcium-activated proteolytic activity could be demonstrated in the perikarya of the giant squid axon, in the axoplasm of the giant axon, and squid retinal fiber but not in nerve terminals, suggesting that the enzyme undergoes autoproteolysis at the latter site.[51]

The calpain-like enzyme of invertebrate axoplasm cleaves *Myxicola* neurofilaments in a characteristic pattern of limited proteolysis.[52,53] Major cleavage sites are located internally within the 150- and 160-kDa neurofilament subunits, separating large soluble fragments (108, 94, and 78 kDa) from the carboxyl terminal ends of the molecules. The remaining portions of the subunits (47, 55, and 70 kDa), comprising the amino termini and the α-helical-rich core elements of the subunits, retain their ability to self-associate and are recovered in the insoluble fraction of the digest. The large solubilized fragments contain most of the phosphorylation sites and are believed to protrude from the neurofilament core. Hence, calpain-like proteolysis appears to cleave off these peripheral portions of the filament. Interestingly, neurofilament fragments identical to those generated by calpain-like digestion are present in *Myxicola* axoplasm, suggestion the enzyme is active in the metabolism of invertebrate neurofilaments.[53]

The calpain-like breakdown of squid neurofilaments also solubilizes portions of the neurofilament subunits that contain the phosphorylation sites and are situated on the periphery of the filaments.[54] At the same time, the neurofilaments remain structurally intact, indicating that the filament-forming core domains in the neurofilament subunits are not disrupted by calcium-dependent proteolysis and are sufficient by themselves to maintain filament structure in the squid. Interestingly, the solubilized peripheral elements that are enriched in phosphorylation sites are also resistant to prolonged calpain-like digestion.

IV. NEUROFILAMENT PROTEOLYSIS IN PATHOLOGICAL STATES

The regulation of calpain activity is critically important in neurofilament metabolism, especially during the prolonged interval of axonal transport when the enzyme is in close proximity to neurofilaments and can be activated by an influx of calcium. It is likely that the enzyme is largely inactive due to the low resting level of free calcium in axoplasm compared to nerve endings.[55] In addition, the coexistence of the potent inhibitor, calpastatin, may limit the inadvertent activation of the enzyme in the axon. It is also possible that calpain activity is regulated by differing phosphorylation states of either neurofilaments[56] or the enzyme.[57,58] Also, the activation of calpain may involve the binding of the enzyme to a specific phospholipid[59,60] or carbohydrate[61] component of the cell that is absent, hidden, or modified within the axon.

Many pathological states appear to result from an increase of free calcium within the axon that triggers a generalized activation of calpain and massive proteolysis of neurofilaments. Since neurofilaments provide the major structural support for the axoplasm, extensive degradation of neurofilaments causes disintegration of the axonal cytoskeleton and instability, collapse, and fragmentation of the axon, as characteristic of the alterations that occur in Wallerian degeneration.[62] The morphological changes of Wallerian degeneration are best depicted with myelin stains that show the collapse of the myelin sheath around disintegrated fragments of axoplasm, a process that can extend along the entire length of the nerve fiber. These changes can be reproduced by incubating excised segments of rat peripheral nerve in calcium-containing solutions.[12] The changes are hastened by increasing the levels of calcium and can be slowed or prevented by reducing or chelating the calcium in the incubation or by the addition of calpain inhibitors.

Wallerian degeneration is the characteristic change that occurs in nerve fibers distal to a nerve transection.[63] The influx of calcium into transected nerve is believed to reflect an altered state of axolemmal permeability. An initial latency period, lasting for about 24 to 48 h, is the interval during which calcium homeostasis can be maintained in spite of an increasing leakiness of the axolemma to calcium. Progressive deterioration of the axolemmal barrier to calcium proceeds due to the inability of the limited reserves in the isolated nerve segments to keep pace with the turnover of membrane or related components. A rise in axoplasmic calcium may also result from the release of internal stores, such as promoted by energy deprivational states.[13]

Activation of calpain with proteolytic destruction of the axonal cytoskeleton appears to be a common endstage pathway of axonal degeneration in many neuropathic conditions, as judged by the similarity of pathological changes to those of Wallerian degeneration. In most instances, the changes are probably due to the abnormal influx of calcium into the axon. It is also possible that additional factors may contribute to the pathology, such as altered states of calpain, of calpastatin, or activational factors (e.g., phospholipids or carbohydrates) in the axon.

V. NEUROFILAMENT PROTEOLYSIS AND NEURONAL METABOLISM

The breakdown of neurofilaments by calpain at nerve endings terminates the elongation of the axonal cytoskeleton at the synapse, thereby establishing and maintaining the length of the axon. Axonal lengths may be further attuned to the precise dimensions of functional circuitries by coupling the activation of calpain to the calcium influxes of neurotransmission. Hence, there is a marked decrease in the density of neurofilaments[64] as well as loss of neurofilament immunoreactivity[65] at nerve endings. Also, the radioactivity that accompanies the proximodistal movement of labeled neurofilament proteins along the axon is rapidly dissipated when the wave of labeled neurofilaments reach the end of the axon.[66] Accumulations of neurofilaments occur in nerve endings treated with leupeptin, a calpain inhibitor.[67] Likewise, neurofilaments also accumulate at inactive synapses, such as produced by lowering the temperatures of poikilotherms.[68,69]

The breakdown of neurofilaments at nerve endings is but one phase in a continuum of neurofilament metabolism within the neuron that is designed to regulate the size and shape of the axon. Somehow the turnover of neurofilaments at nerve endings is matched with the synthesis of neurofilament proteins in the perikarya to produce a steady state of neurofilament metabolism throughout the neuron. The simultaneous decrease of mRNA and proteins to each of the neurofilament proteins following nerve transection[70] indicates that neurofilament expression is signaled by axonal factors that travel retrogradely through the axon. Studies of the developmental expression of neurofilament proteins indicate that these factors arise postnatally,[71] during a period when functional neuronal circuitries become established.

Since markedly differing levels of neurofilament metabolism are established and maintained among individual neurons, it would seem that some sort of feedback signal is needed in order to maintain a steady level of neurofilament metabolism, and to coordinate the levels of neurofilament synthesis with those of neurofilament turnover. Optimally, the feedback arm should accurately reflect the level of neurofilament metabolism in the axon, either directly (e.g., via a breakdown product of neurofilament catabolism) or indirectly (e.g., via an aspect of retrograde vesicular axonal transport that is quantitatively related to axon diameter). Hence, neurofilament fragments from calpain digestion may have special relevance in the regulation of neurofilament metabolism.

Immunoreactivity to neurofilament proteins has been identified in axonal vesicles that undergo retrograde transport.[72] Immunoreactive fragments of neurofilament proteins, identical to those produced by calpain digestion, can be detected at low levels throughout the axon,[42,73] indicating that they are either generated locally or are undergoing translocation within the axon. Most of the calpain-resistant neurofilament fragments in mammalian tissues, like those of invertebrate tissues,[53,54] are derived from the α-helical-rich core domains and the amino-terminal portions of neurofilament proteins, thereby retaining their insolubility in 1% nonionic detergent. It is of interest that similar calpain-resistant fragments of other intermediate filament, e.g., vimentin,[74,75] as well as neurofilament proteins,[76] are known to bind to single-stranded RNA and DNA. Further studies will be needed to identify any neurofilament fragments that are transported retrogradely to the perikarya and that may bind and alter the transcription or translation of neurofilament proteins.

ACKNOWLEDGMENTS

Studies of calpain and neurofilaments in the authors' laboratory was supported by a grant (NS 15722) from the National Institutes of Health.

REFERENCES

1. **Lazarides, E.,** Intermediate filaments as mechanical integrators of cellular space, *Nature,* 283, 1980.
2. **Wuerker, R. B. and Kirkpatrick, J. B.,** Neuronal microtubules, neurofilaments and microfilaments, *Int. Rev. Cytol.,* 33, 45, 1972.
3. **Friede, R. L. and Samorajski, T.,** Axon caliber related to neurofilaments and microtubules in sciatic nerve fibers of rats and mice., *Anta. Rec.,* 167, 379, 1970.
4. **Berthold, C. H.,** Morphology of normal peripheral axons, in *Physiology and Pathobiology of Axons,* Waxman, S. G., Ed., Raven Press, New York, 1978, 3.
5. **Hoffman, P. N., Griffin, J. W., Gold, B. G., and Price, D. L.,** Slowing of neurofilament transport and the radial growth of developing nerve fibers, *J. Neurosci.,* 5, 2920, 1985.
6. **Nukada, H., Dyck, P. J., Low, P. A., Lais, A. C., and Sparks, M. F.,** Axonal caliber and neurofilaments are proportionally decreased in galactose neuropathy, *J. Neuropath. Exp. Neurol.,* 45, 140,1986.
7. **Hoffman, P. N., Griffin, J. W., and Price, D. L.** Control of axonal caliber by neurofilament transport, *J. Cell Biol.,* 99, 705, 1984.
8. **Hoffman, P. N., Thompson, G. W., Griffin, J. W., and Price, D. L.,** Changes in neurofilament transport coincide temporally with alterations in the caliber of axons in regenerating motor fibers, *J. Cell Biol.,* 101, 1332, 1985.
9. **Hoffman, P. N., Cleveland, D. W., Griffin, J. W., Landes, P. W., Cowan, N. J., and Price, D. L.,** Neurofilament gene expression: a major determinant of axonal caliber, *Proc. Nat. Acad. Sci. U.S.A.,* 84, 3472, 1987.
10. **Schlaepfer, W. W.,** Experimental alterations of neurofilaments and neurotubules by calcium and other ions, *Exp. Cell Res.,* 67, 73, 1971.
11. **Schlaepfer, W. W.,** Structural alterations of peripheral nerve induced by the calcium ionophore, A23187, *Brain Res.,* 136, 1, 1977.

12. **Schlaepfer, W. W.,** Calcium-induced degeneration of axoplasm in isolated segments of rat peripheral nerve, *Brain Res.*, 69, 203. 1974.
13. **Schlaepfer, W. W.,** Effects of energy deprivation on Wallerian degeneration in isolated segments of rat peripheral nerve, *Brain Res.*, 78, 71, 1974.
14. **Schlaepfer, W. W. and Hasler, M. B.,** Characterization of the calcium-induced distruption of neurofilaments in rat peripheral nerve, *Brain Res.* 168, 299, 1979.
15. **Schlaepfer, W. W. and Zimmerman, U-J.P.,** Calcium-mediated breakdown of glial filaments and neurofilaments in rat optic nerve and spinal cord, *Neurochem. Res.*, 6, 243, 1981.
16. **Schlaepfer, W. W. and Bunge, R. P.,** The effects of calcium ion concentration on the degeneration of amputated axons in tissue culture, *J. Cell Biol.*, 59, 456, 1973.
17. **Schlaepfer, W. W. and Zimmerman, U-J.P.,** Mechanisms underlying the neuronal response to ischemic injury. Calcium-activated proteolysis of neurofilaments, *Prog. Brain Res.*, 63, 185, 1985.
18. **Schlaepfer, W. W. and Micko, S.,** Calcium-dependent alterations of neurofilament proteins of rat peripheral nerve, *J. Neurochem.*, 32, 211, 1979.
19. **Hisanaga, S. and Hirokawa, N.,** Structure of the peripheral domains of neurofilaments revealed by low angle rotary shadowing, *J. Mol. Biol.*, 202, 297, 1988.
20. **Schlaepfer, W. W. and Freeman, L. A.,** Calcium-dependent degradation of mammalian neurofilaments by soluble tissue factors from rat spinal cord, *Neuroscience*, 5, 2305, 1980.
21. **Tashiro, T. and Ishizaki, Y.,** A calcium-dependent protease selectively degrading the 160,000 M_r component of neurofilaments is associated with the cytoskeletal preparation of the spinal cord and has an endogenous inhibitory factor, *FEBS Lett.*, 141, 41, 1982.
22. **Ishizaki, Y., Tashiro, T., and Kurokawa, M.,** A calcium-activated protease which preferentially degrades the 160 kDa complement of the neurofilament triplet, *Eur. J. Biochem.*, 131, 41, 1983.
23. **Zimmerman, U-J. P. and Schlaepfer, W. W.,** Characterization of brain calcium-activated protease that degrades neurofilament proteins, *Biochemistry*, 21, 3977, 1982.
24. **Zimmerman, U-J. P. and Schlaepfer, W. W.,** Multiple forms of Ca-activated protease from rat brain and muscle, *J. Biol. Chem.*, 259, 3210, 1984.
25. **Banik, N. L., Hogan, E. L., Jenkins, M. G., McDonald, J. K., McAlhaney, W. W., and Sostek, M. B.,** Purification of a calcium-activated neutral proteinase from bovine brain, *Neurochem. Res.*, 8, 1389, 1983.
26. **Malik, M. N., Fenko, M. D., Iqbal, K., and Wisniewski, H. M.,** Purification and characterization of two forms of Ca^{2+}-activated neutral protease from calf brain, *J. Biol. Chem.*, 258, 8955, 1983.
27. **Vitto, A. and Nixon, R. A.,** Calcium-activated neutral proteinase of human brain: subunit structure and enzymatic properties of multiple molecular forms, *J. Neurochem.*, 47, 1039, 1986.
28. **Nilsson, E. and Karlsson, J.-O.,** Characterization of brain calpains, *J. Neurochem.*, 47, 1087, 1986.
29. **Kubota, S., Onaka, T., Murofushi, H., Ohsawa, N., and Takaku, F.,** Purification and characterization of high Ca^{2+}-requiring protease from porcine and bovine brains, *Biochemistry*, 25, 8396, 1986.
30. **Murachi, T.,** Calpain and calpastatin, *Trends Biochem. Sci.*, 8, 167, 1983.
31. **Kamakura, K., Ishiura, S., Susuki, K., Sugita, H., and Toyokura, Y.,** Calcium-activated neutral protease in the peripheral nerve, which requires μM order Ca^{2+}, and its effect on the neurofilament triplet, *J. Neurosci. Res.*, 13, 391, 1985.
32. **Kamakura, K., Ishiura, S., and Sugita, H.,** μ-type calcium-activated neutral protease in the rat peripheral nerve, *J. Neurosci. Res.*, 15, 167, 1986.
33. **Kerlero de Rosbo, N., Carnegie, P. R., and Bernard, C. C. A.,** Quantitative electroimmunoblotting study of the calcium-activated neutral protease in human myelin, *J. Neurochem.*, 47, 1007, 1986.
34. **Banik, N. L., McAlhaney, W. W., and Hogan, E. L.,** Calcium-stimulated proteolysis in myelin: Evidence for a Ca^{2+}-activated neutral proteinase associated with purified myelin of rat CNS, *J. Neurochem.*, 45, 581, 1885.
35. **Banik, N. L., Hogan, E. L., Jenkins, M. G., McDonald, J. K., McAlhaney, W. W., and Sostek, M. B.,** Purification of a calcium-activated neutral proteinase from bovine brain, *Neuchem. Res.*, 8, 1389, 1983.
36. **Gopalakrishna, R. and Barsky, S. H.,** Hydrophobic association of calpains with subcellular organelles. Compartmentalization of calpains and the endogenous inhibitor calpastatin in tissues, *J. Biol. Chem.*, 261, 13936, 1986.
37. **Sato, S. and Miyatake, T.,** Degradation of myelin basic protein by calcium-activated neutral protease (CANP)-like enzyme in myelin and inhibition by E-64 analogue, *Biomed. Res.*, 3, 461, 1982.
38. **Sato, S., Yanagisawa, K., and Miyatake, T.,** Conversion of myelin-associated glycoprotein (MAG) to a smaller derivative by calcium-activated neutral protease (CANP)-like enzyme in myelin and inhibition by E-64 analogue, *Neurochem.Res.*, 9, 629, 1984.
39. **Yanagisawa, K., Sato, S., Amaya, N., and Miyatake, T.,** Degradation of myelin basic protein by calcium-activated neutral protease (CANP) in human brain and inhibition by E-64 analogue, *Neurochem. Res.*, 8, 1285, 1983.

40. **Siman, R., Gall, C., Perlmutter, L. S., Christian, C., Baudry, M., and Lynch, G.,** Distribution of calpain I, an enzyme associated with degenerative activity in rat brain, *Brain Res.*, 347, 399, 1985.
41. **Hamakubo, T., Kannagi, R., Murachi, T., and Matas, A.,** Distribution of calpains I and II in rat brain, *J. Neurosci.*, 6, 3103, 1986.
42. **Schlaepfer, W. W., Lee, C., Lee, V. M-Y., and Zimmerman, U-J. P.,** An immunoblot study of neurofilament degradation *in situ* and during calcium-activated proteolysis, *J. Neurochem.*, 44, 502, 1985.
43. **Nixon, R. A. and Marotta, C. A.,** Degradation of neurofilament proteins by purified human brain cathepsin D, *J. Neurochem.*, 43, 507, 1984.
44. **Banay-Schwartz, M., Dahl, D., Hui, K.-S., and Lajtha, A.,** The breakdown of the individual neurofilament proteins by cathepsin D, *Neurochem. Res.*, 12, 361, 1987.
45. **Susuki, H., Takeda, M. Nakamura, Y., Kato, Y., Tada, K., Hariguchi, S., and Nishimura, T.,** Neurofilament degradation by bovine brain cathepsin D, *Neurosci. Lett.*, 89, 240, 1988.
46. **Chin, T. K., Eagles, P. A. M., and Maggs, A.,** The proteolytic digestion of ox neurofilaments with trypsin and α-chymotrypsin, *Biochem. J.*, 215, 239, 1983.
47. **Gilbert, D. S., Newby, B. J., and Anderton, B. H.,** Neurofilament disguise, destruction and discipline, *Nature*, 256, 586, 1975.
48. **Lasek, R. J., Krishnan, N., and Kaiserman-Abramof, I. R.,** Identification of the subunit proteins of 10-nm neurofilaments isolated from axoplasm of squid and Myxicola giant axons, *J. Cell. Biol.*, 82, 336, 1979.
49. **Pant, H. C., Terakawa, S., and Gainer, H.,** A calcium-activated protease in squid axoplasm, *J. Neurochem.*, 32, 99, 1979.
50. **Pant, H. C. and Gainer, H.,** Properties of a calcium-activated protease in squid axoplasm which selectively degrades neurofilament proteins, *J. Neurobiol.*, 11, 1, 1980.
51. **Pant, H. C., Gallant, P. E., Gould, R., and Gainer, H.,** Distribution of calcium-activity and endogenous substrates in the squid nervous system, *J. Neurosci.*, 2, 1986.
52. **Eagles, P. A. M., Gilbert, D. S., and Maggs, A.,** The polypeptide composition of axoplasm and of neurofilaments from the marine worm, *Myxicola infundibulum, Biochem. J.*, 199, 89, 1981.
53. **Eagles, P. A. M., Gilbert, D. S., and Maggs, A.,** The location of phosphorylation sites and Ca^{2+}-dependent proteolytic cleavage sites on the major neurofilament polypeptides from *Myxicola infundibulum, Biochem., J.*, 199, 101, 1981.
54. **Brown, A. and Eagles, P. A. M.,** Squid neurofilaments. Phosphorylation and Ca^{2+}-dependent proteolysis, *in situ, Biochem. J.*, 239, 191, 1986.
55. **Llinas, R., Blinks, J. R., and Nicholson, C.,** Calcium transient in presynaptic terminal of squid giant synapse: detection with aequorin, *Science*, 176, 1127, 1972.
56. **Pant, H. C.,** Dephosphorylation of neurofilament proteins enhances their susceptibility to degradation by calpain, *Biochem. J.*, 256, 665, 1988.
57. **Zimmerman, U-J. P. and Schlaepfer, W. W.,** Kinase activities associated with calcium-activated neutral proteases, *Biochem. Biophys. Res. Commun.*, 120, 767, 1984.
58. **Zimmerman, U-J. P and Schlaepfer, W. W.,** Characterization of calcium-activated neutral protease (CANP)-associated protein kinase from bovine brain and its phosphorylation of neurofilaments, *Biochem. Biophys. Res. Commun.*, 129, 804, 1985.
59. **Coolican, S. A. and Hathaway, D. R.,** Effect of L-α-phosphatidylinositol on a vascular smooth muscle Ca^{2+}-dependent protease. Reduction of the Ca^{2+} requirement for autolysis, *J. Biol. Chem.*, 259, 11627, 1984.
60. **Imajoh, S., Kawasaki, H., and Suzuki, K.,** The amino-terminal hydrophobic region of the small subunit of calcium-activated neutral protease (CANP) is essential for its activation by phosphatidylinositol, *J. Biochem.*, 99, 1281, 1986.
61. **Zimmerman, U-J. P. and Schlaepfer, W. W.** Calcium-activated proteases (calpains) are carbohydrate binding proteins, *J. Biol. Chem.*, 263, 11609, 1988.
62. **Schlaepfer, W. W.,** Nature of mammalian neurofilaments and their breakdown by calcium, *Prog. Neuropathol.*, 4, 101, 1979.
63. **Schlaepfer, W. W. and Micko, S,** Chemical and structural changes of neurofilaments in transected rat sciatic nerve, *J. Cell Biol.*, 78, 369, 1978.
64. **Palay, S. L.,** The morphology of synapses in the central nervous system, *Exp. Cell Res.*, (Suppl.5), 275, 1958.
65. **Walker, J. H., Boustead, C. M., Witzemann, V., Shaw, G., Weber, K., and Osborn, M.,** Cytoskeletal proteins at the cholinergic synapse: distribution of desmin, actin, fodrin, neurofilaments and tubulin in Torpedo electric organ, *Eur. J. Cell Biol.*, 38, 123, 1985.
66. **Lasek, R. J. and Black, M. M.** How do axons stop growing? Some clues from the metabolism of the proteins in the slow component of axonal transport, in *Mechanisms, Regulation and Special Functions of Protein Synthesis in the Brain*, Roberts, S., Lajtha, A., and Gispen, W. H., Eds., Elsevier-North Holland, Amsterdam, 1977, 161.

67. **Roots, B. I.,** Neurofilament accumulation induced in synapses by leupeptin, *Science,* 221, 971, 1983.

68. **Boycott, B. B., Gray, E. G., and Guillery, R. W.,** Synaptic structure and its alteration with environmental temperature: a study by light and electron microscopy of the central nervous system of lizards, *Proc. Roy. Soc. London, Ser. B.,* 154, 151, 1961.

69. **Bondar, R. L. and Roots, B. I.,** Neurofibrillary changes in goldfish (*Carassius autatus* L.) brain in relation to environmental temperature, *Exp. Brain Res.,* 30, 577, 1977.

70. **Goldstein, M. E., Weiss, S. R., Lazzarini, R. A., Schneidman, P. S., Lees, J. F., and Schlaepfer, W. W.,** mRNA levels of all three neurofilament proteins decline following nerve transection, *Mol. Brain Res.,* 3, 187, 1988.

71. **Schlaepfer, W. W. and Bruce, J.,** Simultaneous up-regulation of neurofilament proteins during the postnatal development of the rat nervous system. *J. Neurosci. Res.,* 25, 39, 1990.

72. **Hollenbeck, P. J. and Bray, D.,** Rapidly transported organelles containing membrane and cytoskeletal components: their relation to axonal growth, *J. Cell Biol.,* 105, 2827, 1986.

73. **Schlaepfer, W. W., Lee, C., Trojanowski, J. Q., and Lee, V.M-Y.,** Persistence of immunoreactive neurofilament protein breakdown products in transected rat sciatic nerve, *J. Neurochem.,* 43, 857, 1984.

74. **Traub, P., Nelson, W. J., Kuhn, S., and Vorgias, C. E.,** The interaction *in vitro* of the intermediate filament protein vimentin with naturally occurring RNAs and DNAs, *J. Biol. Chem.,* 258, 1456, 1983.

75. **Traub, P. and Vorgias, C. E.,** Involvement of the N-terminal polypeptide of vimentin in the formation of intermediate filaments, *J. Cell Sci.,* 63, 43, 1983.

76. **Traub, P., Vorgias, C. E., and Nelson, W. J.,** Interaction *in vitro* of the neurofilament triplet proteins from porcine spinal cord with natural RNA and DNA, *Mol. Biol. Rep.,* 10, 129, 1985.

Chapter 18

PLASTICITY TO PATHOLOGY: BRAIN CALPAINS AS MODIFIERS OF SYNAPTIC STRUCTURE

Peter Seubert and Gary Lynch

TABLE OF CONTENTS

I. INTRODUCTION

Brain calpains are similar, if not identical, to those of other tissues based on a broad spectrum of physical and biochemical criteria. Therefore, the major purpose of this review will be to examine the substrates, functions, and triggering mechanisms of brain calpains during uniquely neural events. It will be argued that the ability of calpain to alter the submembranous cytoskeleton has been adapted to function in the development and alteration of synaptic contacts. Evidence will be discussed that suggests calpain contributes to both normal synaptic restructuring events and particular neuropathological susceptibilities, including aging. Finally, the hypothesis that calpain-mediated proteolysis of the neuronal cytoskeleton is a necessary step in memory formation will be updated.

II. CALPAINS IN BRAIN

A. PHYSICAL AND BIOCHEMICAL CHARACTERIZATION

Considerable progress has been made toward the characterization of brain calpains since the original description of neutral, calcium-activated proteinase activity by Guroff a quarter century ago.[1] Mammalian brain is now known to contain at least two forms of calpain[2-4] along with specific activator[5,6] and inhibitor[7] proteins. Initial confusion as to the subunit composition, relative abundance, molecular weight, and calcium sensitivity of brain calpains was compounded by the use of purification procedures utilizing Ca^{2+}-dependent affinity chromatography[8] which alter all the above properties (discussed in Reference 3).

The best evidence to date is for the existence in brain of a comparatively scarce, high Ca^{2+} sensitivity (μM) form of calpain and a predominant, low Ca^{2+} sensitivity (mM) type (calpain II).[2-4] Immunochemically these proteins resemble the respective calpains isolated from other tissues.[9,10] Physical and enzymatic properties including molecular weights, chromatographic elution profiles, sensitivity to inhibitors, cation specificites, and autolytic behaviors further support the homology of brain calpains with those in other tissues.

Several investigators have postulated the existence of additional calpain-like proteases in brain. Zimmerman and Schlaepfer reported a high molecular weight (154,000) form, which they suggested to be a progenitor of calpains with molecular weights of 96,000 and 76,000, present in brain and muscle.[11] Other investigators have not detected this precursor immunochemically[9,10] and no further characterization of this high molecular weight form exists. Malik et al. have also reported a third form of brain calcium-activated neutral proteinase with properties very similar to those of two other calcium-activated proteases isolated by their procedure, the principal difference being its autolytic behavior.[12] Since Ca^{2+}-dependent casein-Sepharose chromatography was used in this method, and other criteria (peptide mapping, monoclonal antibody reactivity) failed to establish differences among the enzymes, more information will be required to establish whether this represents a modified form of an existing calpain or a separate gene product. Finally, Nixon et al. observed in axons a minor proteolytic activity activated by μM calcium and inhibited by calpastatin but, in contrast to other calpains, relatively insensitive to leupeptin or TLCK, yet inhibited by PMSF. This proteinase was also reported as having a very select substrate specificity.[13] While these properties would suggest a new form of calcium-activated neutral protease, further characterization and purification are required to classify this activity. Thus while several observations suggest the existence of additional calpain-like proteins in brain, definitive proof is lacking.*

* *Note added in proof*: Yoshihara et al. (*J. Biol. Chem.*, 265, 5809, 1990) have recently described a calcium-activated neutral protease in ratbrain whose N-terminal sequence, inhibitor profile, substrate preference, and chromatographic behavior are unlike the known calpains.

In the discussions which follow it should also be remembered that calpain modulatory proteins have also been demonstrated in brain. Calpastatin, the high molecular weight (300 kDa) calpain-specific inhibitor is found in brain at levels in excess of calpain I.[7,14] In addition, reports from independent laboratories have indicated the presence of a calpain-specific activator protein associated primarily with cytoskeletal elements.[5,6] This small (15 to 20 kDa) protein stimulates calpains I and II, isolated from various tissues, 10 to 25-fold. Thus far this activator has only been reported in brain; however, there is no reason to assume such a restricted distribution. The subcellular localizations of these two proteins will obviously be helpful guides in determining the loci of calpain activity.

B. DISTRIBUTION OF BRAIN CALPAINS

1. Among Vertebrate Classes

Studies by Baudry et al. examined the calcium-dependent and -independent proteolytic activities present at neutral pH in the soluble fraction of brain homogenates from several classes of vertebrates.[15] Birds and mammals had by far the largest amounts of calpain (calpain I and II) activity and relatively low amounts of Ca^{2+}-indpendent proteolysis. Amphibian and reptilian brains, on the other hand, contained much lower levels of calpain activity but relatively high amounts of Ca^{2+}-independent activity. Fish brain had indetectable levels of either type. Other investigators have confirmed the presence of calpain activity in the amphibian nervous system as well as demonstrating its activity toward rapidly transported axonal proteins.[16] The lack of calpain detected in fish is puzzling in light of the presence of calpain in invertebrate tissues (see Pintër, Chapter 9, and Mykles and Skinner, Chapter 10) as well as in the axoplasm of giant squid[17] and marine worms.[18]

The second major finding of the above studies is that calpain specific activities are negatively correlated with brain size of both mammals and birds. Maximum lifespan also correlates well with brain size. An exception to this relation is the order Chiroptera (bats) which have surprisingly long lifespans. Interestingly, total calpain activity in Chiroptera is unusually low. A link between calpain and the stability of neurons has been suggested.[19]

2. Regional and Developmental Variation within the Rat Brain

The distribution of soluble calpain activities, measured at 30 μM and 500 μM Ca^{2+}, was determined in several brain regions of various aged rats by Simonson et al.[20] In the adult brain, levels of calpain II activity varied ten-fold with rhombencephalic structures containing the highest and telencephalic structures the lowest levels of calpain activity. An exception to this trend was the olfactory bulb, which contained a level of activity comparable to the hindbrain regions. Calpain I activity was comparatively low throughout the brain and was less prevalent than calpain II in all regions.

Calpain levels also vary remarkably during development. A dramatic reduction (80%) in calpain activity (both I and II) occurs in the forebrain during the first postnatal week. Calpain I activity in the hindbrain also drops dramatically during this period. However, the level of calpain II in the hindbrain remains essentially constant into adulthood. The variations in development are suggestive of a role for calpain in synapse maturation. Regional variations of calpain activity may be related to the relative stability of the neurons in a particular brain region, an idea discussed in more detail elsewhere.[19,20]

3. Celluluar and Subcellular Distributions

Some controversy exists as to the precise localizations of the calpains within brain cells. Hamakubo et al. found calpain II to be most prominent in neuronal cell bodies, myelinated axons and glial cells, while calpain I was primarily localized to neuronal cell bodies and proximal dendrites.[9] No evidence for either calpain in distal portions of dendrites was obtained. In contrast, Siman et al.[10] and Perlmutter et al.[21] detected calpain I in distal dendrites (including spines and postsynaptic densities), axons, and glial cells. At present the reasons for these discrepancies are unclear but the considerable methodological differences are a likely possibility. A second approach toward answering the subcellular distribution question was taken by Baudry et al. who analyzed the calpain I and II activities in different subcellular fractions.[22] In agreement with Hamakubo, calpain II activity was predominantly localized in the cytoplasmic fraction with lesser amounts in the myelin and synaptosomal fractions. Several other investigations have also demonstrated the presence of calpain II in myelin.[23-25] Calpain I, on the other hand, was predominantly localized in the synaptosomal and myelin isolates, suggesting the enzyme is in the finer processes of axons, dendrites, and glial processes.[22] In the following sections further evidence for calpain function in postsynaptic regions will be presented.

III. BRAIN SUBSTRATES OF CALPAINS

A. STRUCTURAL ELEMENTS

Ultrastructural studies of severed axons demonstrated a calcium-dependent process which disrupted the microtubule and neurofilament systems[26] (see Schlaepfer and Zimmerman, Chapter 17). Subsequent investigations have revealed that the proteins in and associated with these filamentous systems are excellent substrates for calpains.

In characterizing the conditions promoting microtubule assembly, Sandoval and Weber[27] noted that an irreversible loss of the ability to assemble occurred if calcium ions were present in crude extracts, an effect not observed using purified microtubule proteins. This alteration was found to be due to Ca^{2+}-dependent proteolysis of the microtubule-associated proteins (MAPs); purification of the microtubules removed the calpain present in crude extract. The presence of calpain-derived MAP fragments in various cell types suggests this interaction occurs *in vivo*.[28] Recent studies have further shown that the susceptibility of MAPs and tubulin to calpain is reduced by assembly into microtubules.[29] Injection of excitotoxic amino acids has been shown to activate calpain and cause the loss of MAP2 in brain, suggesting this interaction occurs under degenerative conditions.[30]

A role for calpain in the disruption of neurofilaments has also been supported by the work of several laboratories. Gilbert et al. described a Ca^{2+}-dependent protease in *Myxicola* (a marine worm) axoplasm which cleaved neurofilament proteins.[18] This activity has since been characterized in squid axoplasm,[17] rat optic nerve,[31] and calf brain.[32] Lasek and Hoffman proposed that calpain in terminals limits axon elongation through calcium fluxes occurring during action potentials.[33] The accumulation of neurofilaments in synaptic terminals observed in the presence of leupeptin supports this hypothesis.[34] Leupeptin in axon tips also blocks the anterograde-retrograde conversion of axonally transported vesicles although the protein substrates are as yet unkown.[35]

Proteins associated with the microfilament system have also been shown to be preferred calpain substrates. A high molecular weight doublet observed to be a preferred calpain substrate[36] was later identified as brain spectrin,[37] also known as fodrin.[38] Spectrin is a long (2000Å), flexible, rodlike molecule which indirectly cross-links and attaches actin filaments to the plasma membrane.[39] The predominant form *in situ* is thought to be a heterodimer consisting of $\alpha\beta$ dimers linked head to head. The 235-kDa β subunit exists in at least two isoforms that have preferential distributions in neurons.[40,41] The 240-kDa α subunit appears to be conserved in both isoforms based on 2-D peptide maps.[42] No biochemical evidence for the existence of 240-kDa isoforms is available to date. Immunocytochemically based claims[43] that different α isoforms exist in axons and dendrites are in dispute. Ivy et al. showed that failure to irreversibly fix tissue sections leads to demonstrable washout of spectrin, particularly in dendrites.[44] This point becomes important in considering anatomical substrates for localized calpain activation (see below).

Two points regarding the calpain-mediated proteolysis of spectrin are worth noting: (1)

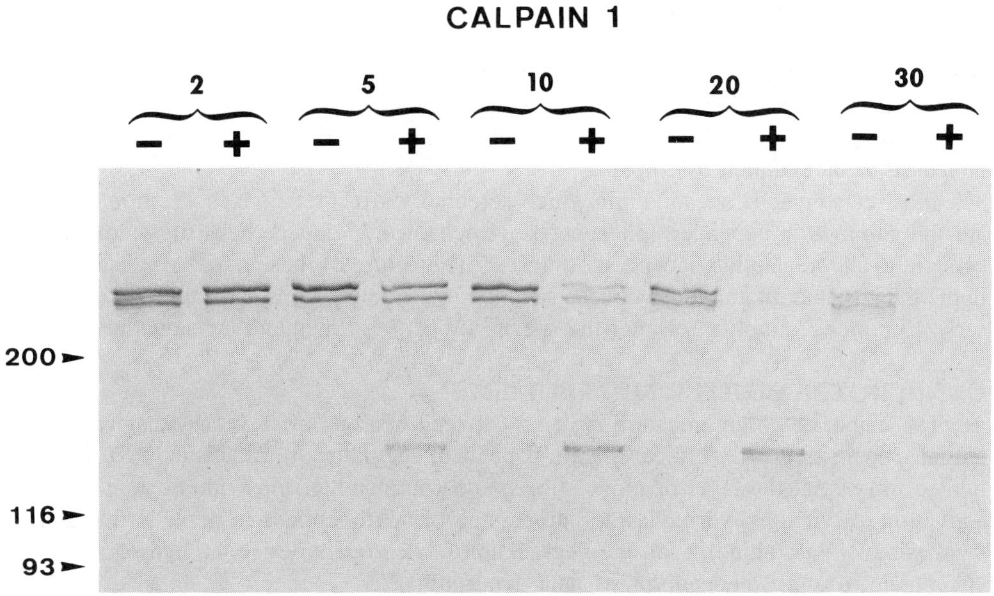

FIGURE 1. Effect of calmodulin on brain spectrin breakdown by calpain I. Spectrin (75 μg/ml) was incubated at 30°C with 100 μ*M* $CaCl_2$, 3 μg/ml calpain I, 20 m*M* Tris-Cl, 5 m*M* mercaptoethanol, 150 m*M* NaCl at pH 7.5. If present, calmodulin was at a concentration of 25 μg/ml (1.5 μ*M*). After the indicated time of incubation (min), aliquots were withdrawn, added to one third volume of 3Xs SDS-PAGE sample buffer, boiled, and subjected to electrophoresis on 3 to 10% gradient gels. Gels were stained with coomassie blue and destained with 7% acetic acid. The breakdown of spectrin is accelerated three- to five-fold by the presence of calmodulin. Calmodulin did not affect the rate at which calpain cleaved the other substrates examined in this study. (From Seubert, P., Baudry, M., Dudek, S., and Lynch, G., *Synapse,* 1, 20, 1987. With permission.)

breakdown products resistant to further cleavage are formed which can be detected on Western blots,[30,45] and (2) proteolysis is accelerated five-fold by the presence of calmodulin (Figure 1).[46] The α subunit contains a highly affinity calmodulin-binding site and cleavage occurs in this vicinity.[47] While calmodulin is known to regulate a variety of other enzymes, this is to our knowledge the first demonstration of a synergistic interaction with the calcium-activated proteases calpain I and II.

Spectrin is found in highly purified postsynaptic densities (psds) along with a calpain-generated spectrin breakdown product.[48] Although the proteolysis most likely occurs during the course of psd preparation, which requires the addition of calcium, this result, in combination with immunocytochemical data, suggests both calpain and spectrin may exist in dendritic spines. As will be discussed below, substantial evidence for the calpain-spectrin interaction exists in response to a wide variety of manipulations (see Sections IV and V).

The substrates for calpain in nonneuronal brain cells also include cytoskeletal components such as glial fibrillary acidic protein,[31,49] which is the primary structural protein of glial filaments, and myelin basic protein.[23,24,50,51] The regulation of calpain activity may then have important consequences in regard to demyelinating diseases.

B. REGULATORY ENZYMES

In addition to altering cytoskeletal proteins, calpain has been suggested to modify the activity of a number of enzymes. This represents a potentially powerful amplification cascade as a number of the suggested substrates are themselves regulatory in nature. To date, little evidence exists for the function of these pathways *in vivo*, which may simply reflect the youth of this field.

The regulation of phosphorylating and dephosphorylating enzymes by calpain has been among the more active areas of inquiry. Before its regulation by Ca^{2+} and lipids was known, protein kinase C was shown to be a kinase which could be irreversibly activated by calpain[52] (see Kishimoto, et al., Chapter 17, for a detailed description). In other tissues, the activity of other kinases including phosphorylase kinase, pyruvate kinase, cAMP-dependent kinase (reviewed in Reference 53) and myosin light chain kinase[54] have also been found to be increased or deregulated by calpain.

Other calpain substrates in brain which potentially affect the phosphorylation of proteins are the calmodulin-dependent phosphatase, calcineurin,[55] and perhaps the calmodulin-dependent cyclic nucleotide phosphodiesterase.[56] The nature of these putative substrates highlights the possible interactions between second messenger systems and suggests calpain may serve to prolong, amplify, or alter the specificity of these regulatory systems in brain.

C. NEUROTRANSMITER METABOLISM

The elaborate communication systems between neurons offer yet another class of potential calpain targets in brain. Calpain-dependent regulatory mechanisms have been postulated to exist at the level of modulation of neurotransmitter biosynthetic enzymes (e.g., activation of tyrosine hydroxylase),[57] processing of neuropeptides in nerve terminals (e.g., synthesis of kyotorphine)[58] and the degradation of neuropeptides (e.g., hydrolysis of Leu-enkephalin, α and β neoendorphins, and dynorphin).[59]

IV. CALPAIN INVOLVEMENT IN NEUROPATHOLOGICAL EVENTS

As described above, calpain is capable of disrupting all the major filament systems as well as modifying the activities of key regulatory enzymes in neuronal and nonneuronal brain cells. These properties, and the known deleterious effects of elevated intracellular calcium levels on cellular viability,[60] suggest a likely role for calpain in pathological conditions.

A. IN DEMYELINATION

Several investigators have suggested the involvement of calpain in demyelinating diseases such as multiple sclerosis and experimental allergic encephalomyelitis.[23-25,50,51] In quaking mouse brain, a CNS myelin deficiency mutation, total calcium levels are elevated and the fraction of calpain in myelin is low while the cytosolic calpain activity is correspondingly increased.[61] Quaking mice also demonstrate abnormal rates of axonal proteolysis of neuronal proteins,[25] suggesting both neuronal and glial proteolytic events are affected in this disorder.

B. FOLLOWING DEAFFERENTATION

The destructive actions of calpain in severed axons, described in detail elsewhere in this volume (Schlaepfer and Zimmerman, Chapter 17), were the first demonstration of calpain function under pathological conditions. Is the damaged axon the only site of calpain activation? Toward answering this question we have undertaken a series of studies involving lesions of the entorhinal cortex. This procedure destroys 90% of the axonal inputs to the outer 2/3 of the dendritic field of the dentate gyrus granule cell neurons; thus a well-defined zone experiencing denervation is produced.[62] The dentate cells are not destroyed by this procedure and, in fact, new axonal growth into the denervated zone occurs several days later.[63,64] Ultrastructural studies of the effects which deafferentation has on dentate granule cells reveal an interesting fact—the psds are removed along with the presynaptic terminal, i.e., unopposed psds are not commonly found following the lesion.[65] Large-scale dendritic cytoskeletal modification must obviously be occurring during this period. Mitochondria in this denervated area also experience changes in that they buffer calcium less effectively.[66]

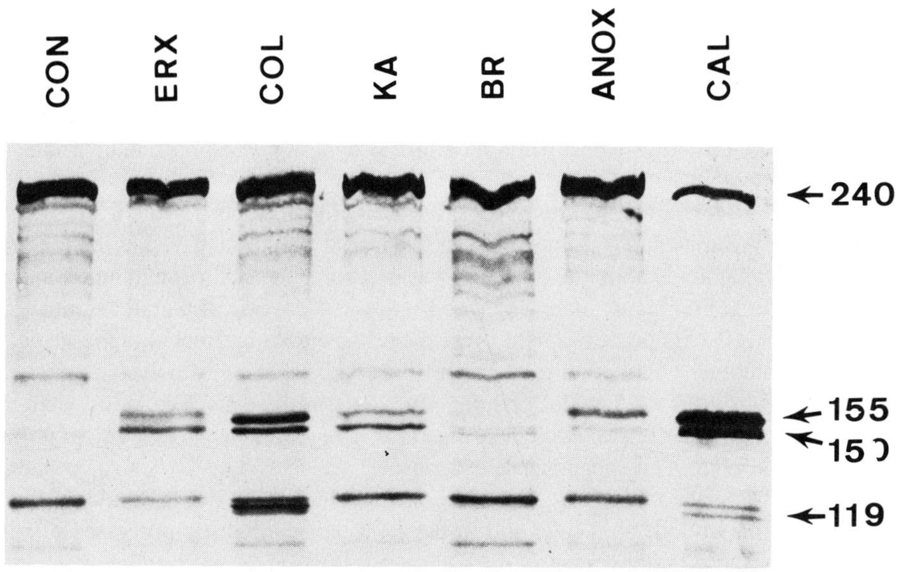

FIGURE 2. Spectrin breakdown occurs following many neural insults. Protein samples (10 to 15 μg/protein) were separated by electrophoresis in 3 to 10% SDS-polyacrylamide gradient gels, transferred to nitrocellulose sheets and probed with affinity-purified anti-spectrin antibodies. The molecular mass in kDa of spectrin (240) and calpain-derived breakdown products (155, 150, 119) is indicated. CON, control rat hippocampus; ERX; rat dentate gyrus, 2 d following entorhinal cortex lesion; COL, rat dentate gyrus 2 d following intrahippocampal injection of 15 μg of colchicine; KA, CA3 region of rat hippocampus 1 d after i.p. injection of 15 mg/kg kainic acid; BR, hippocampus of a 12 d-old Brindled mouse (an hereditary Cu^{2+} metabolic disorder); ANOX, CA1 of gerbil hippocampus 15 min after bilateral clamping of the carotid arteries for 10 min; CAL, crude synaptosomal brain membranes incubated with 4 μg/ml calpain I for 10 min 30° with 1 m*M* Cal_2.

Thus both the trigger (elevated Ca^{2+}) and expected effects (cytoskeletal restructuring) of calpain activation are observed following lesion.

As mentioned above, proteolysis of brain spectrin by calpain results in stable breakdown products. Since spectrin is both a preferred calpain substrate and a component of psds, it seemed reasonable to search for the presence of these spectrin fragments as an indication of past calpain activity. Using antispectrin antibodies to probe Western-blotted samples, we observed that spectrin proteolysis in the dentate gyrus dramatically increases following lesion of the entorhinal cortex (see Figure 2).[45] The extent of this process is reduced by prior infusion of leupeptin. Although axons also contain spectrin, which would be expected to be degraded following lesion, several lines of evidence suggest that proteolysis is also occurring in dendrites. Spectrin breakdown was detectable within 4 h post-lesion, well before deterioration of terminals and synapses occurs. During the period when maximal breakdown was observed (24 to 48 h), no decrease in intact spectrin was detected, suggesting ongoing spectrin synthesis, which implies a dendritic localization. These results, together with the ultrastructural effects noted following lesion and the presence of spectrin and calpain in dendritic spines and psds, indicate calpain may play an active role in modifying dendritic elements.

C. IN RESPONSE TO NEUROTOXINS

Further evidence for the operation of calpain in dendrites was obtained from studies of the effects of colchicine injected into hippocampus. Dentate gyrus granule cells are particularly vulnerable to this microtubule-disrupting drug,[67] which has also been reported to disrupt organelles involved in calcium sequestration.[68] Colchicine injection was found to induce spectrin breakdown (see Figure 2).[69] Subdissection of the dendritic layer from the cell body layer of the dentate gyrus revealed most of the proteolysis was occurring in the dendrites. Furthermore, infusion of leupeptin again reduced the level of breakdown, while chloroquine, a nonspecific lysosomal inhibitor, increased breakdown. Thus, the data are consistent with the proteolysis occurring in dendrites by a nonlysosomal, leupeptin-sensitive process.

More direct evidence for the involvement of calpain in spectrin proteolysis is found in the experiments of Siman and Noszek[30] who showed that several hours following the injection of excitotoxic amino acids, calpain I levels decreased (presumably through autolysis), while the levels of spectrin fragments increased (see Figure 2).[30] Levels of a glial calpain substrate (GFAP) and the 200-kDa neurofilament protein both seemed unaffected, supporting a dendritic site of proteolysis. Although it is impossible to say if this proteolysis is triggered by synaptic events or by the collapse of cellular function, results discussed below demonstrate that the activation of one class of excitatory amino acid receptors triggers spectrin proteolysis within minutes in electrically responsive neurons.

The studies described in this section demonstrate that brain possess calcium-activated proteolytic systems in both glial cells and in the axonal and dendritic processes of neurons. The activation of these systems is most clearly seen following severe trauma, e.g., axotomy or neurotoxin injection. Much valuable information is gained by these studies regarding substrates and subcellular localization. A weakness of toxicological injection studies is that any treatment which eventually collapses the energy-dependent calcium buffering systems of the cell will presumably activate calpain. For example, brain tissue which is not suitably stabilized within minutes of death shows a large elevation in spectrin breakdown.[100] Thus while a tremendous capacity for proteolysis exists, physiological triggering of this process is more difficult to demonstrate. Lesion-induced proteolysis preceded obvious ultrastructural signs of degeneration and was argued to be occurring at least partially in viable cells, suggesting a role in the restructuring of dendritic processes. In the following section we will argue that a similar process, occurring on a smaller scale in response to synaptic events, is critical in the biochemistry of memory formation.

V. CALPAIN AND MEMORY—AN UPDATE

More than 5 years have passed since Lynch and Baudry proposed a new and still controversial hypothesis regarding the biochemistry of memory.[70] Briefly summarized, they postulated that high-frequency synaptic activity caused the postsynaptic activation of calpain, resulting in spectrin degradation and ultimately the expression of an increased number of glutamate receptors. Although some specific details have required revision, the essential involvement of calpain in memory formation has been strengthened by several recent studies.

A. *N*-METHYL-D-ASPARTATE (NMDA) RECEPTOR ACTIVATION IS INVOLVED IN MEMORIAL PROCESSES

Brain possesses a subclass of glutamate receptors, the NMDA receptor, which has interesting property of being a voltage-sensitive calcium channel.[71,72] During periods of high frequency synaptic activity, calcium is admitted through this channel, triggering events such that subsequent synaptic responses in these previously stimulated synapses are enhanced. This persisting enhancement, thought to be related to memory, is known as long-term potentiation (LTP). Calcium is reasoned to be the trigger since intracellular chelation of calcium blocks LTP induction.[73,74] Blockage of the NMDA receptors results in both suppression of LTP induction[75-77] and impairment of the acquisition of memories for certain tasks by rats.[78,79]

B. NMDA RECEPTOR ACTIVATION TRIGGERS CALPAIN-MEDIATED PROTEOLYSIS

When hippocampal slices are exposed to NMDA, proteolysis of spectrin is triggered.[80] This effect is rapid (within minutes), requires extracellular calcium, is not reproduced by potassium depolarization, and is blocked by an NMDA receptor antagonist. The spectrin breakdown product formed is identical to that initially produced by calpain. Under the conditions of this study, excitatory postsynaptic potentials recovered after NMDA washout, suggesting that the proteolysis is occurring in viable neurons.

It should also be noted that NMDA receptors are thought to mediate the deleterious effects of several neuropathological events including hypoglycemia[81] and ischemia.[82] Brief periods of transient ischemia cause the delayed death of certain hippocampal neurons days later.[83,84] Events occurring during or shortly after the ischemia are thought to predispose the vulnerable neurons toward death. NMDA receptor antagonists, administered during the crucial initiating phase, prevent the delayed neuronal loss.[85-87]

In studies presented elsewhere,[88] transient ischemia was found to produce two phases of spectrin proteolysis in the vulnerable region of hippocampus. The first occurred within minutes of the insult (see Figure 2), followed days later by a second, more massive phase, during the terminal stage of neuronal death. The neuroprotective NMDA antagonist MK-801 was found to reduce both proteolytic phases. We suggested the initial NMDA-receptor linked proteolysis in synaptic regions (MK-801-sensitive component) may play an important role in the eventual demise of these neurons days later. The ischemic data highlight the delicate balance which must exist in brain, namely, that while calpain-mediated proteolysis may be necessary for synaptic plasticity, exaggerated activation of the mechanism could underlie certain pathologies.[19]

C. CALPAIN AND HIGH FREQUENCY SYNAPTIC ACTIVITY: SIMILAR SUBSTRATES?

Ultrastructurally the induction of LTP is accompanied by morphological changes in the shape of certain dendritic spines.[89-91] As stated above, calpain is capable of altering all the major cytoskeletal structural systems. Calpain has also been linked to morphological changes in platelets,[92] erythrocytes[93] and hepatocytes.[94] Calpain may in fact be the enzyme most ideally suited to irreversibly alter the submembranous structures of cells.

Biochemically, high frequency stimulation has been found to increase glutamate "binding" in synaptic membranes,[95] an effect reproduced by calpain activation in nonstimulated membranes.[36] Although it was subsequently demonstrated that the increased "binding" represented a sequestration process,[96] the fact remains that a biochemical correlate of LTP can be reproduced by calpain activation, although the functional significance is less clear.

D. INHIBITION OF CALPAIN BLOCKS LTP AND MEMORY FORMATION

The studies discussed above demonstrated that NMDA receptor activation is linked to calpain activity. The question arises as to whether this is related to the formation of memories and/or synaptic potentiation or if this is simply a coincidental biochemical event. To approach this question, rats were chronically administered leupeptin, via osmotic mini-pumps, into the ventricle. Although leupeptin is not a specific calpain inhibitor, this procedure is sufficient to suppress calpain activity[45,69] and has no obvious deleterious effects at the dosages employed.[97] Leupeptin was found to block memory acquisition in certain behavoral tasks[97] as well as to prevent the induction of LTP, both *in vivo*[98] and in hippocampal slices.[99] While the general acceptance of this hypothesis will undoubtedly require studies using a battery of more specific calpain inhibitors and further characterization of the functional consequences of calpain activation in neurons, the progress to date is encouraging.

REFERENCES

1. **Guroff, G.,** A neutral, calcium-activated proteinase from the soluble fraction of rat brain, *J. Biol. Chem.,* 239, 149, 1964.
2. **Siman, R., Baudry, M., and Lynch, G.,** Purification from synaptosomal plasma membranes of calpain I, a thiol protease activated by micromolar calcium concentrations., *J. Neurochem.,* 41, 950, 1983.
3. **Kubota, S., Onaka, T., Murofushi, H., Ohsawa, N., and Takaku, F.,** Purification and characterization of high Ca^{2+}-requiring neutral proteases from porcine and bovine brains. *Biochemistry,* 25, 8396, 1986.
4. **Nilsson, E. and Karlsson, J.-O.,** Characterization of brain calpains, *J. Neurochem.,* 47, 1086, 1986.
5. **DeMartino, G. N. and Blumenthal, D. K.,** Identification and partial purification of a factor that stimulates calcium-dependent proteases, *Biochemistry,* 21, 4297, 1982.
6. **Takeyama, Y., Nakanishi, H., Uratsuji, Y., Kishimoto, A., Nishizuka, Y.,** A calcium-protease activator associated with brain microsomal-insoluble emements, *FEBS Lett.,* 194, 110, 1986.
7. **Nishiura, I., Tanaka, K., and Murachi, T.,** A high molecular weight inhibitor of Ca^{2+}-dependent neutral protease in rat brain, *Experientia,* 35, 1006, 1979.
8. **Malik, M. N., Fenko, M. D., Iqbal, K., and Wisniewski, H. M.,** Purification and characterization of two forms of Ca^{2+}-activated neutral protease from calf brain, *J. Biol. Chem.,* 258, 8955, 1983.
9. **Hamakubo, T., Kannagi, R., Murachi, T., and Matus, A.,** Distribution of calpains I and II in rat brain, *J. Neurosci.,* 6, 3103, 1986.
10. **Siman, R., Gall, C., Perlmutter, L. S., Christian, C., Baudry, M., and Lynch, G.,** Distribution of calpain I, an enzyme associated with degenerative activity in rat brain, *Brain Res.,* 347, 399, 1985.
11. **Zimmerman, U.-J. P. and Schaepfer, W. W.,** Multiple forms of Ca-activated protease from rat brain and muscle, *J. Biol. Chem.,* 259, 3210, 1984.
12. **Malik, M. N., Fenko, M. D., Sheikh, A. M., Kascsak, R. J., Tonna-DeMasi, M. S., and Wisniewski, H. M.,** Third form of calcium-activated neutral proteinase from calf brain: purification, partial characterization and comparison with other forms, *Biochim. Biophys. Acta,* 916, 135, 1987.
13. **Nixon, R. A., Quackenbush, R., and Vitto, A.,** Multiple calcium-activated neutral proteinases (CANP) in mouse retinal ganglion cell neurons: specificities for endogenous neuronal substrates and comparsion to purified brain CANP, *J. Neurosci.,* 6, 1252, 1986.
14. **Murachi, T.,** Calpain and calpastatin, *Trends Biochem. Sci.,* 8, 167, 1983.
15. **Baudry, M., Simonson, L., Dubrin, R., and Lynch, G.,** A comparative study of soluble calcium-dependent proteolytic activity in brain, *J. Neurobiol.,* 17, 15, 1986.
16. **Zalewska, T., Kanje, M., and Edstrom, A.,** A calcium-activated neutral protease in the frog nervous system which degrades rapidly transported axonal proteins, *Brain Res.,* 381, 58, 1986.
17. **Pant, H. C. and Gainer, H.,** Properties of a calcium-activated protease in squid axoplasm which selectively degrades neurofilament proteins, *J. Neurobiol.,* 11, 1, 1980.
18. **Gilbert, D. S., Newby, B. J., and Anderton, B. H.,** Neurofilament disguise, destruction and discipline, *Nature,* 265, 586, 1975.
19. **Lynch, G., Larson, J., and Baudry, M.** Proteases, neuronal stability, and brain aging: a hypothesies, in *Treatment Development Strategies for Alzheimer's Disease,* Crook, T., Bartus, R., Ferris, S., and Gershon, S., Eds., Mark Powley, Madison, CT, 1986, 119.
20. **Simonson, L., Baudry, M., Siman, R., and Lynch, G.,** Regional distribution of soluble calcium-activated proteinase activity in neonatal and adult rat brain, *Brain Res.,* 327, 153, 1985.
21. **Perlmutter, L. S., Siman, R., Gall, C., Seubert, P., Baudry, M., and Lynch, G.,** The ultrastructural localization of calcium-activated protease "calpain" in rat brain, *Synapse,* 2, 79, 1988.
22. **Baudry, M., Dubrin, R., and Lynch, G.,** Subcellular compartmentalization of calcium-dependent and calcium-independent proteases in brain, *Synapse,* 1, 506, 1987.
23. **Banik, N. L., McAlhaney, W.W., and Hogan, E. L.,** Calcium stimulated proteolysis in myelin: evidence for a Ca^{2+}-activated neutral proteinase associated with purified myelin of rat CNS, *J. Neurochem.,* 45, 581, 1985.
24. **DeRosbo, N. K., Carnegie, P. R., and Bernard, C. C. A.,** Quantitative electro immunoblotting study of the calcium-activated neutral proteinase in human myelin, *J. Neurochem.,* 47, 1007, 1986.
25. **Nixon, R. A.,** Increased axonal proteolysis in myelin-deficient mutant mice, *Science,* 215, 999, 1982.
26. **Schlaepfer, W. W. and Bunge, R. P.** Effects of calcium ion concentration and the degeneration of amputated axons in tissue culture, *J. Cell Biol.,* 59, 456, 1973.
27. **Sandoval, I. V. and Weber, K.,** Calcium-induced inactivation of microtuble formation in brain extracts, *Eur. J. Biochem.,* 92, 463, 1978.
28. **Sato, C., Nishizawa, K., Nakayama, T., Nakamura, H., Yoshimura, N., Takano, E., and Murachi, T.,** Rapid proteolysis of brain MAP-1 related cytoskeleton-associated 350 kd protein by purified calpain, *Cell Struc. Func.,* 11, 253, 1986.

29. **Billger, M., Wallin, M., and Karlsson, J.-O.,** Proteolysis of tubulin and microtubule-associated proteins 1 and 2 by calpain I and II. Difference in sensitivity of assembled and disassembled microtubules, *Cell Calcium,* 9, 33, 1988.
30. **Siman, R. and Noszek, J. C.,** Excitatory amino acids activate calpain I and induce structural protein breakdown in vivo, *Neuron,* 1, 279, 1988.
31. **Schlaepfer, W. W. and Zimmerman, U.-J. P.,** Calcium-mediated breakdown of glial filaments and neurofilaments in rat optic nerve and spinal cord, *Neurochem. Res.,* 6, 243, 1981.
32. **Malik, M. N., Meyers, L. A., Iqbal, K., Sheikh, A. M., Scotto, L., and Wisniewski, H. M.,** Calcium-activated proteolysis of fibrous proteins in central nervous system, *Life Sci.,* 29, 795, 1981.
33. **Lasek, R. J. and Hoffman, P. N.,** The neuronal cytoskeleton, axonal transport and axonal growth, in *Cell Motility,* Goldman, R., Pollard, T., and Rosenbaum, J., Eds., Cold Spring Harbor Laboratory, New York, 1976, 1021.
34. **Roots, B. I.,** Neurofilament accumulation induced in synapses by leupeptin, *Science,* 221, 971, 1983.
35. **Sahenk, Z. and Lasek, R. J.,** Inhibition of proteolysis blocks anterograde-retrograde conversion of axonally transported vesicles, *Brain Res.,* 460, 199, 1988.
36. **Baudry, M., Bundman, M. C., Smith, E. K., and Lynch, G. S.,** Micromolar calcium stimulates proteolysis and glutamate binding in rat brain synaptic membranes, *Science,* 212, 937, 1981.
37. **Siman, R. Baudry, M., and Lynch, G.,** Brain fodrin: substrate for calpain I, an endogenous calcium-activated protease, *Proc. Natl. Acad. Sci. U.S.A.,* 81, 3572, 1984.
38. **Levine, J. and Willard, M.** Fodrin: axonally transported polypeptides associated with the internal periphery of many cells, *J. Cell Biol.,* 90, 631, 1981.
39. **Davis, J. and Bennett, V.,** Brain spectrin: isolation of subunits and formation of hybrids with erythrocyte spectrin subunits, *J. Biol. Chem.,* 258, 7757, 1983.
40. **Lazarides, E. and Nelson, W. J.,** Erythrocyte and brain forms of spectrin in cerebellum: distinct membrane-cytoskeletal domains in neurons, *Science,* 220, 1295, 1983.
41. **Lazarides, E. and Nelson, W. J.,** Erythrocyte form of spectrin in cerebellum: appearance at a specific stage in the terminal differentiation of neurons, *Science,* 222, 931, 1983.
42. **Riederer, B. M., Lopresti, L. L., Krebs, K. E., Zagon, I. S., and Goodman, S. R,** Brain spectrin (240/235) and brain spectrin (240/235E): conservation of structure and location within mammalian neural tissue, *Brain Res. Bull.,* 21, 607, 1988.
43. **Riederer, B. M., Zagon, I. S., and Goodman, S. R.,** Brain spectrin (240/235) and brain spectrin (240/235E): two distinct spectrin subtypes with different locations within mammalian neural cells, *J. Cell. Biol.,* 102, 2088, 1986.
44. **Ivy, G. O., Seubert, P., Baudry, M., and Lynch, G.** Presence of brain spectrin in dendrites of mammalian brain: technical factors involved in innunocytochemical detection, *Synapse,* 2, 329, 1988.
45. **Seubert, P., Ivy, G., Larson, J., Lee, J., Shahi, K., Baudry, M., and Lynch, G.,** Lesions of entorhinal cortex produce a calpain-mediated degradation of brain spectrin in dentate gyrus. I. Biochemical studies, *Brain Res.,* 459, 226, 1988.
46. **Seubert, P. Baudry, M., Dudek, S., and Lynch, G.,** Calmodulin stimulates the degradation of brain spectrin by calpain, *Synapse,* 1, 20, 1987.
47. **Harris, A. S., Croall, D. E., and Morrow, J. S.,** The calmodulin-binding site in α-fodrin is near the calcium-dependent protease-I cleavage site, *J. Biol. Chem.,* 263, 15754, 1988.
48. **Carlin, R. K., Bartelt, D. C., and Siekevitz, P.,** Identification of fodrin as a major calmodulin-binding protein in postsynaptic density preparations, *J. Cell Biol.,* 96, 443, 1983.
49. **DeArmond, S. J., Fajardo, M., Naughton, S. A., and Eng, L.,** Degradation of glial fibrillary acidic protein by a calcium-dependent proteinase: an electroblot study, *Brain Res.,* 262, 275, 1983.
50. **Singh, I. and Singh, A. K,** Degradation of myelin proteins by brain endogenous neutral protease, *Neurosci. Lett.,* 39, 77, 1983.
51. **Sato, S. and Miyatake, T.,** Degradation of myelin basic protein by calcium-activated neutral protease (CANP)-like enzyme in myelin and inhibition by E-64 analogue, *Biomed. Res.,* 3, 461, 1982.
52. **Inoue, M., Kishimoto, A., Takai, Y., and Nishizuka, Y.,** Studies on a cyclic nucleotide-independent protein kinase and its proenzyme in mammalian tissues, *J. Biol. Chem.,* 252, 7610, 1977.
53. **Suzuki, K., Imajoh, S., Emori, Y., Kawasaki, H., Mirami, Y., and Ohno, S.,** Calcium-activated neutral protease and its endogenous inhibitor, *FEBS Lett.,* 220, 271, 1987.
54. **Ito, M., Tanaka, T. L., Nunoki, K., Hidaka, H., and Suzuki, K.,** The Ca^{2+}-activated protease (calpain) modulates Ca^{2+}/calmodulin dependent activity of smooth muscle myosin light chain kinase, *Biochem. Biophys. Res. Commun.,* 145, 1321, 1987.
55. **Tallant, E. A., Brumley, L. M., and Wallace, R. W.,** Activation of a calmodulin-dependent phosphatase by a Ca^{2+}-dependent protease, *Biochemistry,* 27, 2205, 1988.
56. **Kincaid, R. L., Stith-Coleman, I. E., and Vaughan, M.** Proteolytic activation of calmodulin-dependent cyclic nucleotide phosphodiesterase, *J. Biol. Chem.,* 260, 9009, 1985.

57. **Togari, A., Ichikawa, S., and Nagatsu, T.,** Activation of tyrosine hydroxylase by Ca^{2+}-dependent neutral protease, calpain, *Biochem. Biophys. Res. Commun.*, 134, 749, 1986.
58. **Yoshihara, Y., Ueda, H., Imajoh, S., Takagi, H., and Satoh, M.,** Calcium-activated neutral protease (CANP), a putative processing enzyme of the neuropeptide, kyotorphin, in the brain, *Biochem. Biophys. Res. Commun.*, 155, 546, 1988.
59. **Hirao, T., Hara, K., and Takahashi, K.,** Degradation of neuropeptides of calcium-activated neutral protease, *J. Biochem.*, 94, 2071, 1983.
60. **Farber, J. L.** The role of calcium in cell death, *Life Sci.*, 29, 1289, 1981.
61. **Banik, N. L., Chakrabarti, A. K., Gantt, G., and Hogan, E. L.,** Distribution of calcium-activated neutral proteinase activity in quaking mouse brain: a subcellular study, *Brain Res.*, 435, 57, 1987.
62. **Matthews, D. A., Cotman, C., and Lynch, G.,** An electron microscopic study of lesion-induced synaptogenesis in the dentate gyrus of the adult rat. I. Magnitude and time course of degeneration, *Brain Res.*, 115, 1, 1976.
63. **Lynch, G., Matthews, D., Mosko, S., Parks, T., and Cotman, W.,** Induced acetylcholinesterase layer in rat dentate gyrus following entorhinal lesions, *Brain Res.*, 42, 311, 1972.
64. **Lynch, G., Deadwyler, S., and Cotman, C.,** Postlesion axonal growth produces permanent functional connections, *Science*, 180, 1364, 1973.
65. **McWilliams, J. R. and Lynch, G. S.,** Terminal proliferation in the partially deafferented dentate gyrus. Time courses for the appearance and removal of degeneration and the replacement of lost terminals, *J. Comp. Neurol.*, 187, 191, 1979.
66. **Baudry, M. Fuchs, J., Kessler, M., Arst, D., and Lynch, G.,** Entorhinal cortex lesions induce a decrease in calcium transport in hippocampal mitochondria, *Science*, 216, 411, 1982.
67. **Goldschmidt, R. B. and Steward, O.,** Preferential neurotoxicity of colchicine for granule cells of the dentate gyrus of the adult rat, *Proc. Natl. Acad. Sci. U.S.A.*, 77, 3047, 1980.
68. **Hindelang-Gertner, C., Stoeckel, M. E., Porte, A., and Stutinsky, F.,** Colchicine effects on neurosecretory neurons and other hypothalamic and hypophysial cells, with special reference to changes in the cytoplasmic membranes, *Cell Tiss. Res.*, 170, 17, 1976.
69. **Seubert, P., Nakagawa, Y., Ivy, G., Vanderklish, P., Baudry, M., and Lynch, G.** Intrahippocampal colchicine injection results in spectrin proteolysis, *Neuroscience*, 31, 195, 1989.
70. **Lynch, G. and Baudry, M.,** The biochemistry of memory: a new and specific hypothesis, *Science*, 224, 1057, 1984.
71. **MacDermott, A. B., Mayer, M. L., Westbrook, G. L., Smith, S. J., and Barker, J. L.,** NMDA-receptor activation increases cytoplasmic calcium concentration in cultured spinal cord neurones, *Nature*, 321, 519, 1986.
72. **Mayer, M. L. and Westbrook, G. L.,** Permeation and block of *N*-methyl-D-aspartic acid receptor channels by divalent cations in mouse cultured central neurones, *J. Physiol. (London)*, 394, 501, 1987.
73. **Lynch, G., Larson, J., Kelso, S., Barrionuevo, G., and Schottler, F.,** Intracellular injections of EGTA block induction of hippocampal long-term potentiation, *Nature*,, 305, 719, 1983.
74. **Malenka, R. C., Kauer, J. A., Zucker, R. S., and Nicoll, R. A.,** Postsynaptic calcium is sufficient for potentiation of hippocampal synaptic transmission, *Science*, 242, 81, 1988.
75. **Collingridge, G. L., Kehl, S. J., and McLennan, H.,** Excitatory amino acids in synaptic transmission in the Schaffer-collatoral-commissural pathway of the rat hippocampus, *J. Physiol. (London)*, 334, 33, 1983.
76. **Harris, E. W., Ganong, A. H., and Cotman, C. W.,** Long-term potentiation in the hippocampus involves activation of *N*-methyl-D-asparate receptors, *Brain Res.*, 323, 132, 1984.
77. **Larson, J. and Lynch, G.,** Role of *N*-methyl-D-aspartate receptors in the induction of synaptic potentiation by burst stimulation patterned after the hippocampal theta rhythm, *Brain Res.*, 411, 111, 1988.
78. **Morris, R. G. M., Anderson, E., Lynch, G., and Baudry, M.,** Selective impairment of learning and blockade of long-term potentiation by an *N*-methyl-D-aspartate receptor antagonist, AP-5, *Nature*, 319, 774, 1986.
79. **Staubli, U., Thibault, O., DiLorenzo, M., and Lynch, G.,** Antagonism of NMDA receptors impairs acquisition but not retention of olfactory memory, *Behav. Neurosci.*, 103, 54, 1989.
80. **Seubert, P., Larson, J., Oliver, M., Jung, M. W., Baudry, M., and Lynch, G.,** Stimulation of NMDA receptors induces proteolysis of brain spectrin in hippocampal slices, *Brain Res.*, 460, 189, 1988.
81. **Wieloch, T.,** Hypoglycemia-induced neuronal damage prevented by an *N*-methyl-D-aspartate antagonist, *Science*, 230, 681, 1985.
82. **Simon, R. P., Swan, J. H., Griffiths, T., and Meldrum, B. S.,** Bockade of *N*-methyl-D-aspartate receptors may protect against ischemic damage in the brain, *Science*, 226, 850, 1984.
83. **Kirino, T.,** Delayed neuronal death in the gerbil hippocampus following ischaemia, *Brain Res.*, 239, 57, 1982.
84. **Pulsinelli, W. A., Brierley, J. P., and Plum, F.,** Temporal profile of neuronal damage in a model of transient forebrain ischemia, *Ann. Neurol.*, 11, 49, 1982.

85. **Boast, C. A., Gerhardt, S. C., Pastor, G., Lehmann, J., Etienne, P. E., and Liebman, J. M.,** The *N*-methyl-D-aspartate antagonists CGS 19755 and CPP reduce ischemia brain damage in gerbils, *Brain Res.*, 442, 345, 1988.
86. **Gill, R. Foster, A. C., and Woodruff, G. N.,** MK-801 is neuroprotective in gerbils when administered during the post-ischaemic period, *Neurosci.*, 25, 847, 1988.
87. **Rothman, S.** Synaptic release of excitatory amino acid neurotransmitter mediates anoxic neuronal death. *J. Neurosci.*, 4, 1884, 1984.
88. **Seubert, P., Lee, K., and Lynch, G.,** Ischemia triggers NMDA receptor-linked cytoskeletal proteolysis in hippocampus, *Brain Res.*, 492, 366, 1989.
89. **Lee, K. Schottler, F., Oliver, M., and Lynch, G.,** Brief bursts of high-frequency stimulation produce two types of structural change in rat hippocampus, *J. Neurophys.*, 44, 247, 1980.
90. **Lee, K. Oliver, M., Schottler, F., and Lynch, G.,** Electron microscopic studies of brain slices: The effects of high frequency stimulation on dendritic ultrastructure, in *Electrical Activity in Isolated Mammalian C.N.S. Preparation,* Kerkut, G. and Wheal, H. V., Eds., Academic Press, N.Y., 1981, 189.
91. **Chang, F. L. F. and Greenough, W. T.,** Transient and enduring morphological correlates of synaptic activity and efficacy change in the rat hippocampal slice, *Brain Res.*, 309, 35, 1984.
92. **Fox, J. E. B., Reynolds, C. C., and Phillips, D. R.,** Calcium-dependent proteolysis occurs during platelet aggregation, *J. Biol. Chem.*, 258, 9973, 1983.
93. **Siman, R., Baudry, M., and Lynch, G.,** Calcium-activated proteases as possible mediators of synaptic plasticity, in *Synaptic Function,* Edelman, G., Gall, W., and Cowan, W. M., Eds., John Wiley & Sons, 1987, 519.
94. **Nicoterra, P., Hartzel, P., Davis, G., and Orrenius, S.,** The formation of plasma membrane blebs in heptocytes exposed to agents that increase cytosolic Ca^{2+} is mediated by the activation of a nonlysosomal proteolytic system, *FEBS Lett.*, 209, 139, 1986.
95. **Lynch, G., Halpain, S., and Baudry, M.,** Effects of high-frequency synaptic stimulation on glutamate receptor binding with a modified *in vitro* hippocampal slice preparation, *Brain Res.*, 244, 101, 1982.
96. **Kessler, M., Petersen, G., Vu, H. M., Baudry, M., and Lynch, G.,** L-phenylalanyl-L-glutamate-stimulated, chloride-dependent glutamate binding represents glutamate sequestration mediated by an exchange system, *J. Neurochem.*, 48, 1191, 1987.
97. **Staubli, U., Baudry, M., and Lynch, G.,** Leupeptin, a thiol-proteinase inhibitor, causes a selective impairment of spatial maze performance in rats, *Behav. Neural Biol.*, 40, 58, 1984.
98. **Staubli, U., Larson, J. Thibault, O., Baudry, M., and Lynch, G.,** Chronic administration of a thiol-proteinase inhibitor blocks long-term potentiation of synaptic responses, *Brain Res.*, 444, 153, 1988.
99. **Oliver, M., Baudry, M., and Lynch, G.,** The protease inhibitor leupeptin interferes with the development of LTP in hippocampal slices, *Brain Res.*, 505, 233, 1940.
100. **Seubert, P. and Lynch, G.,** Unpublished observations.

Chapter 19

CALPAINS IN LENS AND CATARACT

T. R. Shearer and L. L. David

TABLE OF CONTENTS

I. INTRODUCTION

Calpains are present in varying concentrations among tissues of the eye. Relatively high amounts have been detected by immunohistochemistry in corneal epithelium, lens epithelium, inner limiting membrane of the retina, and small localized parts of iris (probably muscle).[1] Aqueous, vitreous, and lens nucleus contained very low to nondetectable levels of calpain antigen. In lens,[2,3] retina,[4] and corneal epitheilum,[5] calpain has also been measured enzymatically. Thus, the presence of calpain in eye tissues has been well documented. The present challenge is to determine the biological functions and control of calpain in ocular tissues. Since the bulk of the work on calpain in eye to date has been on lens, the remainder of this review will concern lens.

II. CALPAIN IN LENTICULAR TISSUES

A. IMMUNOHISTOCHEMICAL DISTRIBUTION

Because of the nature of lens, special techniques were used to provide histologic sections for immunologic study of calpain. Lenses were first fixed in methacarn before sectioning. Sections were incubated with polyclonal antibody which cross-reacted with both calpains I and II.[1] Calpain localization was visualized by the ABC immunoperoxidase method with diaminobenzidine as the chromogen. Results indicated the highest concentration of calpain to be in lens epithelium (Figure 1). Calpain in the epithelial cells appeared cytosolic, with lower concentrations of antigen found in differentiating epithelial cells at the equator. In the superficial cortex, calpain staining sharply outlined the periphery of newly elongated fibers, indicating that calpain was probably associated with the plasma membrane. Thus, lenticular calpain seems to become associated with plasma membranes as the cells differentiate into fiber cells. Calpain in the deep cortex and outer nucleus was again more concentrated in the cytosol and diffuse, and not obviously associated with plasma membrane. No reaction to calpain antibody was observed in the central lens nucleus.

B. BIOCHEMICAL PROPERTIES OF LENTICULAR CALPAIN

1. Demonstration of Calpains in Lens Tissues

Soluble protein isolated from the cortical regions of two pairs of lenses obtained from 64- and 74-year-old human donors was applied to an HPLC-DEAE column.[6] A predominant peak of calcium-dependent proteolytic activity eluted at 300 m*M* NaCl (Figure 2). Immunologic reactivity with an antibody against calpain in a direct ELISA coincided with the proteolytic activity. Proteolytic activity eluted at approximately the same position using soluble protein from rat lenses, but was present at over 10 times the specific activity found in human lens. A calcium activation curve for rat lens calpain showed the proteolytic activity in this peak to be fully activated at 400 μ*M* calcium. These data indicated that this activity is calpain II. Elution position of calpastatin in lens is indicated by the open bar and *. The HPLC-DEAE column is useful for purification of calpains from small amounts of tissue, such as that obtained from lens.

While not observable in human lenses, a smaller peak of antigenic reaction for calpain eluted at 180 m*M* NaCl during HPLC-DEAE separation of rat lens proteins. Since our polyclonal antibody recognizes both calpains I and II, calpain I probably elutes in this peak. Calpain I activity was not measured in these experiments due to coelution with calpastatin. However, the smaller peak of antigenic reaction for calpain I indicated that calpain II is the predominate form of calpain in lens. This was confirmed in a study on bovine lenses.[2] The activity of calpain I in bovine lens cortex was approximately 4% the activity of calpain II. Although the level of calpain I is decidedly lower than calpain II in lens, the presence of calpain I may be important for proteolytic activity under physiological conditions where only μ*M* calcium is present and available for activation of calpain.[7]

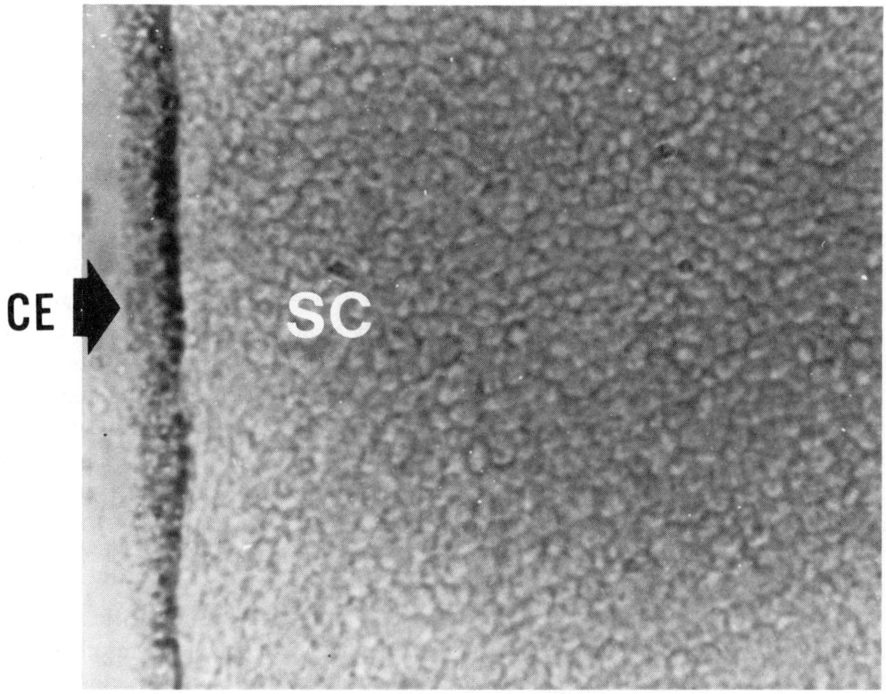

FIGURE 1. Immunoperoxidase staining for calpain in a lens section from an 18-d old rat (400 × original magnification), using antibody against calpains. Darker staining areas indicate the location of calpain antigen, which was highest in epithelium. CE = central epithelium, SC = superficial cortex.

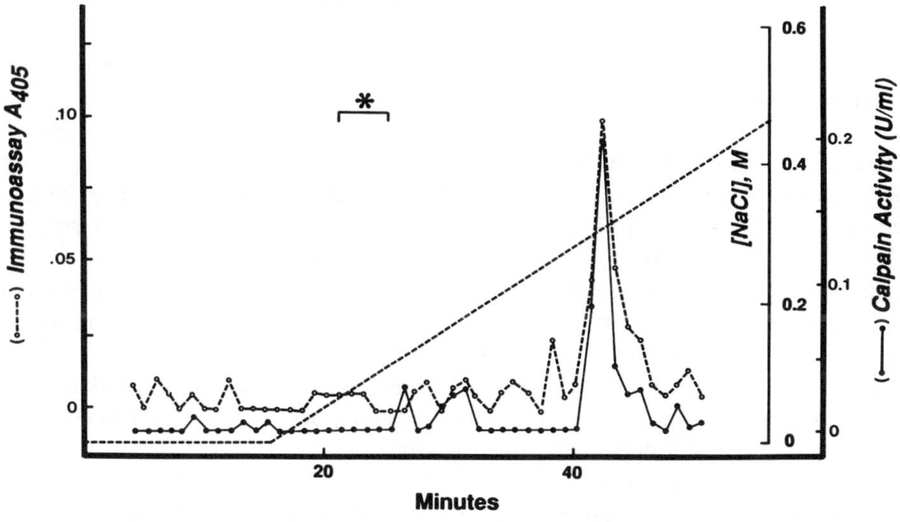

FIGURE 2. Separation of the water soluble fraction from lens cortex of 64- and 74-year old human donors on a 7.5 × 75 mm TSK DEAE 5-PW high pressure liquid chromatography column. Fractions were assayed for calcium-dependent proteolytic activity (●), and for immunoreactivity against calpain antibody (○). One unit of calpain activity was the amount of calpain that caused the release of one μg acid-soluble fragments per minute. Open bar and * indicates the elution position of calpastatin in lens.

2. Distribution of Calpain II in Lens

The relative distribution of calpain II enzyme activity in various regions of the bovine lens was determined to be epithelium ≫ cortex, nucleus = 0[2]. This is similar to the immunohistologic distribution of calpain in rat lens.[1] The rate of protein turnover in the lens is one of the slowest of any tissue, and the central region of the lens contains some of the oldest proteins in the body.[8] Recent studies[6,9] have measured calpain enzyme activity in nuclei (central region) of both rodent and human lenses. Lenses from young animals contained similar specific activities of calpain II in cortex and nucleus, but with aging the activity of calpain II decreased precipitously in the lens nucleus. This was likely the result of an age-related loss of calpain. We hypothesize that calpain concentration decreased more slowly in the outer regions of the lens because new lens fiber cells containing newly synthesized calpain were added to the outer regions of the lens, while no new synthesis of calpain occurred in the inner fibers. The lens nucleus may be an excellent tissue to study posttranslational changes in calpain during aging. While calpain II decreased in the lens nucleus with aging, it was still detectable in the lens nucleus of aged animals.[6,9] This result is important, because we hypothesize that calpain activation may be a cause of cataracts appearing in the nucleus of both young and aged lenses.

3. Purification of Calpain II from Lens

Calpain II has been purified from both bovine and rat lenses.[3,10] The purification schemes of lens calpain II from bovine and rat lens were typical of schemes used in other tissues in that the first two steps were DEAE chromatography followed by gel filtration. Bovine lens calpain II was then purified to apparent homogeneity using two further chromatography steps consisting of blue Sepharose and another DEAE step. Purfication of rat lens calpain II differed in that it was purified following gel filtration by a single chromatography step on a hydrophobic interaction column. While this did not provide as pure a preparation as did the four-step purification of bovine lens calpain II, a three-step purification was necessary because of the smaller amount of starting material from rat lens. Bovine and rat lens calpain II preparations had native molecular weights of approximately 110 to 120 K, contained two subunits of approximately 80 and 30 kDa, and required calcium concentrations in the near m*M* range for full activity. Thus, calpain II from lens is very similar to calpain II isolated from various other tissues.[11]

4. Substrate Specifity of Lens Calpain II

Purified lens calpains have been tested for endogenous substrate specifity. Bovine and rat lens calpain II both caused limited degradation of α-crystallin.[3,10] The preferred sites on which bovine lens calpain II cleaves α-crystallins were subsequently determined.[12] Cleavage at preferred sites caused the removal of 5 residues from the C-terminal end of α-crystallin B chain (cleavage between Thr^{170}-Ala^{171}), and 10 residues from the C-terminal end of α-crystallin A chain (cleavage between Arg^{183}-Glu^{184}).

Neither bovine lens nor rat lens calpain II degraded γ-crystallin polypeptides.[3,10] However, several polypeptides found in β-crystallin aggregates were partially degraded by rat lens calpain II.[3] Like the degradation of α-crystallin polypeptides, calpain II removed no more than several thousand molecular weight from these approximately 23 to 30 kDa polypeptides. In contrast, bovine calpain II did not degrade β-crystallins from the same tissue. The reasons for the resistance of bovine β-crystallin to cleavage is unknown and may be due to actual differences in bovine calpain II specificity, or it may be due to structural differences between rat and bovine β-crystallins.

Lens cytoskeletal proteins and membrane proteins also act as calpain II substrates. Bovine and rat lens calpain II degraded the cytoskeletal proteins actin and vimentin.[3,13,14] Unlike crystallin hydrolysis, degradation of actin and vimentin was not limited. Extended incubation of these proteins with calpain II caused the appearance of material running at the dye front during SDS-PAGE.[14] Rat lens calpain II also caused the extensive degradation of a 66-kDa instrinsic membrane polypeptide, and the partial degradation of the 26-kDa major intrinsic membrane protein.[3] At present, α-crystallin A and B chains remain the only lens polypeptides where calpain cleavage sites have been determined.

5. Regulation of Lens Calpain by Calpastatin

As in other tissues, calpain in lenses may be regulated by the presence of the endogenous inhibitor calpastatin. The calpastatin to calpain ratio in rat lens is low enough to detect calpain activity in crude lens homogenates.[15] However, in bovine, and especially human lens, the amount of calpastatin exceeds calpain,[2,6] and thus calpastatin must first be removed before carrying out the enzymatic assay of calpain. On HPLC-DEAE chromatography lens calpastatin eluted at approximately 100 m*M* NaCl.[6,9]

In contrast to calpain, calpastatin in lens does not decrease with age.[6] In one type of cataract, selenite overdose cataract, calpastatin levels actually increase per milligram of soluble protein, possibly due to insolubilization of noncalpastatin proteins.[16] Thus, any proposed mechanism of action for calpain in lens must take into account the appreciable amounts of this stable endogenous inhibitor. One recent theory adopted from other tissues[17] is that calpain escapes action of calpastatin by binding to lens plasma membrane.

6. Binding of Calpain to Lens Membrane

Recent evidence confirms the immunohistochemical findings that calpain may be associated with lens plasma membranes. The water-insoluble fraction of rate lens was fractionated by SDS-PAGE, electrotransferred to PVDF membrane, and incubated with antibody to calpain. Results showed the 80-kDa catalytic subunit of calpain in the water-insoluble fraction. This association with the water-insoluble fraction is promoted by calcium, but is not totally dependent on calcium, because the association could be demonstrated even by first homogenizing the lens in buffer containing EGTA. Further extraction of the water-soluble fraction with urea did not remove all the calpain, and indeed fiber membrane fractions contained calpain antigen.

During aging, the amount of calpain associated with the water-insoluble fraction from lens cortex actually increased, even though calpain in the water-soluble fraction showed the age-related decrease described above.[9] Two important questions remain concerning this membrane-bound calpain. First, what is the chemical nature of the binding of calpain to lens membrane? Second, is the membrane-bound calpain enzymatically active?

Since membrane components such as phosphatidylinositol are known to lower calcium requirements,[18] we hypothesize that association of calpain with lens membranes has at least two important consequences: (1) The concentration of calcium required for activation of calpain is decreased. (2) Membranes serve as sites for calpain-induced hydrolysis of lens proteins.

III. FUNCTION OF LENS CALPAIN

A. PHYSIOLOGICAL CONDITIONS

1. Possible Role in Differentiation and Maturation

Almost nothing is known about the function of calpain in lens under normal physiological conditions. The high concentrations of calpain in the epithelium, and the apparent association of calpain with the plasma membrane in differentiating fiber cells in outer cortex suggest a role for calpain in normal terminal fiber cell differentiation and maturation. During this period, one could speculate that proteolysis by calpain would be useful for degradation of organelles and cytoskeletal elements. Indeed, cytoskeletal elements are lost during aging,[20] and purified lens calpain II rapidly degrades actin and vimentin.[14]

2. Removal of Damaged Lens Proteins

Another important postulated role for calpain and other lens proteases is removal of damaged lens proteins.[21] Proteins within the lens show a very slow rate of turnover. However, lens proteins are exposed to damaging agents, such as UV radiation, throughout life of the host. Lens proteases including calpain may help rid the lens of damaged proteins by shortening the proteins to peptides and amino acids which can diffuse or be transported out of the lens. Although this seems a logical function for a cytosolic endopeptidase existing within lens (the bulk of the lens contains no lysosomes), no direct evidence yet exists to prove this is a function of lens calpain.

B. CATARACTOUS CONDITIONS

1. Calpain Activation as a Possible Cause of Cataract

Most of the literature concerning the function of calpain in lens is concerned with its role in cataract. Cataract is any opacity occurring in the lens, and it is the universal pathological response to a variety of insults to the lens. Agents or factors associated with cataractogenesis include increasing age (''senile'' cataract), UV radiation, diabetes, hereditary diseases, drugs such as steroids, and toxic chemicals. Cataracts can be located in various regions of the lens such as in the nucleus, cortex, or subcapsular area.

Many cataracts show increases in water-insoluble protein, high molecular weight aggregates and calcium. A number of experimental models exist, and especially in rodent models, increased calcium and proteolysis has been repeatedly documented.[3] From such observations, the hypothesis has arisen that activation of calpain may be part of the mechanism producing the pathology of cataract. If this proves true for some cataracts in man, amelioration of cataract development by application of inhibitors of calpain would have enormous clinical and economic potential. Cataract is the third leading cause of blindness in the U.S. Every year 1 million cases of cataract are diagnosed. The U.S. National Institutes of Health has estimated that if the progression of cataract could be slowed so that surgery could be delayed for 10 years, 608 million dollars could be saved in medical costs annually (1973 dollars).[22] This is in addition to the reduction of human suffering and loss of work by amelioration of cataract.

While examples of possible calpain-induced proteolysis are found in many experiemental cataracts in rodents, the best-characterized models exhibiting evidence of calpain activation are the selenite overdose rat cataract[23] and the Nakano Mouse cataract[24] described below.

2. Selenite Overdose Cataract

Selenite cataract is produced by a single subcutaneous injection of an overdose of sodium selenite at 30 μMol Na_2Se0_3/kg b.w. to suckling rat pups. Within 3 to 5 d a very dense nuclear cataract develops. Calpain is believed to be involved in the mechanism of this cataract for the following reasons:

1. Calcium levels were elevated before formation of selenite nuclear cataract.[15] Furthermore, free ionic calcium is elevated to 104 μ*M* calcium,[7] which is above the minimum level previously shown to initiate activation of calpain II in rat lens.[3]
2. A number of lens proteins including α-, β_H-, and β_L-crystallins, cytoskeletal proteins, and membrane proteins MIP26 and MP66, were hydrolyzed in the nuclear region during cataract formation.[15] Proteolysis was also observed in lenses with selenite cataract induced by incubation with 25 μ*M* sodium selenite.[25] Of particular interest is the fact that incubation of purified rat lens calpain with purified lens crystallins, as described earlier, produces proteolysis very similar to the limited proteolysis observed *in vitro* during cataract formation.[3]
3. A third reason for believing that calpain may be active in selenite cataract is that indirect indicators of activation of calpain were observed in selenite cataract. For

example, significant calpain II activity is present in lens before selenite cataract.[7] However, calpain II activity decreased rapidly after cataract formation, probably due to autoproteolysis following activation of calpain II. Increased amounts of calpain degradation products were observed during selenite cataract formation.[16]

4. Production of cataract *in vitro* by incubation of lenses either with selenite or Ca ionophore is prevented by addition of calpain inhibitor E64 to the culture medium. Proteolysis of lens crystallin polypeptides in lenses was also prevented by calpain inhibitor.[26]
5. Most radioactivity in lens after injection of radioactive selenite was found in the outer layers of the lens.[27] This led to the idea that the initial damage caused by selenite may be largely confined to the lens epithelial cells. In support of this, the earliest change observed so far in selenite cataract was suppression of mitoses, observed in the epithelium only 2 h after selenite injection.[28] The specific insult caused by selenite is unknown. However, selenite is a potent oxidizer of sulfhydryl groups.[29] Oxidation of sulfhydryls increases the permeability of lens membranes to ions.[30] This has lead to the hypothesis that selenite oxidizes critical sulfhydryl groups in lens epithelial cells, causing increased permeability of Ca^{2+} ion. This may lead to the accumulation of Ca^{2+} in lens fibers of the nucleus, and calpain activation.

3. Nakano Mouse Cataract

Parallel studies in the Nakano mouse cataract have also led to the conclusion that calpain is activated in this hereditary cataract:[24]

1. Calcium levels were increased before onset of the pinpoint nuclear opacity at the beginning of the Nakano mouse cataract.
2. Significant levels of calpain II were present at beginning of opacity and these decreased after opacity had formed. Like the selenite cataract, this suggested that calpain II was activated, then lost activity due to autolysis.
3. As in the selenite cataract, partial degradation of α- and β-crystallin polypeptides took place during cataract formation. While α-crystallin A chain did not undergo cleavage, incubation of α-crystallin B chain with calpain II *in vitro* produced partially degraded B chains similar to those produced *in vivo* during cataract formation.

4. A Proposed Mechanism of Calpain-Induced Cataract

Figure 3 shows a proposed mechanism for calpain activation in lens and cataract formation.

1. As a result of physical trauma, accumulation of damage with aging, insult by toxic substances, or hereditary factors; calcium homeostasis of the lens is altered, leading to pathological elevation of intracellular calcium.
2. Large intracellular elevations of calcium activate two pools of calpain, one bound to membrane, the other cytosolic. The sulfhydryl at the active site of calpain becomes more accessible following binding of calcium. Membrane-bound calpain may be activated at low calcium concentration due to the presence of phosphatidyl inositol (PI). While not shown in the diagram, activated calpain may also undergo autolysis, which could also lower its calcium requirements.
3. Some membrane-bound calpain is not inhibited by calpastatin. In contrast, cytosolic calpain is inhibited by calpastatin. However, this inhibition may be only partial in the lenses of some species, due to a high ratio of calpain to calpastatin.
4. Activated calpain at the membrane surface and cytosolic calpain not inhibited by calpastatin cause limited proteolysis of lens proteins. These partially degraded lens proteins cause light scatter and lead to opacity.

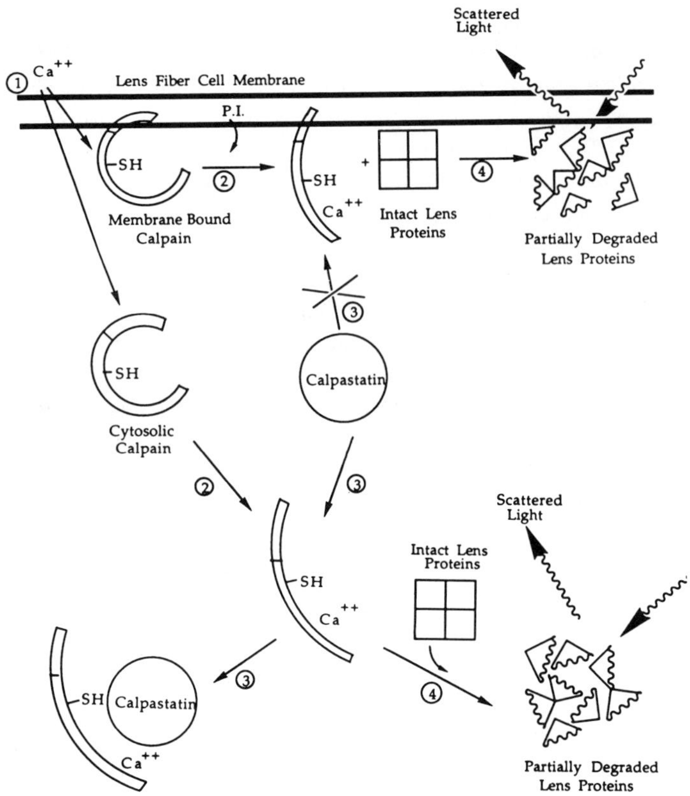

FIGURE 3. A proposed mechanism for activation of lens calpain and formation of cataract. See text for details.

IV. FUTURE RESEARCH NEEDS

1. More conclusive evidence for calpain activation in the lens may be provided by comparing the sites where calpain cleaves lens crystallin polypeptides *in vitro* to the sites where lens crystallins are cleaved *in vivo* during cataract formation.
2. Definitive proof of involvement of calpain in cataract formation as well as affirmation of possible clinical relevance would be obtained by prevention of cataract by application of calpain inhibitors to the whole animal.

ACKNOWLEDGMENTS

The authors would like to thank Dr. Ruth Anderson for providing immunohistochemical results, and making many useful suggestions. The experiments and preparation of this manuscript were partially supported by NIH grants EY-03600 and EY-05786 and by a travel fellowship to T. R. S. from The Japan Society for the Promotion of Science. Special thanks to the late Dr. Takashi Murachi at Kyoto University Hospital and the helpful personnel in his laboratory who graciously hosted Dr. Shearer during the time this manuscript was written.

REFERENCES

1. **Anderson, R. S. and Shearer, T. R.,** Role of calpain in tissue differentiation in rat eye, *Invest. Ophthalmol. Vis. Sci.,* 31(4), 157(abstr.), 1990.
2. **Yoshida, H., Murachi, T., and Tsukahara, I.,** Distribution of calpain I, calpain II, and calpastatin in bovine lens, *Invest. Ophthalmol. Vis. Sci.,* 26, 953, 1985.
3. **David, L. L. and Shearer, T. R.,** Purification of calpain II from rat lens and determination of endogenous substrates, *Exp. Eye. Res.,* 42, 227, 1986.
4. **Tsung, Pi-Kwang and Lombardini, J. B.,** Identification of low Ca and high calcium requiring neutral proteases in rat retina, *Exp. Eye Res.,* 41, 97, 1985.
5. **Shearer, T. R., Azuma, M., David, L. L., Yamagata, Y., and Murachi, T.,** Calpain and calpastatin in rabbit corneal epithelium, *Curr. Eye Res.,* 9(1), 39, 1990.
6. **David, L. L., Varnum, M. D., Lampi, K. J., and Shearer, T. R.,** Calpain II in human lens, *Invest. Ophthalmol. Vis. Sci.,* 30, 269, 1989.
7. **Hightower, K. R., David, L. L., and Shearer, T. R.,** Regional distribution of free calcium in selenite cataract: relation to calpain II, *Invest. Ophthalmol. Vis. Sci.,* 28, 1702, 1987.
8. **Masters, P. M., Bada, J. L., and Zigler, J. S., Jr.,** Aspartic acid racemization in heavy molecular weight crystallins and water-insoluble protein from normal human lenses and cataracts, *Proc. Natl. Acad. Sci. U.S.A.,* 75, 1204, 1978.
9. **Varnum, M. D., David, L. L., and Shearer, T. R.,** Age-related changes in calpain II and calpastatin in rat lens, *Exp. Eye Res.,* submitted.
10. **Yoshida, H., Murachi, T., and Tsukahara, I.,** Limited proteolysis of bovine lens α-crystallin by calpain, a Ca^{2+}-dependent cystein proteinase, isolated from the same tissue, *Biochim. Biophys. Acta,* 798, 252, 1984.
11. **Murachi, T.,** Intracellular Ca^{2+} protease and its inhibitor protein: calpain and calpstatin, in *Calcium and Cell Function,* Vol.4, Academic Press, New York, 1983, 378.
12. **Yoshida, H., Yumoto, N., Tsukahara, I., and Murachi, T.,** The degradation of α-crystallin at its carboxyl-terminal portion by calpain in bovine lens, *Invest. Ophthalmol. Vis. Sci.,* 27, 1269, 1986.
13. **Roy, D., Cheisa, R., and Spector, A.,** Lens calcium-activated proteinase: degradation of vimentin, *Biochem. Biophys. Res. Commun.,* 116, 204, 1983.
14. **Yoshida, H., Murachi, T., and Tsukahara, I.** Degradation of actin and vimentin by calpain II, Ca^{2+}-dependent cysteine proteinase, in bovine lens, *FEBS Lett.,* 170, 259, 1984.
15. **David, L. L. and Shearer, T. R.,** Calcium-activated proteolysis in lens during selenite cataractogenesis, *Invest. Ophthalmol. Vis. Sci.,* 25, 1275, 1984.
16. **David, L. L., Varnum, M. D., and Shearer, T. R.,** unpublished.
17. **Gopalakrishna, R. and Barsky, S. H.,** Hydrophobic association of calpains with subcellular organelles, *J. Biol. Chem.,* 261, 13936, 1986.
18. **Mellgren, R. L.,** Calcium-dependent proteases: an enzyme system active at cellular membranes, *FASEB J.,* 1, 110, 1987.
19. **Imajoh, S., Kawasaki, H., and Suzuki, K.,** The amino-terminal hydrophobic retion of the small subunit of calcium-activated neutral protease (CANP) is essential for its activation by phosphoinositol, *J. Biochem.,* 99, 1281, 1986.
20. **Ellis, M. and Maisel, H.,** Cytoskeletal proteins of the aging human lens, *Curr. Eye Res.,* 3, 369, 1984.
21. **Taylor, A. and Davies, K. J. A.,** Protein oxidation and loss of protease activity may lead to cataract formation in the aged lens, *Free Rad. Biol. Med.,* 3, 371, 1987.
22. Report of the Cataract Panel, in *Vision Research: a National Plan,* Vol. 2, (Part 3), NIH Pub. No. 83-2473, 1983, 1.
23. **Shearer, T. R., David, L. L., and Anderson, R. S.,** Selenite cataract: a review, *Curr. Eye Res.,* 6, 289, 1987.
24. **Yoshida, H., Murachi, T., and Tsukahara, I.,** Age-related changes of calpain II and α-crystallin in lens of hereditary cataract (Nakano) mouse, *Curr. Eye Res.,* 4, 983, 1985.
25. **Hess, J. L., Troxclair, A., Calvin, H., and Bunce, G. E.,** *In vitro* cataract formation in rat lens caused by sodium selenite, *Invest. Ophthalmol. Vis. Sci.,* 29 (Suppl.), (Abstr.), 28, 1988.
26. **Yoshida, H., Saito, I., Veda, K., Honda, Y., and Murachi, T.,** Calpain-catalyzed degradation of lens proteins and its regulation, Proc. Int. Soc. Eye Res., 8th ICER V, 84 (Abstr.), 1988.
27. **Shearer, T. R., Anderson, R. S., and Britton, J. L.,** Uptake and distribution of radioactive selenium in cataractous rat lens, *Curr. Eye Res.,* 2, 561, 1983.
28. **Anderson, R. S., Shearer, T. R., and Claycomb, C. K.,** Selenite-induced epithelial damage and cortical cataract, *Curr. Eye Res.,* 5, 53, 1986.
29. **Tsen, C. C. and Tappel, A. L.,** Catalytic oxidation of glutathione and other sulfhydryl compounds by selenite, *J. Biol. Chem.,* 233, 1230, 1958.

30. **Gandolfi, S. A., Tagliavini, J., Belpoliti, M., Duncan, G., and Maraini, G.,** Oxidative cross-linking of fodrin parallels a membrane conductance increase in the mammalian lens, *Curr. Eye Res.*, 7, 747, 1988.

Index

INDEX

C

D

H

I

N

Q

R

S

U

V

W

X

Z

G.T 19 No to 75 02/01/06

Box Nos — 1093 1094

24 24

Total Books - 48

Total Boxes — 2